FORGOTTEN EXPERTS

STANFORD **OTTOMAN WORLD** SERIES

FORGOTTEN EXPERTS

ASTROLOGERS, SCIENCE, AND AUTHORITY IN THE OTTOMAN EMPIRE, 1450–1600

A. Tunç Şen

STANFORD UNIVERSITY PRESS
Stanford, California

Stanford University Press
Stanford, California

Library of Congress Cataloging-in-Publication Data
Names: Şen, A. Tunç, author.
Title: Forgotten experts : astrologers, science, and authority in the Ottoman Empire, 1450-1600 / A. Tunç Şen.
Description: Stanford, California : Stanford University Press, [2025] | Series: Stanford Ottoman world series | Includes bibliographical references and index.
Identifiers: LCCN 2024047782 | ISBN 9781503642652 (cloth) | ISBN 9781503643017 (epub)
Subjects: LCSH: Astrology, Turkish—History. | Astronomy—Turkey—History. | Turkey—History—1453-1683. | Turkey—History—Ottoman Empire, 1288-1918.
Classification: LCC BF1708.8.T9 S46 2025 | DDC 133.509561—dc23/eng/20241231
LC record available at https://lccn.loc.gov/2024047782

Cover design: Lindy Kasler
Cover art: Seyyid Lokman, *Zubdat al-tavārīkh*, 1585–1590,
405 mm × 277 mm × 53 mm. Chester Beatty Library T. 414, 16b.

The authorized representative in the EU for product safety and compliance is: Mare Nostrum Group B.V. | Mauritskade 21D | 1091 GC Amsterdam | The Netherlands | Email address: gpsr@mare-nostrum.co.uk | KVK chamber of commerce number: 96249943

Printed and bound by CPI Group (UK) Ltd, Croydon, CR0 4YY

In memory of my father, Tayyar Şen

For Sümeyye and Devin

CONTENTS

List of Illustrations ix

Note on Transliteration xi

Acknowledgments xiii

Introduction **Empire of Experts, Experts of Empire** 1

1 ***Munajjims*' Expertise** 39

2 **Persianate Foundations** 79

3 **Heavenly Patronage** 113

4 **Fortunes Turned** 153

5 **Occult Rivalries** 188

Conclusion **Istanbul Observatory and the Aftermath** 225

Notes 239

Bibliography 281

Index 319

ILLUSTRATIONS

FIGURE 1 The Süleymaniye Mosque complex 3

FIGURE 2 Western Asia in the second half of the fifteenth century 7

FIGURE 3 The birth scene of Alexander the Great 50

FIGURE 4 The horoscope page of ʿAbdurrahman's almanac 59

FIGURE 5 An Ottoman astrologer observing the 1577 comet 73

FIGURE 6 The birth horoscope of Sultan Mehmed II 101

FIGURE 7 Ownership records by Khitabi and his student Yusuf b. ʿÖmer 109

FIGURE 8 Sultan Bayezid II 117

FIGURE 9 Sultan Bayezid II's personal copy of a *majmua* 136

FIGURE 10 Ownership statements by Müʾeyyedzade and Riyazi 147

FIGURE 11 Yusuf b. ʿÖmer's almanac for 1511 182

FIGURE 12 Yusuf b. ʿÖmer's almanac for 1517 182

FIGURE 13 Riyazi's horoscope for the construction of the Süleymaniye Mosque 201

FIGURE 14 Taqi al-Din's observatory in Istanbul 227

FIGURE 15 Instruments at the Istanbul observatory 237

NOTE ON TRANSLITERATION

Words in Arabic script that appear in the body of the text are transliterated using a simplified version of the *IJMES* transliteration system. For readability, no diacritics or macrons are included in the body of the text, except for *ʿayn* and *hamza*, which are represented by ʿ and ʾ, respectively. However, in the Notes and Bibliography, words, names, and book titles are fully transliterated.

Terms that have entered English usage are not transliterated; they are written according to the *IJMES* Word List without being italicized (e.g., vizier, madrasa, waqf, shaykh al-Islam, ʿulamaʾ, fatwa, etc.). For Arabic-origin words that also appear in Persian and Ottoman Turkish, the Arabic renderings are preferred in the main body of the text. Hence, *munajjim*, not *müneccim*; *taqwim*, not *takvim* or *taqvim*. Since the words are fully transliterated in the Notes and Bibliography, readers should not find it inconsistent when they encounter different versions of the same word, such as *taqwīm* (Arabic), *taqvīm* (Persian), and *taḳvīm* (Ottoman Turkish). Each is transliterated according to the transliteration conventions of the language in which the quoted text is written.

Regarding the transliteration of individual names, those whose lives unfolded predominantly in the Turkish-speaking Ottoman realm are transliterated using Turkish conventions, while others are transliterated according to their own conventions. Thus, *Taşköprülüzade*, not *Tashkuprizada*; *Müʾeyyedzade* not *Muʾayyadzada*. However, it is not always easy to determine the linguistic context in which an individual flourished. This is particularly relevant for the period examined in the book (i.e., from the mid-fifteenth to the late sixteenth century) and the actors studied (i.e., émigré scientists). The constant movement of learned individuals from one geographic context to another, where cultural and linguistic boundaries were more fluid than they are today, presents a significant challenge. Can we categorize *Qadizada al-Rumi* or *ʿAli Qushji* as Ottoman astral experts, when the cultural and political concept of being "Ottoman" was only in the making and when these individuals gained prominence outside of the Ottoman context? Bearing this challenge in mind, in cases where the name, sobriquet, and background of the individual suggest their primary linguistic context, I follow the transliteration conventions of that language. Thus, *Taqi al-Din*, not *Takiyüddin*; *Salman-i ʿAjam*, not *Selman-ı ʿAcem.*

Place names are rendered in their English equivalents whenever possible, and dates are given according to the Common Era unless otherwise noted.

ACKNOWLEDGMENTS

Thank you for your interest in reading this book, which took many years to come to fruition. Over these years—marked by the joy of raising a newborn, the grief of losing immediate family members, the shock of a global pandemic, and phases of existential questioning—I have accrued countless debts to numerous people who offered their time, guidance, and support. While a few lines of thanks cannot fully repay them, I offer my heartfelt gratitude to everyone who has helped me on this journey.

Contrary to the conventions of acknowledgments, where the tribute to one's family is often reserved for the end, I want to begin by acknowledging my wife and fellow traveler in life, Sümeyye Yar. If you are holding this book in your hands, it is thanks to her unwavering support, tremendous vision, patience, and enormous sacrifices that made its completion possible. Our dearest son, Devin Ulaş, may have taken up much of the time that could have been spent finishing this book earlier, but nothing has been more important to us than watching him grow in good health and experiencing the joy he has brought to our lives.

My amazing colleagues, students, and staff in Columbia University's History Department helped me smoothly transition into a new job at a new institution and significantly broadened my intellectual horizons as a historian. To Manan Ahmed Asif, Pamela Smith, and Rashid Khalidi, whose generous sharing of their time and knowledge has greatly enriched my development as a historian, I am endlessly appreciative. Together with them,

Adam Kosto, Mark Mazower, Marwa Elshakry, and Rhiannon Stephens kindly read and provided feedback on draft chapters, which spurred me to refine my approach and sharpen my narrative. I am honored and privileged to be their colleague. Caterina Pizzigoni, Catherine Evtuhov, Elisheva Carlebach, Gregory Mann, Marc Van De Mieroop, Michael Stanislawski, Neslihan Şenocak, Pablo Piccato, Susan Pedersen, and Zeynep Çelik offered guidance whenever I needed it. I thank them for everything I have learned from them. The members of our "junior faculty" crew— Ali Karjoo-Ravary, Alma Steingart, Hannah Farber, James Stafford, Sailakshmi Ramgopal, and Yana Skorobogatov—created a mental fun zone where we shouldered together the challenges of being on the "track." Students enrolled in the courses I have taught at Columbia, from contemporary civilization and dissertation seminars to lectures on Ottoman history and seminars on occult sciences or manuscripts, have contributed to this book in various ways with their stimulating intellectual engagement, but I would like to specifically acknowledge Yasemin Akçagüner for reading portions of this work and raising insightful questions. I also thank my colleagues from other departments at Columbia, including Brinkley Messick, Gil Eyal, Mana Kia, Najam Haider, Pier Mattia Tommasino, Seth Kimmel, and Wael Hallaq for their intellectually enriching conversations. The members of the outstanding administrative team at Fayerweather Hall, especially Lawino Lurum, Michael Adan, Patricia Morel, Patrick McMorrow, and Theo Cutler assisted me with administrative matters and housekeeping tasks. I thank them for their professionalism and readiness to help.

The journey of this book began in Chicago. Cornell H. Fleischer (1950–2023) was the main reason I went there. His historical and philological erudition, intellectual and ethical guidance, and genuine care for his students, friends, and peers have been a true model for me. It saddens me that he could not see this project realized as a book, though every part of this text carries an echo of our endless conversations. In Chicago, I learned immensely from the philological and historical precision of Hakan Karateke, John Woods, and the late Franklin Lewis, as well as from the warmth and sharp insights of Ahmed El Shamsy, Fred Donner, Holly Shissler, Orit Bashkin, and Persis Berlekamp. I also thank Kay Heikkinen, Muhammad Eissa, and Saeed Ghahremani for spending many hours to introduce me to the beauty and challenges of Persian and classical Arabic. I am likewise grateful for the companionship of friends and colleagues in and from Chicago, including Abdurrahman Atçıl, Alidost Numan, Amir Toft, Basil Salem, Carlos Grenier, Christopher Markiewicz, Emin Lelić, Evrim Binbaş, Ferenc Csirkés, Judith Pfeiffer, Kaya Şahin, Mehmetcan Akpınar, Mohamad Ballan, Nazlı İpek Hüner, Nikolay Antov, Nükhet Varlık, Rao Mohsin Ali Noor, Sinem Eryılmaz, Tolga Cora, Toygun Altıntaş, William Walsh, and Zahit Atçıl. I thank them all for their contributions to my scholarly development.

The year I spent in Leiden was not the most pleasant in memory, owing to the family medical emergencies that occurred then, but my conversations there with Ahab Bdaiwi, Cumhur Bekar, Erik-Jan Zürcher, Gabrielle van den Berg, Hasan Çolak, Jan Schmidt, Jeroen Duindam, and Petra Sijpesteijn helped me think more deeply and widely about this project.

Jamil Ragep and Sonja Brentjes closely read several draft chapters and provided astute comments and critiques that saved me from making blatant mistakes and pushed me to rearticulate some of my arguments in the book. Benno van Dalen, David King, Ekmeleddin İhsanoğlu, Gülru Necipoğlu, İhsan Fazlıoğlu, Julio Samsó, and Serpil Bağcı generously responded to my inquiries, whether related to specific procedures and phenomena in the history of the astral sciences or to the history of art and architecture. Ali Yaycıoğlu, Baki Tezcan, and Robert Morrison kindly participated in a manuscript workshop that I held at Columbia University, together with Adam Kosto, Marwa Elshakry, Pamela Smith, Rhiannon Stephens, and Yana Skorobogatov. They offered invaluable detailed feedback and encouragement. Hasan Karataş and Hasan Umut sent extensive notes about my work, which I occasionally revisited to sharpen my arguments. Boris Liebrenz, Gaye Danışan, Mehmet Arıkan, and Menderes Velioğlu graciously shared some of their findings from their unpublished research. İsmail Emre Pamuk assisted in locating and browsing several registers in the state archives. The anonymous reviewers at Stanford University Press provided incredibly constructive feedback, helping me articulate some arguments more forcefully and avoid several mistakes. I am deeply grateful to all for their time, labor, and attention.

I presented parts of this work at various venues, for which I am grateful to my colleagues for their invitations and to the audiences for their critical engagement. Matthew Melvin-Koushki cordially invited me to several of his panels and conferences on Islamicate occult sciences. The "Science and Craft" conference at the Warburg Institute, organized by Godefroid de Callataÿ and Liana Saif; the "Ilm wa-Amal" conference at Stanford University, hosted by Ali Yaycıoğlu and Tuna Artun; the "Current Trends in the History of Science in Muslim Societies" workshop, led by Justin Stearns and Nahyan Fancy at NYU; the "Ottoman Science from a Global Perspective" conference organized by Harun Küçük and Oscar Aguirre-Mandujano at UPenn; the "Global Shapes of Knowledge: Towards a History of Research" conference organized by Peter Miller and Vera Keller at the University of Chicago's Institute on the Formation of Knowledge; and the two separate workshops of the New York History of Science Society, one hosted by Meira Gold and the other by Sean O'Neil, were particularly valuable for refining my arguments before a critical audience. I would also like to thank Michael Reynolds at Princeton; Cemal Kafadar and Jun Akiba at Harvard; Lawrence Principe and Lijing Jiang at Johns Hopkins; the conveners of the Mongol Zoominar series, Jonathan Brack, Michal Biran, and Qiao Yang; Nikolay Antov at the University of Arkansas; Günhan Börekçi at the former Şehir University in Istanbul; Cengiz Özdemir and Ozan Sağsöz at Medyascope TV; Derya Gürses Tarbuck at Açık Radyo; and Luís Ribeiro, the host of the *Ad Astra* podcast series, for giving me the opportunity to present portions of my work and test my arguments.

Parts of Chapter 3 were adapted from my previously published article, "Reading the Stars at the Ottoman Court: Bāyezīd II (r. 886/1481–918/1512) and His Celestial Interests," in *Arabica* 64 (2017): 557–608. I am grateful to the journal's editors and to Brill's acquisitions editor, Abdurraouf Oueslati, for granting permission to reproduce and adapt this material.

My special gratitude goes to Marinos Sariyannis, who included me, along with a team of amazing scholars, in his multiyear ERC (European Research Council) project, GHOST (Geographies and Histories of the Ottoman Supernatural Tradition). The workshops, conferences, and conversations we had throughout this project were crucial in shaping the scope and methodology of this book.

In addition to funding from the ERC, the generous support of the Social Science Research Council and the American Research Institute in Turkey granted me the privilege of conducting extensive research in manuscript libraries and archives in Turkey and Europe. Columbia University's Department of History and the Graduate School for Arts and Sciences provided substantial material support for my research and writing, including the History Department's Career Development awards and Manuscript Workshop funds, the Lenfest Junior Faculty Grant, and the Hettleman Junior

Faculty Summer Grant. I am indebted to the staff of all the research sites I visited in Turkey and Europe, as listed in the Bibliography, but I would like to specifically acknowledge Tahsin Tahaoğlu for his tremendous help in granting me access to manuscripts at the Kandilli Observatory Library.

I cannot express enough gratitude to William Walsh (*aziz Bill*), who was always ready to offer his tremendous linguistic skills, rhetorical prowess, and historical knowledge to help correct and refine my writing. Two different writing groups—one with Ayşe Baltacıoğlu-Brammer, Emin Lelić, Kameliya Atanasova, and Oscar Aguirre-Mandujano and the other with James Ryan, Madeleine Elfenbein, and Zoe Griffith—could not last long after COVID-19, but the limited sessions we held were still invaluable in helping me identify what was working and what was not in some of my chapters.

Ali Yaycıoğlu and Nükhet Varlık, the editors of the Stanford Ottoman World series, were extremely supportive even before I submitted my book proposal. Kate Wahl, editor-in-chief at Stanford University Press, believed in the book, provided invaluable support, and made critical interventions that strengthened the manuscript. I want to thank them all for taking on this project.

Members of my extended family in Turkey—my mom, Türkan Şen; my elder sisters, Tuba Uztuğ and Tuğgen Erol; my mother-in-law, Fatma Yar; my brothers-in-law, Ferruh Uztuğ, Ali Erol, and Fatih Yar; and my nieces and nephews, Nehir, Ege, Kuzey, and Ada—were incredibly supportive and understanding of the time I had to steal from them while working on or worrying about this book.

This book is dedicated to the cherished memory of my father, Tayyar Şen, now at eternal rest in his small village in Ordu. From my early childhood, he granted me extraordinary freedom, fully trusting my judgments and supporting every decision I made. Without that freedom and trust, I could never have made the abrupt decision fifteen years ago to leave a potentially lucrative career as an economics major and become a historian. It brings me some comfort that he at least saw me complete my dissertation and secure a job. He passed away in 2017. I am sorry that I cannot see him hold this book, flip through its pages without understanding a word, and feel proud of his son.

FORGOTTEN EXPERTS

INTRODUCTION

EMPIRE *of* EXPERTS, EXPERTS *of* EMPIRE

Let us journey back in time to attend an imperial ceremony held in Istanbul in June 1550. We are sure of its year and month but not of its day and hour. We also know those present at the site. Among them are the Ottoman sultan Süleyman the "Magnificent" (r. 1520–66) and the members of his close retinue, an assembly of ruling elites including the chief religious authority of the time and other prominent members of the scholarly establishment (*ʿulamaʾ*). Historians, poets, architects, masons, scribes in the chancery, and common city-dwellers are also present, alongside the court "astrologers" (Ar. *munajjim*, Tur. *müneccim*). This grand ceremony was convened to celebrate the laying of the foundation stones for a new mosque complex that the aging sultan wished to erect atop the city's third hill to memorialize his imperial accomplishments and crown his lasting renown. Named in honor of its benefactor, the Süleymaniye Mosque remains an iconic fixture in Istanbul's skyline as one of the city's majestic landmarks, visited daily by thousands of observing Muslims and annually by millions of tourists. Few contemporary visitors know, however, that the mosque's construction began when the chief religious authority and head of the scholarly establishment, Ebu's-suʿud Efendi (d. 1574), laid the foundations at the precise moment calculated and reported as the most auspicious by court astrologers.[1]

Let us fast forward thirty years and attend another ceremony in the same city in early 1580. Unlike the fanfare heralding the previous imperial construction, this event is marked by a thunderous pandemonium of destruction. The object of this demolition is a newly erected edifice that some contemporary Ottomans deemed "the wellspring of the disruption as the End Times approach."[2] The building, bombarded with cannonballs fired from the seaside by the forces of the grand admiral, is none other than an observatory built only a few years earlier, after the chief court astrologer of the time, Taqi al-Din (d. circa 1585), persuaded the reigning sultan Murad III (r. 1574–95) to generously fund his scientific venture.[3] Yet, in a cruel twist of fate, the very same sultan would issue a decree within a few years, ordering the observatory's destruction. This drastic reversal was preceded by a fatwa (legal opinion) of the Shaykh al-Islam Kadızade Şemseddin (d. 1580), a student of Ebu's-suʿud Efendi, who reportedly warned the sultan in a letter that no country in the world that harbored an observatory and dared to scrutinize celestial spheres had ever prospered.[4] Its destruction not only erased the building from the city's visual topography but also eradicated the observatory from the collective memory of the Ottoman public and literati for centuries to come. It was not until the twentieth century that the observatory and its dramatic end resurfaced as a discursive tool within the prevailing science-versus-religion dichotomy and assumed a central role in grand narratives about the Ottoman decline in science and technology. The shared belief among many modern observers, academics and laypeople alike, is that had those bigoted religious scholars tolerated the observatory and allowed Taqi al-Din to continue his scientific explorations, the Ottoman Empire—and perhaps even the entire Muslim world—might have avoided its ultimate destiny of lagging behind Western/European countries scientifically and technologically from the sixteenth century.[5]

What accounts for this apparent change, where the expertise of a particular class of learned individuals—specifically, court astrologers—was ardently sought and conspicuously honored by imperial and religious authorities in one instance but condemned and repressed in another, just thirty years later? Was this ostensible shift in attitudes toward *munajjims* and their practice linked to the presumed changes in the Ottoman Empire's political culture, social order, and religious orientation from the mid- to the late sixteenth century, which Ottomanists explain as Ottomans' increasing embrace of a stricter Sunni Islamic outlook throughout the sixteenth

FIGURE 1. The Süleymaniye Mosque complex, late sixteenth century. Oxford Bodleian Library Ms. Or. 430, 131a. Courtesy of the Bodleian Library, Oxford.

century?[6] Or should we instead seek the roots of this shift primarily in the varying dispositions and intellectual inclinations of different individual historical actors, who, of course, were not unaffected by the sociocultural dynamics and intellectual currents of their times? Perhaps there was not a marked contradiction at all, given that *munajjims*, as practitioners of an ambivalent "science" (*ʿilm*; *episteme*) or "craft" (*sinaʾat*; *techne*) that was perceived as both useful and controversial at the same time, were compelled to practice their profession on such shaky ground. If this indeed was the true state of affairs, then what did it mean for its aspirants to acquire mastery and pursue a career in a field they understood to be both denounced and, paradoxically, in demand?

Using these questions as a springboard and diving into the words and worlds of *munajjims* from the Ottoman court in the tenth century of the Islamic calendar—which spans from the late fifteenth to the late sixteenth century of the Common Era—*Forgotten Experts* narrates the untold story of a group of scientific experts whose epistemic authority and socio-economic standing constantly ebbed and flowed, often with the ever-changing tastes and priorities of decision-makers. Contrary to prevailing scholarly and popular convictions about *munajjims*, which either portray them as unlettered fortune-tellers and charlatans relaying spiritual utterances about the future or grant them supreme prophetic authority, casting them as key political players and primary producers of millenarian and apocalyptic discourse (notably significant in the tenth-century Islamicate context, as this period marked the end of the first Hijri millennium), *Forgotten Experts* suggests that we should neither ascribe more significance and prestige to them than they held in their time nor undervalue the competence they possessed. We need to examine the ambivalence inherent in their expertise that led to the ebbing and flowing tides of appeal to their service.

Although ambivalence seems to be an unchanging, universal, and hence ahistorical characteristic of astrological practice, taking the long sixteenth century in the Ottoman courtly context as a case study helps trace those ups and downs more sharply. *Forgotten Experts* argues that while the reign of Sultan Süleyman, known for the prevalence of millenarian and apocalyptic discourse bolstered by the use of various occult sciences, was not particularly auspicious for *munajjims*, the earlier decades under Bayezid II (r. 1481–1512), often considered a devout figure aloof from science and innovative ideas, were indeed their heyday. Bayezid II's reign is of special significance, as it marked the appearance of a permanent corps of astrologers under royal patronage, a structure that remained in operation for over four more centuries, playing a pivotal role in institutionalizing the transmission of the corpus of *munajjims*' expertise across generations.

The chronological span covered in *Forgotten Experts* was chosen as follows. The 1450s are conventionally cited as the beginning of Ottoman imperial formation. While plans for and the desire to create a formidable empire with a robust administrative structure and staff trace back to the founding generations of the Ottoman polity in the fourteenth century, this vision only began to materialize in practice from the mid-fifteenth century, particularly after the epochal conquest of Constantinople—which marked the end of the (Eastern) Roman Empire, or its passage to a new set of (Muslim) Caesars.[7] The century following the 1450s is typically viewed as a period during which the Ottoman polity transformed from a relatively minor regional power to a dominant global force. This transformation was evident in the expansion of its territories, population, and administrative personnel, and the challenges of governance that came with them, especially after the gradual integration of extensive territories, stretching from Bosnia to Basra.[8] The decision to choose the closing years of the sixteenth century as the other end of the chronological span is more arbitrary, but it is convenient and pertinent to conclude it with the construction and subsequent destruction in 1580 of the Istanbul observatory, as well as the end of the first Islamic millennium by 1592.

Just as the 1450s were not year zero for the presence of *munajjims* at the Ottoman court, the destruction of the observatory in 1580 was not the end of their story. *Munajjims* practiced their science both before and after our long sixteenth century. Indeed, many more sources are available from later centuries, as the unit of court astrologers continued to operate until the official dissolution of the Ottoman Empire in the aftermath of World War I. While these sources could have no doubt enriched the narrative, this enrichment would have been more a matter of quantity rather than quality. In other words, the sources about *munajjims* from the fifteenth and sixteenth centuries are already richly adequate to reconstruct their story, which is best characterized by ambivalence inherent in the acquisition, application, and reception of their expertise. The more than century-long sketch of it (from 1450 to 1600) further enables us to locate their story within the specific context of the Islamic millennium and test certain scholarly convictions as to the growing influence of millenarian and apocalyptic currents fueled by the use of various occult practices over the long sixteenth century.

Ambivalence permeates nearly every aspect of *munajjims*' expertise, from their learning to their practice. It was a kind of expertise both transmitted and omitted, prized and stigmatized. *Munajjims* demonstrated technical proficiency in the mathematical sciences, astronomical knowledge, and astrological techniques but were often criticized, even caricatured—both in the medieval and early modern periods and in contemporary times—as simpletons lacking reliable bodies of knowledge. The knowledge they needed to acquire for their profession, especially as related to astrology, was text-centered and in open circulation, but it was not commonly taught within the formal educational settings of madrasas. *Munajjims*' services in the form of calculating auspicious hours, casting horoscopes, annually crafting almanacs, or providing on-site astrological counsel appealed to both royals and the public, but it was also met with skepticism from various segments of society, even including *munajjims*' self-doubt about the limitations of their science. Last but not least, *munajjims* were not the sole experts in the domain of predicting the future. There were other, sometimes competing actors, so-called occult practitioners or masters of esoteric knowledge, including dream interpreters, geomancers, experts in the science of lettrism (*ʿilm al-huruf*), and mystics claiming to possess some mantic power, whose authority relied upon distinct bodies of knowledge and hence occasionally came into conflict with *munajjims*' expertise.

While there is a certain convenience in using blanket terms for intelligibility and accessibility, relying too heavily on catch-all phrases such as the occult or esoteric sciences or categorizing *munajjims* simply as yet another group of occultists risks missing the important nuances between the scope and purpose of the respective expertise of different practitioners. The framework of expertise thus serves as a useful analytical lens for discerning the distinctions between different types, registers, and actors of knowledge. This perspective enables us to pinpoint how, where, when, and why the varied authority claims of different practitioners of the divinatory sciences surfaced and came into conflict.

No doubt there were *munajjims* before and after the sixteenth century and beyond the narrower confine of the Ottoman imperial court, but no other Islamicate context provides such extensive and variegated archival

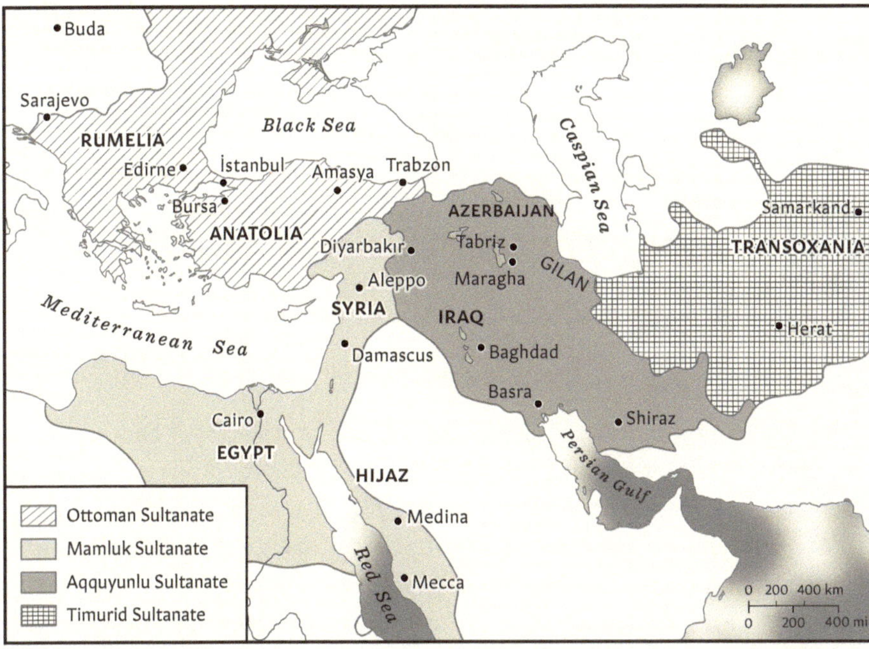

FIGURE 2. Western Asia in the second half of the fifteenth century.

evidence to reconstruct what it meant to be a *munajjim* and how their practice differed from that of other comparable, competing experts. The sheer size of this archival evidence is intrinsically linked to the institutionalization of *munajjims*' expertise through the creation of a permanent unit, even an office, in Ottoman court bureaucracy by the late fifteenth century. This unit was up and running from the late fifteenth to the early twentieth century, and their service was in consistent demand by imperial decision-makers. Yet both the number of *munajjims* employed at any given time and their salaries fluctuated over the years, reflecting the evolving extent of fascination with *munajjims* and their expertise. Contrary to what one might expect, during Süleyman's long, millenarian, and magnificent reign, there was a marked decrease in the number of court *munajjims* staffing the unit, their compensation, and the overall volume and variety of scientific production related to their expertise. This is especially evident when contrasted with the almost equally long term of his grandfather, Bayezid II. The particular chronological focus of *Forgotten Experts*, the period from the mid-fifteenth to the late sixteenth century, limns this waning interest in the expertise of *munajjims* and refines contemporary scholarly views concerning millenarian and apocalyptic currents at the time and the specific role assigned to the occult sciences therein. Here, the framework of expertise once again proves useful, as *munajjims*, owing mostly to the particular constraints embedded in the epistemological foundations of their expertise, rarely took part in the production of millenarian and apocalyptic discourse.

Empire of Experts, Experts of Empire

The wealth of archival documents pertaining to *munajjims*—and indeed similar technical-scientific experts such as physicians, architect-engineers, and others—sets the Ottoman Empire apart from other medieval and early modern Muslim-ruled polities, including the Abbasids, Fatimids, Timurids, Safavids, and Mughals, who also employed court *munajjims* yet left few archival materials behind.[9] This abundance is a direct result of the early modern Ottomans' greater success in utilizing and organizing expertise through a bureaucratic web of office-like entities and positions, as well as in preserving the paper trail over centuries.[10] In the late fifteenth century, as part of the Ottomans' ongoing imperial formation and bureaucratization, a series of administrative units and positions appeared for the first time, as manifested by archival documents and registers. These bodies included the corps of physicians, architect-engineers, and astral experts, among others. Each of these units was led by a practitioner designated as head or chief, denoted in Ottoman bureaucratic correspondence with different combinations of Arabic, Persian, and Turkish words that all had the same meaning of "head" (i.e., *reʾis*, *ser*, and *baş(ı)*, respectively, in Turkish orthography).[11] At first glance, these attendants' employment might appear no different from the typical case of a skilled individual providing consultation and service to a medieval ruler and his extended household.[12] When examined closely, however, a marked drive toward the creation of something far more intricate, impersonal, and public-facing is apparent. Not only was their duty extended from being solely private (i.e., serving only the ruler and his immediate family) to encompassing the public (i.e., addressing the needs of the subjects of the empire whenever required). But the administration of the modus operandi of their profession, across almost the entirety of the vast expanse of the empire, was delegated to the appointed head. This means that the individual occupying the position and hence standing in the status of the chief expert was expected not just to offer regular service in his profession; he was also entrusted with the responsibility of evaluating the merits and qualifications of practitioners within his field and making appointments accordingly.

To take one example, consider the following order dispatched on 5 September 1573 from the Imperial Council to the judge of Istanbul.[13] The order opens with a reference to previous correspondence from the incumbent chief physician, Garsuddinzade Muhyiddin (d. 1581), who had lodged a formal complaint with the Imperial Council. His grievance lay with certain practitioners in the capital and other provinces who, in his view, had been prescribing harmful medicines and toxic purgatives, in clear violation of the established principles of the science of medicine. Following his petition, Garsuddinzade undertook to "test" those practitioners in their crafts.[14] In a later report that he apparently sent to the Imperial Council, he urged that those practitioners whose skills did not meet with his approval should be prohibited from practicing further. The judge, as the city's chief law enforcement authority, was commanded by the Imperial Council to inform those medical practitioners that they would be required to submit to examination and authorization by the chief physician, failing which they would lose their right to practice.

What we see here provides a striking illustration of the significance attributed to a practitioner whom imperial authorities designated, either on merit or at random, as the chief expert within his respective profession. It also exemplifies the deep trust in his expertise shown by both the political decision-makers and the general public. Moreover, the example demonstrates the intricate web of interlinked offices and positions—here, the Imperial Council, the office of the judge of Istanbul, and the chief physician's office. This network facilitated a steady exchange of correspondence, meticulously documented within the early modern Ottoman bureaucracy. Ottoman registers of imperial affairs (*mühimme defterleri*, lit. "registers of important things"), along with the *Ruʾus* records (lit. "Heads" or "Persons," referring to registers of official appointments), abound with similar examples.[15] These documents, both of which started to be systematically compiled from the mid-sixteenth century, with their precursors emerging earlier in the same century, illuminate the imperial court's consistent reliance on the expertise of chief physicians, chief architect-engineers, chief astrologers, or other chief functionaries, whether for appointments of practitioners in their specific domains or for affairs of public interest.

Despite all the relevance of these early modern Ottoman examples, the terms "expert" and "expertise" may still strike one as decidedly modern and potentially anachronistic when applied, prior to the nineteenth century, as historical categories to identify individuals with specialized knowledge and examine their relationships with decision-makers and the general public.[16] Such reservations and concerns are not unfounded, given the evident institutional, infrastructural, and intellectual disparities between societies of the post-nineteenth-century era and those of earlier periods with respect to questions of expertise. For example, the educational institutions and the prevailing epistemological paradigms in earlier centuries rarely advocated for professional specialization. Although books on the taxonomies of the sciences in the medieval Islamicate world classified knowledge and codified distinctions across disciplines with regard to their methodologies and objectives, these different branches of knowledge were often viewed as intimately intertwined facets of a unified episteme, and aspirants of knowledge were encouraged to expand their intellectual breadth to the level of a polymath instead of specializing in a single discipline.[17] Moreover, save for a few well-documented instances, such as those observed in medieval and early modern China, the notion of regular, centralized examinations designed to recruit professionals and meet the needs of administrative divisions was largely absent in the medieval and early modern eras.[18] One must also bear in mind such legitimate questions about whether short- and long-term policy-making based on expert projections ever existed in medieval and early modern times or whether it is plausible to speak of a public that actively sought experts' opinions during a period when the means and media of mass communication were limited or missing altogether. Last but certainly not least, even a cursory foray into the conceptual history of these terms in Anglophone literature, based on the testimony of reliable historical etymology dictionaries, reveals that the earliest usage of "expert" as a noun, signifying "a person regarded or consulted as an authority on account of special skill, training, or knowledge," only dates back to the first half of the nineteenth century.[19]

The very same etymology dictionaries also disclose, however, that the word "expert" was used as an adjective much earlier, with the earliest documented reference in English dating back to the fourteenth century, to mark those who were "trained by the experience of practice; skilled and skillful."[20] If we move away from an Anglophone-centric perspective and open ourselves to exploring concepts prevalent in other literary cultures, we may have a more representative sample of terms related to experts and expertise. Indeed, there is no shortage of terms from the medieval Islamicate lexicon that point to categories bearing a striking resemblance to expertise or experts. Consider, for instance, such Arabic- and Persian-derived Ottoman Turkish words and phrases as *maharet* (denoting special skills in a particular field; *mahir* [sg.] and *mehere* [pl.] as those possessing such skills), *hazık* (signifying somebody able to skillfully perform a task, often used with exclusive reference to physicians), *habir* or *ehl-i hibre* (a person or people, respectively, informed in a particular area; Ar. *ahl al-khibra*), *üstad* (master; Per. *ostad*), and *ehl-i vukuf* (people with discernment; Ar. *ahl al-wuquf*), which were frequently employed by historical actors in a wide array of texts and disciplines.[21] If these terms do not encapsulate the notion of experts and expertise, what exactly do they signify?

The underlying contention of this book is that, notwithstanding the strong modern connotations associated with the terms "expertise" and "expert," they deserve to be used in regard to medieval and early modern contexts, not even merely as heuristic devices and analytical tools but also as authentic categories of historical actors. Scholars of early modern Ottoman history have already deployed these concepts, albeit in a casual fashion, in a broad range of studies, from environmental history and the history of science to empire formation and military history. For instance, environmental historians of early modern Ottoman and Middle Eastern societies have made references to "experts"—those local, knowledgeable people upon whose skills and *metis* the Ottoman central and provincial authorities depended when confronted with matters related to irrigation and the extraction of natural resources.[22] Historians of medicine too have employed these concepts when discussing, for example, the gradual shift of "expertise over plague" from the realm of religious authorities to that of state agencies in the sixteenth-century Ottoman world.[23] One likewise encounters references to "military" or "artillery" experts or "expert[ise] in navigation and cartography" in studies devoted to early modern Ottoman military history and the production of geographical knowledge harnessed in Ottoman territorial expansion.[24] The appeal to the views of experts in qadi courts appears frequently in Ottoman legal sources.[25] The literature on early modern Ottoman bureaucratization and state formation also evinces a heightened sense of "bureaucratic expertise" from the sixteenth century onward, manifest primarily in the specialized division of offices and the codification of protocols regarding the appointment and promotion, though not always the training, of individuals.[26] The terms "expert" and "expertise" further feature in scholarship pertaining to the social history of artisans and craftspeople, who were in possession of a specific type of practical or "tacit knowledge" deemed beneficial for both the state and society.[27] Despite their references to experts and expertise, these studies seldom delve into the more abstract questions regarding what constituted expertise, who were considered experts and by whom, what personal strategies and institutional mechanisms individuals could leverage to attain and assert their putative expertise, and how these individuals established, negotiated, and disputed their relationships with decision-makers as well as other experts within their particular field or relevant area.

Was the early modern Ottoman Empire an empire of experts? The studies mentioned above hint that it was. Some other works go even further, asserting that the early modern Ottoman Empire, in particular the "first" one from the mid-fifteenth to the late sixteenth century, functioned as a meritocratic state, rationally entrusting offices to those who deserved them based on their expertise and qualifications.[28] Otherwise, it would be difficult to fathom, as staunch advocates of the Ottoman meritocracy thesis contend, how the Ottoman ruling elite managed to establish one of the most formidable and enduring world empires of the day, a polity that remained resilient even when monumental changes from the early seventeenth century in ecological, economical, and societal terms threatened its foundations.[29]

The romanticized belief in the early modern Ottoman Empire as a meritocratic polity stems in part from Ottoman sources, particularly the political tracts and advice literature authored by disillusioned Ottoman literati of the late sixteenth and seventeenth centuries. Faced with the combined adverse effects on the Ottoman social and political order of inflation, environmental catastrophes, rapid population movements, and stagnation in territorial expansion, these learned figures looked back on their past (i.e., the period from the mid-fifteenth through the late sixteenth century) through an undeniably nostalgic lens.[30] Their disillusionment, spurred mainly by personal career concerns, was articulated within a discourse of distress concerning the present and future trajectory of their empire. Figures such as Mustafa ʿAli (d. 1600) professed that in the most glorious and judicious times of Ottoman rule—for him, this was the reign of Mehmed II (r. 1444–46; 1451–81)—merit was the sole criterion the Ottoman ruling elite considered.[31] In the view of these literati, key positions had been assigned to those who demonstrated their aptitude and expertise for the tasks at hand. For many of these "Ottoman observers of Ottoman decline," who were obsessed with the ideal of justice and order, these notions meant nothing more than preserving existing social hierarchies by keeping every individual in the position they deserved.[32] They maintained that from the late sixteenth century onward, this presumed meritocratic culture began to dissolve, and favoritism and nepotism became the norm and disrupted the order.

Notwithstanding the subjective and ever-changing assessment of early modern Ottoman observers concerning their empire's stance toward merit and expertise, the Ottoman Empire, much like its early modern counterparts across Eurasia, was a polity teeming with experts.[33] But who, one might ask, were these experts of the empire? They were individuals whose erudition, technical skills, and experiential knowledge were sought after by decision-makers, even when this know-how, as in the case of *munajjims*' expertise, was not always held in the highest regard or commonly taught within formal institutional settings. These experts were not specialists in the modern sense of the word, nor did they qualify as professional scientists. Many actors had or claimed to possess a level of erudition that was not confined to a single field. Indeed, it is this versatility and polymathic character that enabled them to diversify and enhance their claims of authority and expertise as circumstances dictated.

The era primarily under scrutiny in this book—namely, the long sixteenth century, stretching from the mid-fifteenth to the turn of the seventeenth century—was a time when imperial bureaucracies were emerging worldwide. It was a period when the idea and practice of governmental ministries, along with their attendant offices and agencies, were on the horizon, and the demand for competent individuals to assist in the governance of imperial affairs was steadily mounting.[34] These affairs ranged from securing military organization and the management of taxation and finances to the execution of architectural projects and provision of public healthcare. All of these areas and more were burgeoning spatially, bureaucratically, and conceptually in tandem with the empire's gradual maturation and sophistication. In this respect, the term experts employed here encapsulates a broad spectrum of learned practitioners and craftspeople, from physicians, architect-engineers, and navigators to artillerymen, scholar-bureaucrats, scientists of stars, and others. Each of these individuals, within their peculiar roles, capacities, and expertise, contributed to the diverse tapestry of early modern Ottoman governance.

In the relative absence of institutional and impersonal means to certify one's expertise in a given field during the period under study, the recognition and endorsement of imperial authorities through the distribution of positions, titles, and honors assumed crucial importance. After all, expertise was—and still is—more about the recognition one received than an inherent trait one possessed.[35] In that regard, early modern imperial authorities and decision-makers were the creators of experts, inasmuch as they elevated particular individuals to the expert status once they capitalized on their skills and assigned them specific roles and titles. At a time when certificate-issuing institutions, licensing mechanisms, centralized examinations, and print media that could publicly affirm one's mastery in a given field were limited, the recognition by imperial authorities was the primary if not the sole way to affirm an individual's erudition and expertise in a particular domain.

The nature of the relationship between empires and experts was symbiotic. While empires relied on the services of skilled individuals, the experts sought validation and recognition from the imperial authorities to authenticate their expertise. This recognition was ultimately conferred by the sultan, grand vizier, or another high-ranking official with executive authority. There were undoubtedly intermediaries, or brokers of expertise, as one might say, who vouched for the individual's skills and usefulness—which might already have been the word on the street—and introduced them to the ultimate decision-makers. The eventual endorsement was formalized through the issuing of an official "certificate" (*berat*), detailing the individual's unique qualifications within their realm of expertise.[36] Unfortunately, the number of surviving certificates that must have been given to individuals with a demonstrated ability or expertise in a specific art or science is rather sparse. Still, the copy of the *berat* apparently given to Taqi al-Din before the establishment of the observatory, identifying him as "the master of the masters of astronomy and star[s], the pinnacle of the experts of philosophy and almanacs, the knower of the decrees of solar and lunar motions," is a precious specimen that discloses the Ottoman vocabulary of early modern expertise in the astral sciences.[37] Notwithstanding the dearth of extant certificates, abundant references to the practice of distributing *berats* and certifying one's expertise in a given field are available in other historical sources. These include chronicles, biographical dictionaries, and manuals of epistolography that contain model letters or copies of authentic correspondence, all of which testify to the act of credentialing expertise.[38]

The question remains whether these official appointments and nominations were made purely on objective grounds, a query that holds relevance even today. After all, appointments were typically made when the decision-makers deemed a candidate suitable for the position, based either on their own assessment, on some form of an internal review, or on their reliance upon the views of others they trusted.[39] We cannot ignore the influence of personal connections, kinship, and an overall sense of familiarity. However, it is not my intention here to argue that early modern Ottoman imperial authorities were paragons of impartiality in assessing expertise and assigning positions in a perfectly meritocratic manner. Rather, my assertion is that the notion of expertise in the early modern Ottoman context and elsewhere is embodied in the very act of decision-makers creating positions and assigning them to individuals whom they perceived to be capable. By doing so, they validated their expertise and thus generated experts.

Adopting expertise as both an analytical framework and the actors' category thus promises a wealth of insights into the intricate social relations established and maintained between the men of knowledge and the men of rule—or between science and governance—in the early modern Ottoman imperial context. By capturing the crucial nuances and variations between different kinds, registers, and agents of knowledge, the concept of expertise stands as an exceptionally useful lens through which we can identify rivalries and trace competing assertions of authority among different actors. Such competition might surface within a single field of knowledge or profession, with its experts locked in a race for prestige and patronage. In the context of the early modern Ottoman court and its peculiar way of managing expertise, this contest often culminated in the pursuit of attaining the status of chief practitioner. Beyond rivalries within professions, conflicts of authority also emerged between separate yet intertwined fields of knowledge. *Munajjims*, for instance, are typically believed to possess unrivaled authority in predicting the future. However, they often found the domain of prediction contested by other practitioners with expertise in different forms of occult knowledge, each desiring to exercise its own predictive mastery. While modern scholars studying the occult sciences and divinatory arts tend to group practitioners of these fields under a single term and flatten the important distinctions between them, there were considerable variations and instances of competition among various prognosticators. This contention is particularly discernible in the discord between the experts of letter divination and *munajjims*, a topic set to receive a more in-depth examination in Chapter 5.

Munajjims and Their Forgotten Expertise

Munajjims were those astral experts whose specialized knowledge and skills in tracking, timing, predicting, and interpreting celestial movements were sought after by early modern Ottoman imperial authorities and the public. Much like physicians applying their medical expertise to treat diseases or architect-engineers using their theoretical and experiential knowledge for construction projects, *munajjims* provided a wide range of astrological and astronomical services valued by imperial decision-makers. For example, at the onset of the new solar year in March—that is, *Nowruz* (Tur. Nevruz; lit. "the new day")—*munajjims* submitted annual almanacs (Ar. *taqwim*, Tur. *takvim*) that included general astrological prognostications, comprehensive calendrical information, and monthly astrological advice. They also routinely reported astrologically auspicious hours for commencing planned events, be they military campaigns, construction projects, or ceremonies. On occasion, *munajjims* cast birth horoscopes, primarily for the members of the royal family, interpreting the indications of the planetary configurations at the exact moment of birth. It is also almost certain that *munajjims* provided verbal astrological guidance when called upon, specifically during military campaigns or other forms of travel, though capturing such unrecorded counsel is inherently impossible to document. Many *munajjims* also operated as timekeepers (Ar. *muwaqqit*, Tur. *muvakkit*), housed in imperial mosque complexes where they undertook time-related duties such as determining the five daily prayer times or identifying the first day of the month.

Whether compiling annual almanacs, reporting auspicious hours, casting horoscopes, or serving as timekeepers, *munajjims*' peculiar expertise in tracking and interpreting time based on celestial configurations was deemed so indispensable that an official position, even an office, was created within the emerging bureaucratic machinery of the Ottoman court in the late fifteenth century. Extant Ottoman bureaucratic records reveal this court unit's discernible appearance by the late fifteenth century, presided over by a chief practitioner with the title "head *munajjim*" (*müneccim-başı* in Turkish; also known by corresponding Persian and Arabic calques as *ser-müneccim* and *reʾis el-müneccimin*, respectively). This organizational framework helped routinize their service and recruitment throughout the following four centuries, though from the late sixteenth century onward, the chief physician's office took charge of overseeing the appointments of court *munajjims*, hinting at a decline in *munajjims*' standing and professional status. The number of *munajjims* in the unit at a given time varied, sometimes with as many as six and as few as one. While it is true that the unit often consisted of the son(s) and student(s) of the incumbent chief *munajjim*, its members did not always operate on collegial terms.

Munajjims hold a peculiar place in histories of science, considering their attention to mathematical and astronomical precision in computing time, their commitment to the use of instruments and a concomitant desire for technological advances, and their reliance on extensive data about astronomical and astrological variables—collected through computational and empirical methods—before making projections and predictions. Yet despite the undeniably scientific nature of their expertise, *munajjims* remain notably absent from the narratives of science both in the Ottoman context and the broader Islamicate world. Several misconceptions and underestimations concerning the nature and scope of their expertise have stood in the way of their recognition as significant historical actors of scientific practice and noteworthy subjects for historiographical investigation. These misapprehensions and misappreciations about *munajjims* became particularly entrenched in the aftermath of the nineteenth century, when evolving definitions of science and scientist—the latter known to have been coined as a term only in the 1830s—began to draw a stark distinction between what is scientific and what is not, and between who qualifies as a scientist and who does not.[40] *Munajjims* were ultimately relegated to the status of fortune-tellers, crystal-gazers, and newspaper horoscope columnists.[41] While this skewed perception solidified irrevocably in modern times, it is crucial not to forget the medieval and early modern discursive conventions, ubiquitous among the accounts by traditional ʿulamaʾ and Sufi-minded individuals, where *munajjims* were frequently cast as charlatans, devoid of any serious foundation by the epistemic standards of their era. A substantial body of polemics against *munajjims* and their craft, much like the anti-astrology polemics of medieval and early modern Europe, was produced and circulated within the Islamicate and Ottoman domains.[42] Yet focusing solely on the polemics and viewpoints of outsiders—that is, those who were not *munajjims*—runs the risk of overlooking *munajjims*' expertise and their often critical self-reflections about it.

Forgotten Experts, therefore, attempts to restore the agency of *munajjims* and explores their vast paper trail that has been largely overlooked both by modern scholars of Ottoman history and those specializing in histories of science in the Islamicate world. This rich array of sources includes the technical pieces composed and consumed by *munajjims*: annual almanacs, reports on the calculation of auspicious hours, birth horoscopes (*zayicha*), and various astronomical handbooks with tables (*zij*), as well as astrological textbooks. Some of these technical sources even contain *munajjims*' autobiographical sketches of their lives and tribulations that shed light on their personal challenges in and perspectives on their profession. Such details allow us to reconstruct how a skilled but periodically disparaged class of learned people perceived their sciences, selves, and others.

Available sources about *munajjims* and their science are not limited to these sorely unexplored technical texts. There is also an equally wealthy archival treasure of documents and registers providing raw data on the salaries and status of early modern scientific experts, an exceptional Ottoman phenomenon that illustrates the extent and dynamics of these experts' service to the imperial court. Furthermore, the invaluable observations of their contemporaries, spanning a broad spectrum of genres from poetry and theological treatises to chronicles and biographical dictionaries, enrich the gallery of sources. Such an expansive collection of materials by and about *munajjims* offers crucial firsthand insights into "science [and scientist] in action" in the sixteenth-century Ottoman imperial context.[43] Thereby, we can address the types of questions posed at a history of science symposium held over sixty years ago at Oxford University:

> Who were the people taking part in scientific activity?—What were their numbers, education, social position, means of livelihood, personal motives and opportunities, means of communication, institutions? What critical audience was there to be convinced by, use, transmit, develop, revise or reject their conclusions? What social pressures were there within the scientific community itself to affect the consensus of opinion in favor of the old or the new?[44]

Rescuing *munajjims* from oblivion is of paramount importance to a variety of intersecting historiographical fields and inquiries. Since the main protagonists of this book—that is, *munajjims*, their patrons, and their contemporaries who generated discourse about them and their expertise—hail from the fifteenth- and sixteenth-century Ottoman world, this book is primarily and intrinsically grounded in the domain of Ottoman history. But even within the broader canvas of Ottoman studies, *Forgotten Experts* seeks to strike different chords, ranging from the understudied histories of science in the Ottoman realm to the scholarship that grapples with questions concerning imperial formation and bureaucratization during the long sixteenth century.[45] The pronounced focus of this book on the history of science, and more specifically on the history of astrology and astrologers, along with their social, political, and cultural ramifications, also places this work in conversation with parallel trajectories in other regional specializations, such as late medieval and early modern European and Islamicate history. This conversation, however, should not be perceived as a contrived comparison merely to attract the attention of Europeanists and scholars of Islamic studies. Instead, the aim is to demonstrate how and why Ottoman practitioners of science and their experiences, as extracted from the rich textual corpus of *munajjims*, should be considered an integral and inextricable part of the global narratives of late medieval and early modern science, which have often ignored Ottoman episodes and agents of scientific practice.[46]

Science as the Ottomans' Achilles Heel

The marginalization of *munajjims* in modern scholarship cannot be fully comprehended without revisiting and recognizing the relative disregard accorded to the history of science as a subdiscipline within Ottoman studies. In the broader context of modern Ottoman historiography, which has stretched well over a century, the history of science has rarely found itself at center stage, despite the gravity of the science-versus-religion discourse in the prevailing narratives of Ottoman-Turkish modernization.[47] The first half of the twentieth century was marked by an overriding emphasis on political and military history, which, drawing primarily on narrative sources, shaped the main contours of debates surrounding the various pivotal periods of the empire: its rise to prominence in the fifteenth and sixteenth centuries, its ensuing crises from the seventeenth through the nineteenth centuries, and its eventual demise following World War I. The 1950s and 1960s provided greater access to state archives, and the trove of documents they contained instigated a shift toward social and economic history. Subsequently, social and economic history have never lost their commanding appeal and dominance in the scholarly trends of Ottoman historiography, but the linguistic turn of the 1980s ushered in a resurgence of interest in narrative sources, this time from a more critical cultural-history perspective.[48] Amid these evolving academic trends, the history of science has largely stayed on the periphery, scarcely influencing major debates in modern Ottoman historiography except when it was weaponized to fuel contentious discussions about the infamous Ottoman decline, failed efforts toward modernization, and the complex interplay between science and religion.

One of the primary reasons for the marginalization of the history of science within Ottoman historiography is the widely shared presumption about the shortage of Ottoman scientific enthusiasm, curiosity, and achievements.[49] All too often, science came to be viewed as the Ottomans' Achilles heel, the softest spot of the otherwise glorious early modern empire. When compared to their formidable military strength or elaborate administrative organization, which genuinely impressed some early modern European observers, science has seldom been regarded as the Ottomans' standout characteristic.[50] Even for some, including pioneering historians of Ottoman scientific endeavors such as Adnan Adıvar, the early modern Ottomans' military achievements and administrative resilience were not due to but rather in spite of their scientific vigor. This common perception often acted as a deterrent, discouraging many from paying attention to scientific materials produced or disseminated in the Ottoman world, under the assumption that an investigation would hold little to no significance, given the

Ottomans' presumed lack of contribution to the genesis of modern science and technology.

Two particular narratives, prevalent among both scholarly circles and lay readers, have been instrumental in reinforcing these dismissive attitudes and assumptions about the Ottoman scientific patrimony—or its perceived absence. The first narrative pertains to the so-called Golden Age of Islam, which is believed to have taken place from the eighth through twelfth centuries. This narrative maintains that during this period, the exact sciences thrived with the support of enlightened caliphs, leading to the emergence of prominent scientists such as Ibn Sina (d. 1037), Biruni (d. after 1050), and Ibn al-Haytham (d. circa 1040).[51] Still guiding both the learned and lay conversations about the scientific triumphs of the medieval Muslim world today, the Golden Age narrative regrettably implies that the period after its designated time frame, including the time under Ottoman rule, was not conducive to science and scientists.[52] The second grand narrative is that of the Scientific Revolution, which is commonly considered to have created, from the sixteenth century, an insurmountable divide between the West and the rest of the world, including the Ottomans.[53] Both narratives have deeply plagued the ways in which Ottoman scientific enterprises and actors have been treated by scholars and nonspecialists alike. Measured against the acknowledged scientific achievements of the earlier Islamicate and the early modern Western/European domains, Ottoman engagement with science appeared remarkably inadequate and lackluster, often evoking notions of backwardness, belatedness, and decline. These comparative assessments, precipitated by a fixation on seeking—and failing to find—discoveries, inventions, innovators, and contributions to the formation of the modern sciences, have either resulted in a complete disinterest in the Ottomans' scientific past, leading to its sidelining in Ottoman studies, or triggered apologetic and counter-triumphalist narratives.

The notion of decline still looms large in the study of early modern Ottoman intellectual history and scientific enterprises. Despite the vigorous critique the decline paradigm has undergone over the past few decades by several generations of Ottomanists, as well as the proliferation of studies on postclassical Islamicate intellectual history that have challenged the Golden Age of Islam narrative, the concept of decline continues to shape scholarly debates. Nearly every new study in the fields of intellectual history and the history of science is expected to grapple with the question and propose an alternative analytical and chronological framework. The robust critiques of the paradigm in recent decades by historians specializing in political, social, economic, or military history have demonstrated that the concept of decline fails to explain the complex Ottoman realities in the political, social, economic, or military arenas. They suggest that a better analytical framework is needed to portray the remarkable resilience of the empire from the late sixteenth through the twentieth century more accurately.[54] However, it remains unclear how well the proposed alternatives—such as transformation, adaptation, or the idea of "the second Ottoman Empire"—apply to early modern Ottoman intellectual history and the history of science.

Some of the most recent scholarly examinations of early modern Ottoman science and intellectual life perfectly exemplify the persistent scholarly preoccupation with the notion of decline, albeit approached and interpreted in opposite manners. In his *Islamic Intellectual History in the Seventeenth Century*, for instance, Khaled El-Rouayheb contests the assertions of a decline in the rational sciences during the seventeenth century. Instead, by exploring works in logic and rational theology produced at the time, especially those written by Kurdish scholars, he posits that the period witnessed "the reinvigoration of the study of the rational and philosophical sciences."[55] Despite its substantial contribution toward challenging widespread beliefs about the stagnation and decline in Ottoman intellectual affairs from the late sixteenth century, El-Rouayheb's book, which did not explore the astronomical, mathematical, or other technical-scientific literature, leaves unanswered why the intellectual vigor he so substantially documents within the specific context of logic and rational theology should be taken as a sign of revitalization in other, more practice-oriented areas of science and technology.

At the opposite end stands Harun Küçük's *Science without Leisure: Practical Naturalism in Istanbul, 1660–1732*. Küçük thinks there was a decline in his designated time frame and space, but instead of subscribing to outmoded cultural-civilizational claims for its explanation, he attributes it to "the decline of teaching salaries and decline in the material conditions of scholarship."[56] He argues that what he perceives as the absence of theoretical astronomy, theoretical medicine, and natural philosophy in the late seventeenth and early eighteenth centuries resulted from the collapse of the salaries madrasas offered their instructors, which was an inevitable consequence of the pervasive inflationary trends beginning in the late sixteenth century. His provocative argument notwithstanding, Küçük does not address in greater depth why we should consider madrasas as exact counterparts to early modern or contemporary European universities for the teaching and research of the physical and mathematical sciences, or how we might know that madrasa scholars would have pursued different and more "theoretical scientific" interests had they received higher salaries.

The narrative of the perceived dearth and decline of Ottoman scientific enterprises—triggered by the unfavorable comparisons with both earlier Islamicate and early modern European counterparts—has indeed profoundly influenced modern scholarship on the sciences and scientists of the Ottoman world. In an attempt to counterbalance the prevailing presumptions about the negligibility of Ottoman scientific enterprises, some scholars have sought to spotlight those exceptional figures who could measure up to the higher standards set by earlier Muslim or contemporary European scientific geniuses. As part of this strategy, which is all too common in the broader Islamic history-of-science tradition, a few select men of knowledge, who are often dubbed as "individuals well ahead of their times" (*çağını aşan*), have received disproportionate attention.[57] Among these favored figures of modern literature are ʿAli Qushji (d. 1474), a late fifteenth-century mathematician and astronomer;[58] Taqi al-Din, the founder of the Istanbul observatory in the late sixteenth century;[59] Katib Çelebi (d. 1657), a seventeenth-century polymath and encyclopedist with texts and translations on new European knowledge;[60] and İbrahim Müteferrika (d. 1747), who established the first Muslim Ottoman–run printing press in the early eighteenth century and authored several texts on the new sciences.[61] Within this picture featuring only outstanding figures, there has been little curiosity for exploring such supposedly petty practitioners as *munajjims*, the scope of their expertise, and the complex web of their social relations. It is also worth adding that some of these celebrated names, such as ʿAli Qushji, were purposefully labeled as Ottoman scientists even though they were nurtured, educated, and rose to prominence elsewhere, spending only a limited period in Ottoman service.

Another approach, often appearing in tandem with the foregoing narrative prioritizing the exceptional works of extraordinary learned people, is to lay an exclusive emphasis on the technical and textual facets of scientific production, with sparse consideration given to the wider political, social, economic, and cultural contexts within which these sciences were practiced and practitioners operated. While it is undoubtedly commendable, indeed necessary, to comprehend and reflect the technical and philological intricacies of the sciences and practitioners under scrutiny, a myopic focus on technical contents runs the risk of overlooking the broader socioeconomic and politico-cultural mechanisms that shaped the practice of science. At times, the mere act of publishing a scientific treatise on a subject such as mathematics or astronomy is branded as a clear indication of the scientific vitality or even supremacy of the Islamicate or Ottoman case, with the complex technical vocabulary and scientific specifics often left uncontextualized from a social, cultural, or political perspective. What these exclusively textual and technical approaches to the history of science, and specifically to the history of the astral sciences, in the Ottoman and, by extension, the Islamicate context, inevitably lead to is the further insulation and compartmentalization of the field, rendering it an exceedingly specialized domain, essayed primarily by students and scholars with a background or career in science and engineering who have developed an interest in the history of their own profession and are already equipped with a thorough understanding of their field's technical aspects.[62] In parallel, those scholars who identify primarily as historians of Ottoman politics, society, or culture may grow even more reluctant to tread into the terrain of science and thus have avoided the subject. For instance, doyens of modern Ottoman historiography, such as Suraiya Faroqhi and the late Halil İnalcık, who have composed an extraordinary number of scholarly works encompassing virtually every conceivable topic and period of Ottoman history, have written scarcely anything about the sciences and their practitioners in the Ottoman realm.[63]

Despite this rather gloomy perspective on the state of the field, we should acknowledge that many valuable studies and projects have been undertaken in the Ottoman historiography of the sciences that aspire to transcend the narratives of scarcity and decline or, conversely, the glory and triumph of exceptional figures. To take one example, the sustained efforts Ekmeleddin İhsanoğlu has exerted since the 1980s in his capacity as the founder of Istanbul University's History of Science Department and the general director of IRCICA (The Research Centre for Islamic History, Art, and Culture) bear mentioning. İhsanoğlu has endeavored to explore what he defines as "Ottoman science"—encompassing the natural-physical, mathematical, and applied sciences developed and practiced within Ottoman territories—in a more holistic, comparative, and sociological manner.[64] A monumental cataloging project İhsanoğlu initiated and coordinated for many years at IRCICA together with his colleagues, the *History of Ottoman Scientific Literature*, shares some parallels in principle with Joseph Needham's multivolume work *Science and Civilization in China*.[65] The aim was to generate an inventory and bio-bibliographical catalog of extant scientific manuscripts produced in the Ottoman world, albeit not including, regrettably, those generated by Greek, Armenian, Jewish, and other non-Muslim communities of the empire.[66] Amounting today to more than fifteen bio-bibliographical inventories and providing catalog information about thousands of authors and texts written in Islamicate languages, these volumes have become indispensable sources for any student of history interested in the mathematical, astronomical, medical, and other natural and applied sciences theorized and practiced by Ottoman learned and laypeople from the fourteenth through the early twentieth centuries.

The inaugural two volumes of this colossal project, released in 1997, were devoted to astronomy. Although these volumes present precious catalog information about the copies of books composed or used by *munajjims* for astrological purposes, there was a distinct reluctance to acknowledge that so many of the names listed as "astronomers" were indeed *munajjims*, who also undertook astrological pursuits. Almost fifteen years after the publication of these catalogs on Ottoman astronomical materials, a new volume was issued in 2011 to supplement the earlier list of materials and introduce Ottoman astrological lore. In his preface to the new catalog, Ekmeleddin İhsanoğlu tendered a belated apology, explaining the reasons behind their initial disregard for *munajjims* and astrology:

> While we were preparing the first two volumes on astronomy we had chosen to leave aside the works on astrology on account of the latter generally being considered as an "occult science." This was due to our understanding of the sciences at the time: our intention was to exclude astrology from the body of sciences that are based on observation and experiment and to consider it a "pseudo-science." But quite a long time passed since then, and given the maturity presently reached in history of science studies we believe that it was not the right choice and we are compensating for it now as we complete the series.[67]

The reluctance to acknowledge the significance of *munajjims* and explore the textual corpus written by and about them—a sentiment encapsulated succinctly in Ekmeleddin İhsanoğlu's delayed apology—has come to define modern scholarship on the history of Ottoman sciences. Put differently, the history of *munajjims* and their scientific endeavors remains one of the most neglected areas within the already marginalized field of the history of science in Ottoman studies. The marginalization of *munajjims* was not an innocent oversight. On the contrary, *munajjims* and their body of work were deliberately excluded from the narratives of Ottoman sciences, as they were not deemed scientific enough by post-nineteenth-century definitions and standards of sciences and scientists. This debarment is, however, far from being an idiosyncrasy of Ottoman historiography. In other historiographical traditions covering different geographical scopes within the broader discipline of the history of science, even including European and Western historiography, at least for some time, one could observe similar attitudes dismissing astrologers as "wretched subjects."[68] But in the context of non-European historiographical traditions, particularly the Islamic history of science, there are additional layers to consider. These strata involve concerns about and reactions against colonialist, Orientalist, and Eurocentric perspectives that were particularly prone, especially in the nineteenth century, to associate the Islamicate realm—and indeed all non-Western domains—with magic, the occult, superstition, and the overall absence of scientific mentality. Consider, for instance, the following remarks by William Whewell, the inventor of the word "scientist," in his *History of Inductive Sciences* (1837): "We need not attempt to trace the progress of this science [astrology]. It prevailed extensively among the Arabians, as we might expect from the *character of that nation*."[69]

Such deep-seated essentializing views, prevalent in Western and European intellectual milieux from at least the nineteenth century onward, undoubtedly influenced the development of the history of science in the Islamicate context as a discipline in the twentieth century, resulting in considerably less scholarly attention being given to astrologers and their science.[70] This is not to dismiss, of course, the indispensable work done by David Pingree, Edward Kennedy, Bernard Goldstein (all students of Otto Neugebauer), along with Richard Lemay (a student of Lynn Thorndike) and Charles Burnett, who edited some of the major texts of the early Islamic astrological canon or studied the transmission of astrological theories and concepts from pre-Islamic civilizations into the early Islamic realm.[71] However, with their exclusive focus on Arabic astrological sources produced during the so-called classical period of Islamic history, these vital contributions have, perhaps inadvertently, reproduced the Golden Age narrative. Even the encyclopedic works of Manfred Ullman and Fuat Sezgin on major astrological sources in the Islamic past clearly show that astrological materials produced after the twelfth century in languages other than Arabic were not necessarily deemed worthy of examination by twentieth-century historians of science.[72]

Occult Sciences and the Islamic Millennium

In light of nineteenth- and twentieth-century Orientalist legacies, it is partly understandable why the Golden Age narrative was so eagerly embraced by specialists and enthusiasts of Islamic history. The contention was that to highlight the exact scientific accomplishments of brilliant scientists in the medieval Muslim world would help unseat the essentialist convictions of post-nineteenth-century Western scholarly circles, where it was too common, as William Whewell's quotation above exemplifies, to associate the Islamicate world with magic, the occult, superstition, and a general lack of science. Such a reactive approach seeking to emphasize the grand role of Muslim scientists and their contributions to the world's scientific stock also went hand in hand with and was to a certain extent inspired by scholarly calls in the first half of the twentieth century to write the history of science as a truly universal enterprise of all humankind. Pioneered by George Sarton and articulated through his "New Humanism" theory, this perspective sought to underscore the multi-civilizational origins and global character of the modern sciences.[73] Often regarded today as the founder of the history of science as an academic discipline, George Sarton was no ordinary scholar, establishing the first degree-granting institution for the history of science at Harvard University, along with the flagship scholarly association (i.e., History of Science Society) and journal (i.e., *Isis*) in the field. Notwithstanding the global elements of his call, Sarton's triumphalist and progressivist vision was still Eurocentric in nature. For him, the intrinsic value of non-Western scientific pursuits could only be measured against their impact on the development of modern (read Western) science.[74]

Sarton is also notable for his marked disdain for the occult scientific practices of the pre-modern world. In one of his famous book reviews, which later became the subject and target of a noted response by Otto Neugebauer, Sarton criticized the publication of a book on Mandaean astrology, which he characterized as "a wretched collection of omens, debased astrology and miscellaneous nonsense derived from Arabic, Greek, Persian and of all the superstitious flotsam of the Near East."[75] In another work, Sarton described the occult sciences as the "mud of the streets and dirt of the houses," demonstrating the "imbecility of the human mind."[76] For Sarton, modern historical scholarship should not take these materials seriously unless they are instrumentalized to illustrate how such "intellectual delusions" impeded the technological and scientific advancement of humanity.[77]

As a direct consequence of the combined effects of Golden Ageism and Sartonian apprehension, the occult sciences have received in Islamic studies far less attention than they deserve. The vast range of occult scientific sources produced and circulated in the Islamicate world for over a millennium beginning in the eighth century were inadequately explored, possibly because of concerns either that these materials could reinforce and perpetuate Orientalist convictions about the presumed incompatibility of exact, rational sciences with Muslim societies or that they might distort the narrative on the role of Muslims in the generation and accumulation of modern scientific knowledge. Even though several important texts on astrology or alchemy, particularly from the earlier periods of Islamic history, were edited and published through individual initiatives, only in recent decades has a systematic resurgence emerged in Islamic studies across various regional and chronological specializations, recognizing the ubiquitous presence and importance of occult sciences and esoteric practices in the social, political, intellectual, and cultural histories of the Islamicate pasts.[78] This new interest has, in part, been influenced by the proliferation of studies in Western historiography since at least the 1970s, which have come to terms with the undeniable significance of occult sciences and esoteric knowledge practices in medieval and early modern European social, intellectual, political, and cultural contexts.[79]

While no definitive scholarly consensus has been reached on the use of terms such as "occult" versus "esoteric" sciences, and important nuances distinguishing between magic and divination or between prognostication and prophecy are often disregarded, one key conclusion of the growing research on the occult sciences in the Islamicate world is the peculiar significance of the fifteenth- and sixteenth-century Islamicate East for the flourishing of this array of knowledge and its experts, both in royal courtly settings and across informal intellectual networks. This period coincided with the growing ascendancy of messianic, apocalyptic, and millenarian discourse among political and intellectual elites, as well as the less learned masses of post-Mongol Islamicate societies. The era is sometimes even referred to as the "Messianic Age," as it was marked by the presence of mystics and self-declared messiahs with pronounced political ambitions backed up by a broad spectrum of intellectual endeavors, ranging from numerology and metaphysics to astrology and magic.[80] Although Islamic history is replete with periods of intense messianic activity and heightened apocalyptic expectations, the fifteenth- and sixteenth-century chapters of these currents stand apart.[81] This transitory era, coming after the dissolution of the Abbasid caliphal and Chinggisid models of rule—after the fall of the Abbasid capital of Baghdad in 1258 and the eventual fragmentation of the Mongols by the second half of the thirteenth century—and preceding the emergence and consolidation of territorial early modern Muslim empires (of the Ottomans, Safavids, and Mughals), led to a constitutional void that allowed messianic currents to have an even more secure foothold.[82] Furthermore, the imminent arrival of the tenth century of the Hijri calendar by 900 AH (1494–95 CE) and then the impending close of the first millennium by 1000 AH (1591–92 CE) likely triggered, at least among certain segments of society and the learned elites, a flurry of concerns, claims, and considerations pertaining to the End Times.

The heightened currency of messianic and apocalyptic discourse in the fifteenth- and sixteenth-century Islamicate East brought with it an increased appeal to various occult sciences and esoteric practices, to the extent that the articulation of messianic and sacral forms of political claims was shaped by an "occultist arms race."[83] In particular, the science of letters and [divine] names (*ʿilm al-huruf wa'l-asmaʾ*), of which the foremost authority was the Andalusian mystic Ibn al-ʿArabi (d. 1240), as well as the science of political-eschatological prophecies (*ʿilm al-jafr*)—attributed to the prophet Daniel and ʿAli b. Abi Talib—played a pivotal role in expressing apocalyptic expectations and promulgating messianic political assertions.[84] Operating within a nexus of correspondences and affinities linking letters in the alphabet, numbers, the four elements, prophets, cyclical historical eras, the seven planets, twenty-eight lunar mansions, and related phenomena, the adepts of these prophetic sciences contended that they had the authority to perfectly understand these correspondences between the heavenly and terrestrial realms and to predict and prophesy, even manipulate, future events by deciphering the secrets encoded in letters and numbers. In the apocalyptic vision of one of the paramount exponents of the science of letters at the time, ʿAbd al-Rahman al-Bistami (d. 1454)—a prominent figure within an informal, transregional intellectual network stretching across Iran, Egypt, and Anatolia with a shared interest in the science of letters—the age of the world was 7,000 years.[85] Each of its seven millennia was presided over by a prophet, with a corresponding planet, alphabet, and letter unique to it. The current epoch, the last one before the prophesied end of history, was the Lunar Age, overseen by the revelation of Muhammad.

One should ask, however, to what extent the erudition of Bistami or those adept in the science of letters and letter divination aligned with the expertise of the *munajjims*. Would the mere incorporation of astral references and the vocabulary of planetary correspondences into an esoteric understanding of the cosmos render such knowledge equal or close to the *munajjims*' technical domain of astronomy and astrology? Did the intricate technicalities mastered and employed by lettrists and *munajjims* in their respective fields of expertise overlap, or were there notable distinctions in their methodologies, procedures, and objectives? *Forgotten Experts* underscores the necessity of distinguishing the authority and expertise of *munajjims* from that of other esoteric experts. This endeavor, however, demands a comprehensive exploration of *munajjims*' technical writings—a task with which the growing literature on the rise of Islamicate occult sciences and messianic currents in the fifteenth and sixteenth centuries has rarely engaged. For example, in his work on the vocabulary of sacred kingship and the role of messianic claims in the sixteenth-century Islamicate world, more specifically within the Mughal context, Azfar Moin remarks that during this period, "future was as important as the past, divination as important as genealogy, and astrology as valuable as history."[86] Although such a stimulating remark leaves the impression that the book has explored relevant astrological materials produced and circulated during the time under study, readers later realize that no technical work was in fact explored to investigate whether "the science of the millennium," as Moin calls it, truly encouraged individuals to employ sophisticated astrological techniques to predict cosmic changes, as well as whether the writings of *munajjims* genuinely engaged with millenarian, apocalyptic concerns.[87]

Despite easily accepted but sparsely documented presumptions as to *munajjims*' heavy involvement in producing and disseminating a discourse attuned to the larger millenarian debates, the writings of Ottoman court *munajjims* reveal strikingly little evidence about their participation in calculating and interpreting an apocalyptic future. This seemingly surprising case serves as a poignant reminder that the nature and scope of the expertise possessed by *munajjims*, as well as the extent of their forecasts and computations relayed as prognostications and conjectures, differed significantly from those of other types of experts, such as dream interpreters, lettrists, geomancers, and mystics with some arcane power whose future-oriented reflections qualified by contrast as prophecies. In that sense, although the terms "prognostication" and "prophecy" are often used interchangeably and the original terms in Islamicate languages, such as *kihana* (Tur. *kehanet*), do not necessarily imply a conceptual difference, we should analytically distinguish between the two. While both terms deal with the realm of the future and the unknown, prognostication does not carry the same spiritual or moral weight as a prophecy, relying instead on methods of prediction based on visible portents, patterns, and data. The analytical framework of expertise used in this book is, therefore, all the more important, as it enables us to trace these frequently overlooked nuances and discrepancies among different groups of learned individuals, who are all too often hastily considered to be part of a similar or single tradition.

To explore such instances of rivalry among various experts peering into the future, the long sixteenth century serves as an intriguing historical laboratory. The belief in End Times, already a long-standing tradition among Christian, Jewish, and Muslim communities, took on a newfound eschatological significance and momentum during this period that witnessed the epoch-making events of the Fall of Constantinople, the Columbian discoveries, the Christian expansion on medieval Iberia, the Lutheran Reformation, and the imposition of confessional uniformity.[88] The far-reaching political and societal implications of apocalyptic thought and practice reverberated from the western to the eastern ends of the Mediterranean. Plus, the period's alignment with the final century of the first Islamic millennium infused an additional chiliastic urgency within the

Islamicate realm. Millenarian beliefs and apocalyptic rhetoric permeated political and public discourse, triggering sovereigns and their spokespersons to incorporate such elements into their vocabulary of sovereignty. However, this usage was more than just a discursive tool. The fresh surge of apocalyptic thought and practice almost turned into a science unto itself, with a group of experts emerging as authorities to decipher its secrets. Although *munajjims* are all too often viewed as the usual suspects of the "science of the millennium," their surviving textual repertoire from the period suggests otherwise. Exceptions could be and have been found, especially among fifteenth-century almanacs, which, unlike later examples of the genre from the sixteenth century and beyond, contain reverse-chronology tables that date events from Creation to the present, almost like an hourglass flowing to the eventual end. Even though these tables lack explicit references to apocalyptic and millenarian thought, the peculiar arrangement of time in these reverse-chronology tables, cascading from Creation toward an implied end, alludes to, if not decisively documents, an apocalyptic understanding of time.

Structure of the Book

Forgotten Experts zooms in on astral experts from the Ottoman world, hitherto neglected by historians, including specialists in Ottoman history as well as those exploring histories of science in the Islamicate and early modern European worlds. It unearths materials by and about *munajjims* that have been previously unexamined. I must stress, however, that my aim is not simply to transpose the already well-known theme of loyal astrologers at the service of their royal benefactors—a theme extensively visited and illuminated in European historiography—into a non-Western context by adding a commonplace critique of Eurocentrism.[89] Neither is my intent to reiterate the significant role the so-named occult sciences played in medieval or early modern times. While these objectives are neither insignificant nor outdated, my exploration is guided by a slightly different set of inquiries, some of which closely resonate with the questions Martin Mulsow has raised in his work *Prekäres Wissen* (translated into English as *Knowledge Lost*):[90] What, for instance, did the expertise and authority of this distinct class of scientific practitioners represent, particularly in an environment where their epistemic legitimacy was under perpetual critique?

How were their authority and expertise constructed and maintained in the relative absence of institutional structures structuring their practice? In their capacity as experts in a field that ignited heated epistemic controversies, incited faith-based backlash, and largely lacked an institutional framework of its education, how did *munajjims* sustain their craft amid all these scientific, cultural, and societal vulnerabilities? What, indeed, was the specific body of knowledge possessed by *munajjims*, and how was it different from the expertise of other learned individuals and (occult) scientific experts? How and where did they acquire the requisite knowledge and skills to gain recognition as competent *munajjims*? Were those *munajjims* with demonstrated expertise accorded any authority, and if so, whence did it originate? Further still, what was the nature of their relationship with their patrons and the "state," their professional peers and competitors, and the broader society?

Sources utilized in *Forgotten Experts* relate exclusively to courtly practitioners, thus painting a picture that focuses on court *munajjims* and their high-register astrology. This is not to suggest that other forms, registers, and actors of astrological practice did not exist. There were practitioners who might be characterized as "lay" or "street" astrologers.[91] In the extant examples of this kind of astrology, circulating under the titles *ruzname*, *melhame*, and *yıldızname*, no effort was made, for instance, to calculate the degrees of planetary positions mathematically and astronomically or to identify variables based on the exact time and location of the matter in question. Instead, they either relied on age-old traditional knowledge concerning the cyclical patterns of natural phenomena or employed a basic numerological and lettrist framework that associated heavenly objects with certain letters and their numerical values, which were believed to provide deeper insights about one's future.[92] For example, as detailed in Chapter 1, the exact computation of the degree of the ascendant (i.e., the rising point of the ecliptic—the eastern point where the horizon intersects with the ecliptic) at a given time and location was central to any serious astrological endeavor. Only after establishing its precise astronomical degree and identifying other astrological houses and variables could *munajjims* begin their interpretations. The practitioners of lay astrology, however, paid little attention to such elaborate astronomical and mathematical techniques. In one particular case, as prescribed in a relatively popular book (*Taliᶜname* [Book of the ascendant]), the author advises: "If you would like to know the ascendant of a person and predict his or her course and length of life, add the numerical value of the individual's name to that of his or her mother. Then divide the sum by twelve. If the remainder is one, then it means the ascendant is in Aries. If two, it is Taurus, and so forth."[93] Additionally, such texts include explicit references, underlining how their knowledge was qualitatively different from that of *munajjims*.[94] Although these texts have left behind more extant manuscript copies than the court *munajjims*' horoscopes and annual almanacs, thanks to the simplicity of their suggested operations and the accessibility of their language, *Forgotten Experts* pays less attention to these works and practitioners of lay astrology, not just because these works, as they themselves acknowledge, are qualitatively different from *munajjims*' accounts, but also because there are far fewer clues to trace and reconstruct the lives and perspectives of practitioners of lay astrology.

Chapter 1 provides a comprehensive overview of *munajjims*' technical expertise and its scope, as well as the means and sites of its acquisition. This chapter serves two primary objectives: first, to familiarize readers with the technical concepts and content intrinsic to *munajjims*' practice, which will be further explored in subsequent chapters, and second, to highlight the ambivalent reception of their expertise, which was simultaneously demanded and doubted. *Munajjims* and their science were neither explicitly prohibited nor actively promoted in the Ottoman and Islamicate contexts. Frequently met with suspicion by contemporary epistemic authorities, *munajjims*' astrological expertise largely failed to develop an institutional mode of knowledge transmission typical of mainstream madrasa disciplines such as theology, jurisprudence, or logic. Yet despite skepticism from prevailing scholarly circles and the overall absence of an institutional framework for its propagation, the *munajjims*' knowledge was transmitted and put into practice, notably within royal courts and through master-disciple relationships established in less formal settings. It was available to all those eager to master it, yet was undeniably distinct, grounded in both openly circulating texts and not so easily accessible material technology and instruments. This textual mode of dissemination defies its easy characterization as an occult or esoteric wisdom if one considers these labels to denote secretive, clandestine knowledge exclusive to a select few. Nevertheless, in an environment largely lacking alternative institutional support for their learning and brimming with criticism from competing epistemic authorities, the *munajjims*' practice was by and large contingent on the distinct favor of individual benefactors. Therefore, the presence and activities of *munajjims* fluctuated constantly, shaped by the varying levels of personal interest exhibited by individual patrons.

Chapter 2 begins the chronological examination of the social and intellectual lives of *munajjims* in the early modern Ottoman realm. At the heart of the narrative are two practitioners from the late fifteenth and early sixteenth centuries who have remained virtually unknown to modern scholars and readers. The stories of these two men exhibit both shared traits and distinctive qualities that reflect the practice of the astral sciences within the Ottoman sociopolitical order in the making. Flourishing at a time when there was a surge in courtly interest in the celestial sciences, thanks largely to the genuine curiosity of Sultan Bayezid II and his inner circle about this field of expertise, these two figures' writings shed light on their common strategies when jockeying for royal support and rebranding their expertise whenever needed. Unlike later court *munajjims*, who acquired their expertise within the Ottoman domains and wrote primarily in Ottoman Turkish, these two were trained in Iranian lands and composed their works in Persian. This heritage allows us to trace the "Persianate" influence on the formation of indigenous Ottoman cadres and customs of astral practice from the early sixteenth century.[95]

Chapter 3 shifts the focus from *munajjims* to their patrons, delving into the extraordinary scope and far-reaching consequences of the celestial fascination that gripped the court of Bayezid II. This chapter contends that, despite modern portrayals of Bayezid II as a pious and even zealously religious figure who has often been accused of preventing the Ottoman enterprise from becoming a modern, science-and-technology-friendly world power, he was a true champion of science and institutor of a well-functioning imperial structure. His three-decade rule was the seventh heaven for *munajjims*, and this was not at odds with his renowned piety. On the contrary, his commitment to leading a pious life, his pursuit of astral scientific interests, his patronage of astral experts and other learned individuals, and his efforts to help create a rationalized imperial administration were all interrelated facets of his complex profile that measured up to the idealized philosopher-king. But Bayezid II was not the sole impetus for the remarkable fascination with the astral sciences during the time. Several prominent figures from the sultan's inner circle, such as Müʾeyyedzade ʿAbdurrahman (d. 1516) and Mirim Çelebi (d. 1525), also warrant special recognition for their contributions to the creation of an exceptional intellectual milieu that facilitated the influx of astral experts and their knowledge into the Ottoman capital, especially from the post-Timurid Persianate realm as well as from the Mediterranean world in the wake of the 1492 Jewish expulsion. Although the enthusiasm for *munajjims*' expertise was not always maintained to the same degree by Bayezid's successors, the formation of a formal administrative framework for the employment of court *munajjims* and the standardization of their service influenced how the impersonal, institutional use of *munajjims*' expertise was preserved in the Ottoman realm for centuries thereafter.

Chapter 4 examines the change in identity, recognition, compensation, and linguistic preferences of court *munajjims* by the mid-sixteenth century, viewed through the experiences of practitioners from the era of Süleyman the Magnificent. Archival sources document at least six *munajjims* offering their expertise during different decades of Süleyman's extended reign. However, the uneven nature of biographical and autobiographical details about these *munajjims* has compelled focus on one particular figure: ʿAli (d. circa 1587), better known as Riyazi—a name signifying someone adept in the mathematical sciences. Riyazi was among the corps of court *munajjims* from the early 1540s until the mid-1550s, by which time the number of court *munajjims* had gradually decreased to two. Despite Riyazi's continued service into the 1550s, he never ascended to the position of chief *munajjim*. Perhaps driven by unmet ambitions, he left behind a significant body of autobiographical records that candidly express his disillusionment with the inadequate acknowledgment of his expertise. Moreover, Riyazi's extant archives comprise a treasure trove of manuscript records and microscripts—ownership statements, colophons, and a variety of marginalia and paratexts—that provide a window onto the self-reflections of Ottoman court *munajjims* and their articulation of claims of expertise. In order to balance the heavy focus on Riyazi in this chapter, occasional forays are made into the lives, works, and experiences of other astral experts from the period. Together, these diverse accounts from the Süleymanic era reveal a decline in the stature of *munajjims* and their expertise, even as their counsel remained in demand. Rather than taking this decrease in the number and compensation of court *munajjims* under Süleyman's reign as an indication of the so-called Decline in the Ottomans' scientific enthusiasm, a more nuanced perspective would view it in the context of the evolving trends and preferences of those in power regarding their use of experts and fields of expertise. Throughout Süleyman's lengthy reign, there certainly was a royal interest in heavenly matters, but it leaned more into the esoteric and prophetic practice of lettrists than the technical and astronomical knowledge of *munajjims*.

Building on the findings of the previous chapter, Chapter 5 scrutinizes various textual examples of forecasting, some penned by *munajjims* and others by competing actors in the domain of future predictions. The aim here is to address the question of who held the authority and expertise to predict the future in the early modern Ottoman world. While a majority of these examples are derived from the time of Süleyman, references to materials from earlier and later periods are also made, demonstrating the enduring and ubiquitous nature of certain practices as well as the subtle distinctions between them. Süleyman's reign is of particular significance for this chapter, given the heightened currency of prophetic lore and the growing expression of millenarian and apocalyptic concerns at the time. The rising tides of swelling messianic and apocalyptic currents are often thought to have promoted and enhanced the discourse produced by experts of the future, but little consideration has been given to unearthing the possible variations and competitive interactions between those experts. Even though *munajjims* were not frequently involved in the production and circulation of apocalyptic discourses, they still were—with their ability and authority to interpret both imminent and far-reaching futures through heavenly portents—subjected to denunciation and demonization by certain geomancers and prominent Sufi shaykhs. Those geomancers and eminent mystics professed to hold genuine authority over speaking about the future in regard to their privileged and esoteric knowledge of letter divination. By adopting expertise as a framework, this chapter traces and reconstructs the ways in which ostensibly analogous types of experts and branches of knowledge, often categorized under the umbrella terms of "occult sciences" or "esoteric knowledge," were in fact engaged in vigorous competition for social capital and epistemic recognition.

The Conclusion revisits modern debates surrounding the construction and subsequent destruction of the observatory established in Istanbul in the late sixteenth century, with the aim of reassessing the place of *munajjims* both during and after this episode. Like other observatories founded in the medieval Islamicate world or contemporary Europe, *mu-*

najjims' need for greater precision was the primary impetus for its establishment. Contrary to the prevalent narrative, which views the demolition of the Istanbul observatory—sanctioned by a fatwa from the then Shaykh al-Islam—as a decisive defeat of *munajjims* by religious authorities and as a turning point for the infamous Ottoman decline in the pursuit of science, court *munajjims* preserved their status and continued their practice. Their expertise, marked by ambivalence and subject to constant ebbs and flows, endured throughout the centuries, albeit largely forgotten.

ONE

MUNAJJIMS' EXPERTISE

Who were the *munajjims*, and what did their expertise encompass? Translating this term into English poses a challenge. At first glance, "astrologer" might seem an obvious choice, and it is indeed the most common one. For instance, standard modern Arabic dictionaries, such as Hans Wehr's *A Dictionary of Modern Written Arabic*, use "astrologer," referring to the verb root of the word *najjama*, which means "to observe the stars; to predict the future from the stars; to practice astrology."[1] However, the modern pejorative connotations of "astrologer"—conjuring images of crystal-gazers and psychic shops—mask the *munajjims'* competence in the mathematical and astral sciences, which classical Arabic dictionaries capture better. Ibn Manzur's (d. 1311) *Lisan al-Arab* (The tongue of the Arabs) and Firuzabadi's (d. 1414) *al-Qamus al-muhit* (The encompassing ocean), for example, note that the word, derived from *najm* (star), means "those who look at the stars, compute their timings and movements."[2] The *Dehkhoda Dictionary*, drawing on hundreds of classical Persian and Arabic sources, also provides the following: "an individual informed in the science of the stars; compiler of almanacs; one who measures the timings and movements of the stars to understand the conditions of the universe."[3] While these latter sources grasp the astronomical expertise of *munajjims* together with their erudition in astrology, it would not be entirely apt to translate the term merely as "astronomer," as that occludes the *munajjims'* inherent engagement with astrology. After all, in the Islamicate historiography of science, "astronomer" is a somewhat loaded term, often used to strip many "astral experts" from their documented involvement in what has been dismissively characterized as "astrological nonsense."[4] Given these complexities, I prefer to retain the original term *munajjim*, which more accurately represents the main protagonists in this book: experts in the astral sciences who also actively practiced astrology.

Munajjims possessed technical proficiency in mathematics and astronomy, which they applied in their practice of astrology. In this book, those proficient in mathematics and astronomy but lacking evidence of having practiced astrology are not considered *munajjims*. For example, ʿAli Qushji, one of the stellar scientists at the Samarkand observatory in the mid-fifteenth century who relocated to Istanbul in the early 1470s, is not included in the gallery of *munajjims* owing to the lack of evidence regarding his astrological activities.[5] In contrast, Mirim Çelebi, a descendant of ʿAli Qushji who rose to prominence in the late fifteenth- and early sixteenth-century Ottoman intellectual realm as a high-ranking scholar, deserves to be recognized as a *munajjim*, among his other accomplishments, thanks to his notable output in astrology—through his books as well as his students.[6] As a general rule, this book treats as *munajjims* all those learned individuals who identified themselves with this term in their self-proclaimed sobriquets, even if their astrological endeavors are often overlooked by modern historians of science. On the contrary, sorcerers, magicians, and clairvoyants, who lacked documented expertise in mathematics, astronomy, and astrology but were branded as *munajjims* by their contemporaries for derogatory reasons, are excluded because of their apparent knowledge gap compared to the practitioners of astral science.[7]

This chapter provides a broad overview of the technical expertise of *munajjims*, detailing its scope and delineating the means and locations through which it was acquired. This foundational exploration serves two objectives. First, it aims to familiarize readers with the technical concepts and content central to *munajjims*' practice, which will be explored in the separate case studies of subsequent chapters. Second, it underscores the ambivalence inherent in their expertise, which was at once trusted and mistrusted, rewarded and rebutted. With their expertise requiring proficiency in diverse epistemic traditions of medieval Islamicate knowledge systems—namely, the mathematical and natural sciences—*munajjims* had to navigate domains and forms of knowledge that were not always easily reconcilable or highly regarded. Moreover, since *munajjims*' astrological guidance was often seen as violating the belief that only God knows the unknown (*ghayb*), they frequently drew criticism from Sharia-minded scholars and pious mystics, whose intellectual stance was grounded in the knowledge and interpretation of Islamic sources. While *munajjims* were not explicitly banned in the Ottoman and Islamicate contexts—in the absence of an Inquisition-like disciplinary mechanism or anti-divination laws—the strictly astrological aspects of their expertise largely failed to develop an institutional mode of knowledge transmission typical of mainstream madrasa disciplines such as theology, jurisprudence, and logic.

Yet despite skepticism from prevailing learned circles and the shortage of institutional means for the propagation of their astrological expertise, the *munajjims*' knowledge was indeed transmitted, notably within royal courts and through familial ties and master-disciple relationships maintained in less formal settings than madrasas. It is essential to note that the textual mode of the dissemination of their expertise defies its easy characterization as an occult or esoteric wisdom if one understands these labels as denoting secretive, clandestine knowledge founded on intuition and exclusive to a select few.[8] In reality, their knowledge was available to all those eager to master it. Nonetheless, the *munajjims*' *métier* was undeniably distinct, grounded both in openly circulating textual technology such as tables and textbooks and not so easily accessible material technology, such as astronomical instruments. In an environment brimming with criticism from competing epistemic authorities and providing limited institutional support for their learning, the *munajjims*' practice was by and large contingent on the "fond attachment" (Ar. *raghba*, Tur. *rağbet*) of individual benefactors to this type of expertise. Therefore, the presence and influence of *munajjims* fluctuated constantly, depending on the level of personal interest exhibited by individual patrons.

Munajjims' Expertise in Taxonomies of Science

Were *munajjims* merely street fortune-tellers, lacking any sophisticated, scientific foundation? This is the impression one might get from George Saliba's widely quoted article on the role of *munajjims* in medieval Islamic society or from historical sources written by those ʿulamaʾ and mystics slandering *munajjims*, almost to the point of becoming a trope, as simple-minded and malicious soothsayers, violating core tenets of Islamic faith by attributing divine power to celestial objects.[9] Based on this view, it becomes unthinkable to recognize the broader scope of *munajjims*' expertise or to acknowledge that some esteemed figures of the celestial arts in Islamic history were indeed *munajjims*. Numerous historical sources cogently document that the experts in astral sciences adopted this title in their scholarly sobriquets. For instance, Jaghmini (fl. first half of the thirteenth century), renowned for his widely read work on astronomical theory, *al-Mulakhkhas fi al-hay'a al-basita* (The epitome of plain astronomical theory), was known as *ʿaliyy al-munajjim* (sublime among *munajjims*).[10] Likewise, al-Wabkanawi (d. 1320), a notable practitioner from the early fourteenth century and one of the revisers of the Ilkhanid tables, called himself *shams* (the sun of) *al-munajjim*.[11] Mahmud b. Yahya al-Kashi (d. after 1411), possibly the grandfather of Jamshid al-Kashi (d. 1429) at the Samarkand observatory and a reputed practitioner of the astral sciences in the early Timurid realm, signed his autograph as *ʿimad* (the chief [pillar] of) *al-munajjim*.[12]

Despite the noted reluctance of several modern historians of science to associate the term *munajjim* with scientific pursuits, sources composed prior to the late nineteenth and early twentieth century—a period marking the eventual decline of astrology and the crystallization of its disciplinary boundaries with respect to astronomy—show no hesitation in using the term when referring to exact scientific activities. For instance, eighteenth-century Ottoman ambassadors in Europe, including Yirmisekiz Mehmed Çelebi (d. 1728) in Paris and Mustafa Hatti Efendi (d. 1760/1) in Vienna, had no reservations about referring to the Paris and Vienna observatories as "the house of *munajjims*."[13] Even as late as the nineteenth century, Süleyman Sudi Efendi (d. 1896), a high-ranking Ottoman finance bureaucrat with a personal interest in the history of the astral sciences, unhesitatingly titled his book on the universal history of astral experts as *Tabakat-ı Müneccimin*, where he provided brief biographical information on figures such as Biruni, Nasir al-Din al-Tusi (d. 1274), and Qushji.[14] Dictionaries produced at the turn of the twentieth century continued to emphasize the mathematical and astronomical underpinnings of *munajjims*' expertise besides their engagement with astrology. For example, in Şemseddin Sami's (d. 1904) popular dictionary of Ottoman Turkish, the term is defined in twofold fashion: "(1) an individual well versed in the science of the stars, familiar with the movements and configurations of celestial bodies and capable of celestial observations. *Astronome* in French. (2) A fortune-teller with the false belief of drawing inferences from the positions and patterns of stars. *Astrologue* in French."[15]

This capaciousness of the term *munajjim* is not to suggest that during the medieval and early modern periods it denoted a completely ambiguous specialization or that the ambits of astronomy and astrology were considered indistinguishable. In fact, the common conviction that these two disciplines—astronomy and astrology—were indistinct until the modern era is an oversimplification. Both ancient and medieval sources reveal that practitioners of the astral sciences were genuinely aware of and described the important differences between these two domains. For instance, Claudius Ptolemy (d. circa 168), the supreme ancient authority on the study of the heavens in both the medieval Islamicate world and Europe, distinguished conceptually—if not semantically—between astronomy and astrology as early as the second century, addressing them in two distinct works: while his *Almagest* focuses on the mathematical and astronomical aspects of the study of the heavens, his *Tetrabiblos* delves into the practical astrological applications of this knowledge, which he terms "prognostication through astronomy."[16] The conceptual distinction between these different facets of the study of the heavens eventually found its semantic expression in late antiquity, if not earlier. For instance, in his encyclopedic work *Etymologiae*, which reflects the scope of learning in late antique Europe, Isidore of Seville (d. circa 636) elaborates on the differences between *astronomia* and *astrologia*: the former is the science dedicated to understanding the motions of the stars and the rotation of the heavens, while the latter is an art of divination that can easily slip into superstition.[17]

This distinction between the astronomical study and astrological treatment of the heavens became increasingly pronounced in the Islamicate world from at least the ninth century. In the hands of those studying the properties and motions of celestial bodies and striving to demarcate their expertise from that of diviners and fortune-tellers, a new discipline known as *ʿilm al-hayʾa* (lit. "the science of structure or configuration [of the heavens]"), devoted to studying the dimensions, positions, and motions of celestial bodies, was developed that came to draw a strict line between itself and "the science of the decrees of the stars" (*ʿilm ahkam al-nujum*, or sometimes *tanjim*), that is, astrology.[18] Since the subject matter of *hayʾa* is the unchanging realm of the celestial spheres and since it rests upon mathematical demonstrations, it was universally classified among the "mathematical sciences," together with geometry, arithmetic, and music. *Ahkam*, on the other hand, was concerned with making judgments about the influences of specific celestial configurations upon the ever-changing world of generation and corruption. As its main objective was to trace and predict changes in the physical world—based on heavenly signs and portents—it was traditionally located among the "natural sciences," along with medicine and other divinatory arts.

Besides the notions and disciplines of *hayʾa* and *ahkam*, which shaped the general understanding and conceptual framework in the Islamicate world for the study of the heavens for centuries, another term, "the science of the stars" (Ar. *ʿilm al-nujum*, Tur. *ʿilm-i nücum*), was also frequently used. The multiple connotations of this last category in relevant sources—ranging from books on taxonomies of knowledge to astronomical and astrological treatises—occasionally led to certain ambiguities. While some were inclined to equate *ʿilm al-nujum* with *ʿilm al-hayʾa*, the majority linked it more closely to astrology.[19] More often than not, *nujum* encompassed various practical and applied pursuits that entailed both *hayʾa*- and *ahkam*-like elements. Compiling *zijes* (astronomical handbooks with tables) and annual almanacs (*taqwims*), for instance, or using astronomical instruments and conducting celestial observations, along with determining times and directions—notably the five daily prayer times and the orientation of *qibla*, the direction of the Kaʿba (the point toward which Muslims turn in praying)—were often associated with *nujum*.[20]

Notwithstanding these conceptual and semantic distinctions among various aspects and genres of the study of the heavens, those aspiring to become a *munajjim* were expected to excel in all three areas, along with other mathematical and natural sciences. Astral experts often emphasized in their works that without a foundational knowledge of the forms and physical structures (i.e., *hayʾa*) of celestial objects, it was impossible to calculate the motions and positions of these bodies, and without the data gleaned from mathematical operations concerning the motions and positions of the stars, it was impossible to interpret the indications of celestial configurations. Put simply, the notion that *ahkam* is a natural sequel to *hayʾa* and *nujum* was, as Carlo Nallino astutely noted, "common to all the Musalman [*sic*] astrologers and astronomers."[21] For example, Biruni asserts in his *Kitab al-tafhim li-awaʾil sinaʿat al-tanjim* (Book of instruction in the elements of the art of astrology) that to become a *munajjim* one needs to study geometry (*handasa*), arithmetic (*hisab*), the science of structure [of the heavens] (*hayʾa*), and the science of decrees [of the stars] (*ahkam*) in a prescribed sequence.[22] Likewise, another prolific scientist of stars from Shiraz in the first half of the fifteenth century, Rukn al-Amuli (d. after 1455), concisely states in his treatise on astrolabes that the pinnacle of the mathematical sciences—comprising geometry, calculation, and *hayʾa*—is the science of the decrees of the stars (*ahkam*).[23] These examples can easily be multiplied, showing how, in the very eyes of astral experts, these separate disciplines were deemed interconnected and complementary facets of *munajjims*' expertise.[24]

While proficiency in the science of "the structure [of the heavens]" (*hayʾa*) and the science of the "stars" (*nujum*) was essential for any *munajjim*, what was decisive to qualify as a *munajjim* was the knowledge of the "decrees of the stars" (*ahkam*), which is astrology. In other words, to be acknowledged as a *munajjim*, one needed to practice *ahkam*, which should be built upon the necessary theoretical, technical, and mathematical knowledge of celestial phenomena. At its core, the *munajjims*' craft was anchored in the fundamental Hermetic concept and the basic Aristotelian principle: "As above, so below."[25] What this implies is that occurrences in the ever-changing terrestrial world are inherently connected to the unchanging celestial realm. Each unique celestial configuration, at a given time, exerts a specific influence on the sublunary world. These influences were believed to be predicted or retrospectively interpreted through meticulous mathematical and astronomical analysis of celestial bodies and their movements and positions. While the ability to map individual celestial configurations required erudition in the mathematical sciences, the task of determining astrological implications for the natural world also demanded proficiency in the natural sciences.

This Janus-faced nature of *munajjims*' expertise, swinging between the mathematical sciences of the celestial realm and the natural sciences of the physical world, surfaces in the writings of encyclopedists and classifiers of knowledge in the medieval Islamicate world, who sometimes encountered complications in locating *munajjims*' expertise in their taxonomies of the sciences. From Farabi (d. circa 950) and Ibn Sina to Fakhr al-Din al-Razi (d. 1210) and Taşköprülüzade (d. 1561), various authors of books on the classifications of the sciences displayed a certain level of ambiguity in classifying and characterizing the knowledge in which the *munajjims* were experts. For instance, Farabi, in his *Ihsa' al-ʿulum* (Enumeration of sciences), employed the singular category of the astral sciences (*ʿilm al-nujum*) by dividing it into two main categories, reminiscent of Ptolemy's discussion in his *Almagest* and *Tetrabiblos*: the first branch (*ahkam al-nujum*) investigates and interprets the indications of celestial objects for predicting future events, while the second (*ʿilm al-nujum al-taʿlimi*) studies the measurable features of the heavenly objects, including their sizes, motions, and distances from one another.[26] Although Farabi never rejected the fundamental cosmological axiom of celestial causation on the sublunary world, for him the epistemic foundation of the first division was ineluctably inadequate compared to the latter.[27]

Farabi's distinction between the mathematical investigation of heavenly objects and the astrological study of celestial influences on the physical world prefigures the gradual semantic and categorical distinction between *hay'a* and *ahkam*. The decisive shift came with Ibn Sina, who brought relative clarity to the astronomy/astrology debate by placing them in different epistemological categories within his overarching Aristotelian concept of knowledge. For Ibn Sina, *ilm al-hay'a* was an independent discipline whose subject and objectives were primarily defined by the use of mathematical demonstrations; hence, it was grouped under the mathematical sciences.[28] *Ahkam*, however, was classified as one of the branches of the natural sciences, alongside medicine and the divinatory arts.[29] Ibn Sina's model profoundly influenced the later development of taxonomies, including many examples produced in the Persianate, Mamluk, and Ottoman intellectual realms. Even some leading experts of the astral sciences in the postclassical Islamicate world, including Tusi and Qutb al-Din al-Shirazi (d. 1311), referred to the same hierarchical boundaries between *hay'a* and *ahkam*. Tusi, for instance, reiterated in his *Nasirean Ethics* that while *hay'a* is one of the mathematical sciences, the practice that aims at interpreting the knowledge of the heavens in order to predict what will happen in the sublunary world falls into the category of natural science, given that the latter is concerned with matters that are subject to change and corruption.[30] A similar discussion also appears in Shirazi's encyclopedic work *Durrat al-taj* (The crown jewel). While it is worth noting that in his exposition on the mathematical sciences, Shirazi uses *ʿilm al-hay'a* and *ʿilm al-nujum* interchangeably, he clearly underlines that *ahkam al-nujum* is not within the purview of this branch but belongs to the natural sciences, along with medicine, agriculture, physiognomy, and alchemy.[31]

Despite its immense popularity, Ibn Sina's model categorizing *ahkam* as a natural science and *hay'a* as a mathematical science was not the only paradigm in the postclassical period. Fakhr al-Din al-Razi, for instance, in his *Jami' al-'ulum* (Compendium of sciences), where he addressed sixty different sciences, classified *ahkam* as an inextricable part of the mathematical sciences. Unlike the Aristotelian and Avicennan traditions, Razi categorically detached *ahkam* from the group of natural sciences and placed it immediately after the mathematical sciences, including, in his exact order, geometry, algebra, arithmetic, music, and *hay'a*.[32] Razi's preference also found favor among certain scholars and classifiers of knowledge in the Ottoman world, especially in the fifteenth century. Mehmed Şah Fenari (d. 1436), an important member of the influential Fenari intellectual circle of the fifteenth century, wrote a detailed work modeled on Razi's account, in which he grouped *nujum* and *ahkam*, like Razi, together with other mathematical sciences, separate from the natural sciences and divinatory arts.[33] Another example that employs a similar Razian framework, locating *ahkam* among the mathematical sciences, is the Ottoman palace library inventory prepared in 1502–4 by 'Atufi (d. 1541).[34] In this voluminous inventory, 'Atufi, a noted scholar and learned natural philosopher, recorded around 7,200 titles in 5,700 volumes, grouping them in different clusters of disciplines. 'Atufi's gallery on occult sciences, which included dream interpretation, physiognomy, alchemy, geomancy, lettrism, and magic, did not encompass *nujum* and *ahkam* texts; rather, 'Atufi deliberately collected them together with other mathematical sciences such as *hay'a*, arithmetic, geometry, and music.[35]

Munajjims as Masters of Time

As the definitions in classical Arabic dictionaries mentioned before attest, "time is the thing; time is the essential piece of interpretation" for *munajjims*.[36] The preoccupation of *munajjims* with time and its celestial underpinnings has two major implications. First, the reckoning of time—then as now—was intrinsically linked to tracking celestial bodies, specifically the Sun and the Moon, whose cyclical, calculable movements over set intervals determined the span of an hour, day, month, or year. The exact length or the beginning and end of these units of time might slightly differ based on the conventions adopted. For example, while some took the start of a day to be sunset, others chose midday or sunrise. Likewise, preferences differed between using lunar and solar years.[37] In any case, accurately calculating these temporal markers demanded a profound understanding of celestial movements and configurations, a skill the *munajjims* were expected to possess. Their ability to measure time in its smaller and larger units was especially important in a period and setting where mechanical timepieces (i.e., clocks) and mass-consumed calendars were still not common.[38]

Second, and more important, in order to interpret the configuration of the heavens at a precise moment and location—hence practice astrology—a *munajjim* needed to take into consideration the exact moment of the issue at stake and calculate the relative positions of the celestial entities based on the latitude of the present locale. These moments were varied, each guiding and informing the particular focus within the various subfields of astrological practice. For example, to foretell the fortunes of an individual or a political entity, *munajjims* had to ascertain the moment of birth of that person or polity, a practice that falls under the domain of natal astrology or genethlialogy, known in Islamic sources as *mawalid*. Closely related to the *mawalid* genre, anniversary horoscopes of the birth of individuals or the revolutions of solar years (Tur. *tahvil-i sal-i ʿalem*) were routinely prepared to interpret the fortunes of the upcoming (solar) year for a single individual or society in general. The latter formed the essence of the annual almanac genre, which I will explore in more detail below. Similarly, the determination of an auspicious moment to embark on a particular endeavor involved determining or "projecting" the precise time when the astrological indicators related to that task would be most favorable. This sort of ascertainment was the domain of electional astrology (*ikhtiyarat*). The repertoire of activities ranged from everyday matters, such as when to go to a public bath or cast a talisman, to issues concerning imperial policies, such as when to embark upon a military campaign or lay out the construction of a building. Furthermore, if one sought to know whether particular plans or requests would eventually come to fruition, *munajjims* had to discern the celestial alignment at the exact moment the question was posed, which constituted a branch known as horary or interrogational astrology (*masaʾil*).[39] The questions asked by the querent might typically touch on several quotidian aspects of life, including marriage, travel, and household affairs.

In addition to natal, electional, and horary astrology, *munajjims* computed planetary conjunctions (*qiranat*), specifically those of Saturn and Jupiter, which were often interpreted as signs of important political and dynastic changes.[40] Conjunctions vary, with each type associated with a different sublunary transition. For instance, the regular conjunctions of Saturn and Jupiter, recurring about every twenty years in a different sign ("the lesser conjunction"), were believed to indicate a change of ruler. These conjunctions remain in the same astrological triplicity—that is, a group of three zodiac signs belonging to the same element—for an extended period. However, roughly every 240 years, they move into a new triplicity. This type of conjunction ("the middle conjunction") often symbolized a more serious change, such as the emergence of a new dynasty or a nation. Furthermore, the completion of a cycle of shifts through all four triplicities every 960 years ("the greater conjunction") was thought to herald even more sweeping transformations, like the advent of a prophet or the foundation of a new religious dispensation.[41] Conjunctionist astrology was notably widespread in the early Abbasid realm during the eighth and ninth centuries, leading to the emergence of what some modern historians of early Islam term "astrological histories" or "historical astrologies."[42] In these texts, changes in the politico-religious sphere were dated and justified with explicit or implicit references to Jupiter-Saturn conjunctions.

While various branches and genres of astrological practice might employ distinct techniques drawn from different legacies, what is central to all of them is the need to pinpoint the exact celestial configuration at a specific moment in a particular locality. This could be a moment of birth, the beginning of the year's revolution, an auspicious time to embark upon a certain activity, or the precise moment a question was posed to the *munajjim*.[43] To make an astrological judgment about the moment in question, the *munajjim* first had to calculate the degree of the ascendant (*taliʿ*). This represents the rising point of the ecliptic, that is, the eastern point where the horizon intersects with the ecliptic. Moreover, the ascendant stands as the beginning of the first house of the twelve astrological houses, each of which was associated with certain aspects of life. After determining the ascendant, the next steps involved equalizing the cusps of the remaining astrological houses and establishing relevant astrological indicators (*dalaʾil*).[44] The determination of the ascendant was so crucial in the practice of

FIGURE 3. The birth scene of Alexander the Great. From Nizami, *Iskandarnama*. The David Collection, Ms. 74/2006, f. 14a. Courtesy of the David Collection, Copenhagen.

astrological interpretation that in vernacular Turkish, the technical Arabic term *taliʿ* eventually transformed into *talih*, which came to mean fortune and luck.[45]

Munajjims and Their Textual Output

In executing mathematical and astronomical operations related to the positions and movements of celestial objects, *munajjims* may have utilized instruments such as astrolabes, which are vividly documented in the typical visual representations of *munajjims* in the Islamic miniature tradition as well as in poetry.[46] However, on account of the prohibitive costs of owning large astronomical instruments, which practicing *munajjims* occasionally lamented, they predominantly determined planetary positions by relying on astronomical handbooks with tables (*zij*) that were less costly and more widely accessible.[47] While modern scholarship has sometimes undervalued the role of *zijes* in astrological practices, these texts were, in fact, *munajjims*' indispensable tool for making the necessary calculations before their astrological interpretations.[48] Prominent historians of science and astronomy in the medieval Islamicate world, including Edward Kennedy and David King, have already highlighted in their research the specific contents of *zijes* related to natal astrology. However, as they acknowledge, "there is precious little evidence of how these works were used in practice."[49] It would not be far-fetched to compare medieval *zijes* to today's computer programs, allowing for manual calculations of all quantities needed for establishing the horoscope and other astrological indicators. As the foundation of a *munajjim*'s art rested upon the meticulous reckoning of celestial alignments at specific moments for particular locations, *zijes* equipped them with the necessary data and parameters for their calculations. It is, however, difficult for a nonspecialist to fully reconstruct the ways *munajjims* employed *zijes* when they made their computations. While it is obvious that practicing *munajjims* had to consult the tables that arrange the detailed celestial data and parameters for the mean motions of planets in sexagesimal numbers, they do not go into the essential details about the particularities of this operation in their writings.[50] The standard phrase *munajjims* used in their works, *istikhraj* (lit. "extraction"), indicates that they "took out" the data and parameters on the planetary positions from a *zij* and applied it to their compilation of almanacs or horoscopes.

In his detailed study on the mathematics of Islamic astrology, Benno van Dalen elucidates the methods *munajjims* used to compute planetary positions and calculate the specific degrees of the zodiac signs in horoscopes.[51] As van Dalen explicates, calculating the true longitudes of planets is fundamental to all astrological activities. In order to determine the longitudes of planets for a specific moment, *munajjims* consulted the tables in *zijes* that detail the mean motions and equations for each planet. These equations often required that values in sine tables be checked and several trigonometric operations be carried out. Apart from determining the true longitudes of planets, *munajjims* often used the "oblique ascension functions" (*matali*ᶜ *al-buruj*) to calculate the ascendant. While many *zijes* feature tables for these functions, some treatises on astronomical instruments also offer insights into how these oblique ascension functions are calculated as a step toward computing the ascendant. After establishing the ascendant, the other important astrological variables, including the descendant and the remaining two "pivots" (*watad*)—the upper Mid-Heaven (*Medium Coeli*, or literally, the middle of the sky) and the lower Mid-Heaven (*Imum Coeli*)—

are identified. Once all twelve astrological houses are set, the horoscope chart (*zayicha*) is ready for interpretation. Each astrological house relates to certain aspects of life: the first house pertains to an individual's personality and demeanor, the second to material possessions and finances, the third to siblings and short journeys, and so forth. These houses are influenced not only by the zodiac signs passing through them but also by the planets associated with each sign through a complex system encompassing lords/rulers (*sahib*), exaltations (*sharaf*), terms (*hadd*), triplicities (*muthallatha*), and decans (*wajh*).[52] These parameters can also be derived from the astrological tables found in *zijes*.[53]

The content of *zijes* varied considerably, depending on the quality and duration of celestial observations undertaken. Not all *zijes* were the result of extensive research carried out in observatories; therefore, they differ in scope, precision, the number of tables, and the parameters they deploy. The accuracy of the data and precision of the values of the parameters, such as the mean motions of planets, were intimately tied to the depth and breadth of research in observational astronomy that preceded the preparation of tables. The primary motivation for constructing a new observatory, ideally equipped with larger instruments and designed for longer observation time, was to update and correct imprecise celestial data and parameters. While not all practicing *munajjims* demanded or were qualified for new research, the majority genuinely understood that ideally, in order to achieve more precise results from the observations, at least thirty years of consistent observation were required. This time frame was grounded in the understanding that Saturn, the outermost planet in traditional cosmology, took approximately thirty years to complete its full cycle around the ecliptic.[54] Another factor influencing the precision of observations was the quality of the astronomical instruments available at the observation site and the observational skills of the participants. A prevailing sentiment among astral experts was that larger instruments yielded greater precision, as they allowed for more detailed markings on the instrument, measuring smaller parts of a degree such as minutes and even seconds.[55]

These rather abstract remarks are better illustrated in the introductory prose sections of *zijes*, where authors elucidate their motivations and circumstances for conducting new observational research (i.e., *rasad*) and consequently compiling new data sets (i.e., *zij*). For example, in the Ilkhanid tables produced at the Maragha observatory in the mid-thirteenth century,

Tusi explicitly states that one must observe and calculate the positions of celestial objects to gain insight into earthly affairs, such as the security of the country, warfare and peace among rulers, health and diseases, the state of agricultural production and market trends, and weather conditions. Tusi contends that accurate astrological assessments regarding these issues are contingent upon a precise knowledge of celestial positions, which, in turn, necessitates diligent observation of the heavens.[56] After meticulously tracking the motions of celestial objects, the data on computed cycles was inserted in the tables of the *zij*. Using the information and methodologies presented in these tables, almanacs were crafted annually, detailing the positions of celestial entities throughout the year and deriving associated astrological predictions. Tusi expresses his aspiration for their new *zij* to serve as the main reference work for *munajjims* when drafting their annual almanacs and casting horoscopes.[57]

While the compilers of a *zij* were keen for their work to help practicing *munajjims* and provide them with precise data for their astronomical and astrological computations, discrepancies often arose between the calculated times of celestial phenomena, such as eclipses or planetary conjunctions, and their personal observations. This led successive generations of astral experts to undertake even more rigorous observational programs, provided they received financial backing from ruling elites. The construction of observatories in the Islamicate world, especially in the post-thirteenth-century Turko-Mongolian-Persianate context, perfectly illustrate how new, systematic celestial observations were conducted, usually in observatories, to enhance the precision of planetary data *munajjims* needed. For the *Zij-i Ilkhani* in particular, many contemporary and subsequent *munajjims* found errors and inconsistencies between the computed times of celestial phenomena based on Tusi's tables and the actual times when these phenomena were observed firsthand. ʿAli-Shah Bukhari (d. after 1291), a noted *munajjim* from late thirteenth-century Iranian lands, whose Persian astrology textbook *Athmar va ashjar* (Fruits and trees) later became a favorite among Ottoman *munajjims*, detailed in his commentary on Tusi's tables an encounter where several individuals expressed to him their deep dissatisfaction with the Ilkhanid tables. Apparently, when his interlocutors attempted to calculate the true longitude of the Sun to determine the degree of the ascendant, they found "stark discrepancies" between results from the *Zij-i Ilkhani* and older tables.[58] Yet Bukhari showed some sympathy toward the *Zij-i Ilkhani.*

He maintained that Tusi's motivation for initiating an observational enterprise in the first place and gathering various experts was his own awareness of "grave inconsistencies" in the existing tables of his period.[59] To overcome this perennial problem, says Bukhari, Tusi assembled in Maragha an ensemble of esteemed astral experts from different corners of the Islamicate world and amassed cutting-edge astronomical instruments. Regrettably, Tusi did not live long enough to see the completion of the program of observation.[60] As Bukhari emphasizes, conducting a systematic celestial observation and compiling a *zij* are tedious tasks, hinging on several factors, including "financial resources," "physical space," "capable colleagues," "precise instrumentation," and "ample time" to complete different sets of observations with patience.[61] Concluding his account, Bukhari states that every new research endeavor is set to surpass its predecessor in terms of greater precision in instrumentation and longer duration of observation. Given the inconsistencies in the Ilkhanid tables, he says, a new and ideally more meticulous *rasad* should be implemented to enhance *munajjims*' precision in computation and accuracy in prediction.[62]

In the two centuries following the Maragha observatory's establishment in the 1250s, funded by Ilkhanid Hülegü Khan (r. 1256–65), at least two major observatories were founded in the Islamicate East, in addition to numerous other individual observational enterprises.[63] These endeavors aimed to test existing *zijes*, refine the extant planetary data within the tables, and, if possible, compile new astronomical handbooks with more precise data. While the Samarkand observatory, founded in the first half of the fifteenth century by Ulugh Beg and his colleagues, is the most renowned of all, another noteworthy facility was erected in Tabriz at the turn of the fourteenth century with the support of another Ilkhanid ruler, Ghazan Khan (r. 1295–1304). The chief scientific actor in Ghazan Khan's observatory was Shams al-Din Muhammad al-Wabkanawi—also known as *shams al-munajjim*. Wabkanawi eventually compiled his revised handbook, *al-Zij al-muhaqqaq al-Sultani* (The confirmed sultanic handbook of tables), which, he stated, was the culmination of over thirty years of observations in Maragha and Tabriz. In this text, Wabkanawi also highlights inconsistencies between observed celestial phenomena and their respective values derived from the Ilkhanid tables. In one of his examples, he states that in 1285 CE, the Jupiter-Saturn conjunction transpired in the ninth degree of Aquarius. However, there was a discrepancy of about fifteen days between the observed time of this event and its predicted computation based on the *Zij-i Ilkhani*. More important, Wabkanawi emphasizes the profound astrological significance of the conjunction, which, he says, influences earthly affairs, revealing his understanding that imprecision resulting from flawed celestial data might trigger undesirable outcomes. [64]

In addition to *zijes* and treatises on the use of astronomical instruments—often categorized in modern history-of-science literature only as astronomical or mathematical sources—there were textbooks elucidating standard rules and foundational astrological principles that aspiring *munajjims* needed to learn before practicing their art. These books usually covered topics including the presumed characteristics and dispositions of planets, zodiac signs, and specific celestial degrees, and introduced specific techniques in various branches of astrological practice, all tailored to guide the craft of *munajjims*. There were several alternative textbooks and *summae* on astrological principles openly and copiously circulating at the time. While the current state of scholarship on the history of astrology in the medieval and early modern Islamicate world does not allow for a precise count of extant manuscript copies for each or the social background of their readers, the frequent references by early modern Ottoman *munajjims* and the number of titles listed in the Ottoman palace library inventory from the dawn of the sixteenth century provide insights into the key texts and authorities.[65] The list includes:

(i) Nasir al-Din al-Tusi's *Thirty Chapters* (*Si fasl*), also known as the *Epitome of the Knowledge of the Almanacs* (*Mukhtasar dar maʿrifat-i taqvim*): This concise Persian textbook in thirty chapters informed those aspiring to astrological knowledge about sexagesimal numbers, calendar systems, general characteristics of planets, zodiac signs, and the indications of planetary aspects. The text was one of the most widely copied scientific works of the time, documented by the substantial manuscript evidence supporting its widespread use in Anatolia and Iranian lands from the late thirteenth century. It was translated into Turkish as early as the turn of the fifteenth century.[66] The Ottoman palace library inventory of 1502–4 lists fourteen copies of the text, including translations and commentaries.

(ii) *Thamara* (known in Latin as *Centiloquium*): Comprising one hundred astrological aphorisms, this work was attributed to Ptolemy but originally composed in Arabic in the early tenth century by a certain Abu Jaʿfar Ahmad b. Yusuf (d. circa 944). Later translated into Persian by different parties, including Tusi, it was mostly through Tusi's Persian translation that the text reached Ottoman *munajjims*, though some did refer to the original Arabic aphorisms.[67] Serving as a simpler and more accessible rendition of Ptolemy's major astrological work, *Tetrabiblos*, it became a standard textbook for astrological instruction, particularly in the field of birth horoscopes.[68] The Ottoman palace library inventory lists ten copies.

(iii) Kushyar ibn Labban's (d. after 1025) *Compendium of Principles in Astrology* (*Mujmal al-usul fi ahkam al-nujum*), also referred to as the *Introduction to the Craft of Astrology* (*al-Madkhal fi sinaʿat ahkam al-nujum*): A detailed guide echoing Ptolemy's *Tetrabiblos*, it elucidates, in four major chapters, fundamental elements of astrology and methods for practicing its various branches such as birth and electional astrology.[69] The Ottoman palace library inventory lists eight copies.

(iv) Biruni's *Book of Instruction in the Elements of the Art of Astrology* (*Kitab al-tafhim li-awaʾil sinaʿat al-tanjim*): An expansive resource of astronomical and astrological knowledge, available in both Arabic and Persian renditions, it was essential for mastering precise techniques and terminology for the practice of *munajjims*.[70] The Ottoman palace library inventory lists five copies.

(v) Abu Maʿshar (d. circa 886): Although Ottoman *munajjims* frequently cited his name in their annual almanacs and birth horoscopes, specific references to the titles of his renowned treatises, including *General Introduction to Astrology* (*Kitab al-mudkhal al-kabir*), *Book on Religions and Dynasties* (*Kitab al-milal wa'l-duwal*), or *Book of Thousands* (*Kitab al-uluf*) were rather rare. The Ottoman palace library inventory includes six different titles recorded under his name.[71]

(vi) Abu'l-Mahamid b. Masʿud al-Ghaznawi's (fl. late 12th century) *Sufficiency of Learning in the Art of Astrology* (*Kifayat al-taʿlim fi sinaʿat al-tanjim*): This textbook, frequently cited by Ottoman *munajjims*, combines—in the style of Biruni's *Tafhim*—two major bodies of celestial knowledge in Persian, *hayʾa* and *ahkam*, each further subdivided into chapters and subsections. The Ottoman palace library inventory lists two copies.

(vii) ʿAli-Shah Bukhari's *Fruits and Trees in the Decrees of the Stars* (*Athmar va Ashjar dar ahkam-i nujum*): Divided into five chapters, it covers in Persian specific aspects of astrological practice such as the horoscope of the year-transfer or birth horoscope.[72] The Ottoman palace library inventory lists one copy.

(viii) Shahmardan b. Abi'l-Khayr Razi's (d. after 1072) *The Garden of Astrologers* (*Ravzat al-munajjimin*): This comprehensive introduction to astrological rules, composed in Persian, is structured in fifteen chapters and designed for novice learners.[73] The Ottoman palace library inventory lists one copy.

As for those titles on more specialized branches of astrology, such as horary and electional astrology, the *Book of Interrogations* (*Kitab al-masaʾil*) by Yaʿqub b. ʿAli al-Qasrani stands out. Little is reported about Qasrani by his contemporaries, but later sources, including Fakhr al-Din al-Razi, recognize him as an authority from the early Abbasid period, along with Abu Maʿshar.[74] His compilation on horary astrology, which aims at treating a wide array of questions a querent might ask a *munajjim*, seems to have garnered considerable attention in the early modern Ottoman realm. Mirim Çelebi, a prominent astral expert in the early sixteenth-century Ottoman world, recommends Qasrani's work to curious readers in his own treatise on horary astrology.[75] The Ottoman palace library inventory also lists four copies of the text. Besides Qasrani, Fakhr al-Din al-Razi's influential work on electional astrology, *On Sublime Elections in Heavenly Judgments* (*Al-Ikhtiyarat al-ʿalaʾiyya fi ahkam al-samawiyya*), of which Razi produced both Arabic and Persian renditions during his own lifetime, was often cited by *munajjims* in the late fifteenth- and early sixteenth-century Ottoman realm.

Relying on the knowledge accumulated from *zijes*, treatises on astronomical instruments, and textbooks imparting the principles of astrological techniques, *munajjims* crafted annual almanacs (*taqwim*) and cast horoscopes. Of these, almanacs were the quintessential medium for showcasing the *munajjims*' expertise in the astral sciences. Although *taqwims* are often called "calendars," which might sound akin to a mere day-list to a modern reader, it is essential to understand that medieval and early modern *taqwims* were much more than that. They were yearly almanacs that combined astronomical data, calendrical information, and astrological prognostications.[76] Numerous references to the practice of almanac-making from at least the tenth century, and some surviving *taqwims* or fragments of them from pre-Ottoman Anatolia suggest that it was a ubiquitous and popular genre in medieval Islamicate culture.[77] However, unlike other periods and realms of the Islamicate world, for which a dearth of evidence prevents a definitive interpretation of the significance and development of this genre, the early modern Ottoman era offers a substantial amount of *taqwim* texts that enables us to chart the genre's development, even on a yearly basis.

Taqwims were typically crafted by *munajjims* and presented to their patrons, usually rulers, around the time of the solar year transfer, that is, the spring equinox, marking the onset of the new solar year in early March—approximately March 10–11 per the Julian calendar and March 21 under the Gregorian calendar introduced in 1582. In constructing their *taqwim*, the first task of *munajjims* was to calculate the precise degree of the ascendant (*taliᶜ*) at the exact moment the Sun completed its annual cycle and transitioned into the zodiac sign of Aries.[78] It is important to note that according to the dominant, geocentric model of the universe, to which fifteenth- and sixteenth-century Ottoman *munajjims* adhered, the Sun was assumed to revolve around the Earth rather than the converse. Using the ascendant of the new year's commencement, the *munajjim* would then designate the remaining astrological houses and establish other astrological parameters. The specifics of these might differ based on the conventions the *munajjim* followed. After determining the horoscope for the upcoming year, the *munajjim* would then delve into interpreting the fortunes of various societal strata.

These general astrological forecasts for the coming year (Tur. *ahkam-ı külliye* or *ahkam-ı taliᶜ-i sal-i ᶜalem*) invariably begin with the destiny of the reigning sovereign, who was often likened to the Sun—the cornerstone of the cosmos—and to whom the almanac was usually dedicated. The narrative then proceeds to the generalized astrological predictions for six other sets of social groups, each theoretically associated with one of the remaining planets. For instance, following the examination of the sovereign's fate, the discussion moves to the fortunes of viziers and high-ranking states-

FIGURE 4. The horoscope page of ʿAbdurrahman's almanac. Süleymaniye Library, Esad Efendi Ms. 1978, 20a. Reproduced by permission from the Directorate of the Turkish Institution for Manuscripts.

men. Next come predictions for religious scholars and Sufis, followed by military personnel and governors. The fifth section covers the fortunes of women and servants, and the sixth delves into the fate of court personnel, including physicians, poets, astrologers, and scribes. The final section pertains to commoners. Upon completing the forecasts for various social classes, the *munajjim* would offer a series of prognostications about worldly matters, such as wars, diseases, crop yields, and market prices. Adhering to the conventions of the *taqwim* genre, these forecasts never named individuals directly but were couched in a generic narrative structure marked by equivocations and boilerplate.[79] Therefore, no matter how tempting it might be to ascribe an astrological prediction about a particular social group or natural event to a real historical occurrence, one must be cautious about reading these annual predictions as direct references or allusions to specific historical figures and events.

General astrological predictions concerning the fortunes of the upcoming year were then followed, over the next six folios (twelve pages), by the laborious tabulation of astronomical, astrological, and calendric information for each month of the year ahead. These monthly tables were sometimes indexed on the basis of the lunar Islamic calendar and sometimes based on the solar (old) Persian calendar. In any case, they marked the corresponding days in the major chronological systems in use, including the Islamic calendar (Ar. *al-taʾrikh al-ʿArabiyya*), Syrian calendar (Ar. *al-taʾrikh al-Rumiyya*), and Jalali calendar (Ar. *al-taʾrikh al-Jalaliyya*); identified the significant dates of feasts and fasts with additional advice on what to perform and avoid doing on these days; and often recorded in detailed fashion the planetary positions in a given month.[80] Monthly astrological forecasts, based on the horoscope of the beginning of that month, were written at the top of the tables, offering brief indications of what could be expected regarding the fortunes of various social groups and assorted natural phenomena in that particular month. Following these monthly tables came the last section of a typical almanac, wherein the compiler would note the day, hour, and duration of the expected eclipses that year, together with providing relevant astrological commentary. The section on eclipses always came at the end because eclipses were interpreted as bad omens, so it was considered inappropriate to put their astrological interpretation earlier.

As with the *taqwim* genre, the creation of natal charts and other horoscopes produced for determining an auspicious moment for initiating a planned endeavor also required *munajjims* to undertake elaborate mathematical and astronomical operations before providing their astrological interpretations. All extant birth horoscopes from the Timurid and Ottoman periods display their compilers' preoccupation with computing and explaining astronomical and astrological variables before drawing their inferences. For example, in the birth horoscope of the Timurid Mirza Rustam b. ʿUmar-Shaykh (d. 1424/5) from 1419, Yahya—likely the son of the *munajjim* who created the more famous Mirza Iskandar horoscope now kept in the Wellcome Institute Library—first meticulously details the celestial arrangement at the time of Rustam's recorded birth date. This description constitutes more than two-thirds of the entire manuscript—the first sixty-three folios of the eighty-four-folio account. When he finally moves on to relaying astrological interpretations, he emphasizes the importance of having completed precise computations, such as the determination of the zodiacal degree of the ascendant, equalization of all the remaining astrological houses, tabulation of the true longitudes of the seven planets as well as the fixed stars and astrological lots, and designation of the prorogators (e.g., *haylaj, kadkhuda, tasyirat, intihaʾat*).[81] To Yahya and many other *munajjims*, interpreting the decrees of the stars was the pinnacle of extensive mathematical and astronomical calculations.[82] One often finds a similar approach in Ottoman natal horoscopes: For example, in the natal chart produced in the late 1470s to interpret the fortunes of Mehmed II, the *munajjim* dedicated two hundred folios to discussing, in mathematical and astronomical terms, the celestial configuration at Mehmed II's reported birth.[83] Only upon establishing this foundation was he able to begin providing his astrological interpretations, which occupy the final sixty folios of his voluminous account.

Munajjims' Reservations about Their Expertise

Despite the common agreement of the astral experts on the premise that drawing astrological judgments heavily relied on mathematical and astronomical knowledge of the celestial spheres, many also acknowledged the important epistemological differences between the former and the latter. More important, not every astral expert was keen to make astrological judgments, even though they were perfectly capable of calculating astronomical positions and providing their astrological interpretations. For example, a cursory examination of the works of luminaries from the Samarkand mathematical-astronomical school, such as Qadizada al-Rumi (d. after 1440) or ʿAli Qushji, reveals that they left no text documenting their practice of *ahkam*, although their major output based on their systematic celestial observations, the *Zij-i Ulugh Beg*, like many other *zij*es, contains specific sections on horoscopic astrology for the use of practicing *munajjims*.[84] Furthermore, many *munajjims* were acutely aware of the inherent epistemological constraints in the discipline of *ahkam*, which stemmed from its conjectural (*zanni*) and unreliably experiential (*tajribi*) nature, as opposed to demonstrative (*burhani*) and apodictic (*yaqini*) knowledge of other disciplines. Some *munajjims* even expressed their hesitance or unease about practicing astrology.

This hesitation concerning the practice of *ahkam* finds pertinent expression in a pardon letter penned by an anonymous *munajjim* who seems to have served the Ottoman court around the turn of the sixteenth century. Located within the folder of written communications from the era of Bayezid II, this undated letter, composed in Persian, seeks the sultan's sympathy and forgiveness. The author confesses his recent lapse in delivering the annual almanacs with astrological prognostications, offering two reasons for his recent inattentiveness. First, he says, his attention was lately geared more toward medicine than the astral sciences; second, and more remarkably, he avows that as he has aged and perceives death drawing nearer, the task of studying the stars, especially determining their decrees, has become profoundly unsettling.[85] This letter provides an intimate glimpse into the introspection of a *munajjim*, illustrating the contested nature of astrological practice, which could lead even its practitioners to abandon the profession owing most probably to faith-based and existential concerns. The hesitance of this anonymous *munajjim* is no isolated instance: in numerous extant almanacs, *munajjims* often append notes underscoring the epistemological limitations of *ahkam* as a field of expertise.

A recurrent argument raised by various *munajjims* in different almanacs centers on the limitations of human perception in grasping the vastness and diversity of celestial influences upon the sublunary world. Many practicing *munajjims* believed that mere experiential knowledge and observation of celestial patterns were inadequate, given the virtually limitless occasions for which astrological judgments may be drawn.[86] For example, in a *taqwim* produced for the year 1495, an unnamed *munajjim* poses a rhetorical question: "Which being in the world can truly fathom the myriad celestial influences that incessantly impact the terrestrial realm?" This leads him to conclude that "the experts of *ahkam*" are fallible by definition.[87] Yet the foundational doctrines of the discipline established by past luminaries through experience and analogical reasoning (*qiyas*) maintain that terrestrial entities are indeed influenced by heavenly bodies. As the works of Ptolemy and later contributions by Abu Maᶜshar demonstrate, according to this anonymous *munajjim*, celestial configurations provide clues for terrestrial occurrences.[88]

The same arguments appear, with minor modifications, in at least ten different almanacs from the early to mid-sixteenth century. For example, in the almanac for the year 1502, an unnamed *munajjim* refers to a vast array of astrological indicators, including the projections of rays, terminal signs, progressions, astrological lots, and the varying strengths of stars. These indicators are so abundant, says the *munajjim*, that it is nearly impossible to track each with precision and accuracy. Consequently, he argues, the science of *ahkam* can offer only a glimpse into potential future events in the terrestrial realm. Moreover, this discipline relies more on unreliable experiential and speculative (*zanni*) knowledge than on unwavering mathematical proofs and geometrical demonstrations.[89] An underlying empiricism, however, is impossible to fully attain, as the terrestrial effects of celestial configurations are so numerous and the empirical data showing these relationships so inherently fragmented and inconsistent. No one can accumulate the comprehensive experiential knowledge of celestial influences in a single lifetime. Nevertheless, thanks to the experience-based findings of past astral experts and the use of analogical reasoning, predictions about future sublunary events are possible, though these predictions are destined to remain conditional. These internal epistemological constraints of astrology prompted *munajjims* to use discreet language in their forecasts. Typically, they emphasized the probabilistic nature of their craft, frequently employing qualifiers that express prospects and expectancies, such as *mumkin*, *yahtamilu*, *ihtimal-i*, or *umid ast*—when making their predictions.

Equally noteworthy is that, despite frequently facing accusations of astral determinism, primarily from Sharia-oriented and Sufi-minded individuals, *munajjims* consistently refrained from implying any such determinism in their writings.[90] Instead, they typically framed their references to celestial influences within the context of divine will, underscoring the belief that the entire cosmic order operates under God's omnipotent control. The Ottoman astrological corpus from the fifteenth and sixteenth centuries abundantly documents instances where *munajjims* peppered their astrological predictions with clauses, provisions, Qur'anic verses, and hadith (the traditions of the Prophet Muhammad). These include phrases such as "with the permission of God," "God is the most knowing one," "with Him are the keys of the unseen," "No one knows the unseen except God," or "Whatever Allah willed to be, shall be, and whatever Allah did not will to be, shall not be." While these inclusions may be viewed as mere gestures or face-saving attempts to deflect potential charges of heresy and infidelity, a more plausible and charitable reading, in light of the *munajjims*' own cautious remarks about their craft as mentioned above, suggests that most *munajjims* did genuinely reject astral determinism and understand celestial influences as subordinate to God's infinite command. Even if one insists on viewing all these remarks as disingenuous, face-saving attempts rather than sincere expressions of self-doubt or reservations about their expertise, such remarks still testify to the constrained epistemic and social ground upon which *munajjims* had to operate.

The Sites of *Munajjims*' Training

Where and how could individuals with aspirations to become a *munajjim* learn the specialized disciplines and texts essential for their profession? Given the long-standing absence of an observatory in the early modern Ottoman realm, with the exception of Taqi al-Din's short-lived one in the late sixteenth century, were there institutionalized means for vocational training in this field? Did madrasas, the institutes of higher education, cater to aspiring *munajjims*, equipping them with the requisite skills and knowledge? How instrumental were roles such as the timekeeper (*muwaqqit*) at mosques and court astrologers at the palace in preserving and transmitting this specialized knowledge through the ages?

The observatories that were sporadically founded in the medieval and early modern Islamicate domains, especially in the Persianate realm, stood as the key institutional framework for aspiring and established experts in the mathematical-astral sciences, facilitating their assembling and aggregating relevant texts and tools.[91] Beginning with the reign of the Abbasid caliph al-Maʾmun (r. 813–33) and gaining a marked momentum in the post-Mongol era, several observatories sprouted across the central and eastern Islamicate territories. The underlying reasons for the notable enthusiasm in the post-Mongol Islamicate East for building state-of-the-art observatories and bringing together experts in the celestial arts remain unclear. Nevertheless, heightened interactions with Chinese civilization, where the scientific study of the heavens was institutionally promoted, may have influenced the cultivation of similar expertise in the Islamicate East, the vast region stretching from Iran to China.[92] Aydın Sayılı, in his pioneering work on observatories in Islamic history, also posits that Turko-Mongol dominance over the Islamicate East from the thirteenth century might have played a role in fostering celestial investigations. Their enthusiasm may be attributed not only to their pre-Islamic beliefs, centered around the idea of the sky god (*gök Tengri*), but also to the fact that many patron dynasts and scholars in the region were less strictly indoctrinated in the stringent Sunni Islamic traditions that usually frowned upon the practice of astrology.[93]

As previously discussed in relation to the production of *zij*es, the primary objective of constructing a new observatory and conducting systematic observations was to rectify the existing tables that were indispensable for *munajjims*' practice. However, the operations of many observatories were short-lived as the result of inconsistent financial support on the part of ruling elites, unstable political circumstances, and the overall lower social and religious standing of *munajjims*' expertise.[94] While they were in operation, observatories undoubtedly served as a space to facilitate private instruction and apprenticeship between experienced experts and aspiring students of celestial knowledge who were every so often connected by family ties.[95] Some of these observatories were also accompanied by specialized madrasas, where different matters related to the mathematical and astral sciences evidently were taught. Thanks to a few extant letters and teaching certificates (*ijaza*) from key members of the Samarkand observatory in the first half of the fifteenth century, we have concrete evidence of both theoretical and hands-on practical celestial instruction at the Samarkand observatory and its affiliated madrasas.[96] While none of these sources specifically mention the teaching of an exclusively astrological textbook in the classroom, Jamshid al-Kashi's letters to his father reveal that students—including a significant number of *munajjims*—actively utilized *zij*es and addressed problems related to calculating the ascendant (*taliʿ*).[97]

Was instruction in the astral sciences available in other madrasas that were not affiliated with observatories? This question is inherently tied to broader inquiries about the role and scope of madrasas in the medieval and early modern Islamicate landscape, and more specifically about the teaching in madrasas of "rational" sciences (*al-ʿulum al-ʿaqliyya*) and the sciences of the ancients—besides the mainstream, transmitted sciences (*al-ʿulum al-naqliyya*). The publication in 1981 of George Makdisi's *The Rise of Colleges* engendered a new wave of scholarly interest in discerning the extent of pre-modern madrasas.[98] His contention, defining madrasas exclusively as "colleges of law," with a distinctly institutional character and defined curriculum of traditional Islamic sciences, faced criticism from several scholars.[99] The main criticism revolved around Makdisi's leaning toward viewing the madrasa as the sole avenue for learning and education in the medieval Islamicate world. Later critics emphasized that madrasas were not the only site for education and knowledge transfer and that, in the medieval Islamicate context, *personae* were much more important than *loci*.[100] That is to say, what really mattered in terms of the scope of teaching in medieval Islamicate learning was not the name of the school whence an individual graduated but rather those of the scholars under whom one studied certain books and topics chosen at the preference of the instructors.

Makdisi's critics have opened up new vistas for a more nuanced exploration of the role the mathematical and "rational" sciences might have played in the knowledge regimes of the medieval Islamicate world. The revisionist perspective, with its emphasis on the deeply personalized nature of education and the role of individual preferences of specific instructors, implies that the study of the astral sciences might have taken place within madrasas, depending on the intellectual proclivities of individual teachers. However, this viewpoint runs the risk of perpetuating certain long-held scholarly convictions that suggest that genuine interest in the less mainstream, "rational" sciences during the medieval Islamic period was only sporadic and informal, limited to a few inquisitive minds and largely devoid of a solid institutional grounding.[101] In that regard, the shift from Makdisi's excessive emphasis on madrasas as institutions dedicated solely to certain sciences to the revisionists' stress on the personalized nature of learning does not adequately resolve the question of whether a consistent institutional mechanism for disseminating the astral sciences existed in the medieval Islamic world. In other words, the central question posed by Abdelhamid Sabra over thirty years ago still remains to be tackled in the case of *munajjims*' expertise: "How did a significant scientific tradition maintain itself for such a long time largely outside the only stable institution of higher learning in medieval Islam?"[102]

Several important publications in the last two decades have presented compelling evidence, primarily about the post-thirteenth-century Islamicate East, that documents *hayʾa* being taught in madrasas, either as a stand-alone discipline or subsumed within philosophical theology (*kalam*). Given that *hayʾa*, as a discipline, focused intensely on studying the celestial spheres as manifestations of God's creation, it is little wonder that many scholars and students, deeply immersed in theology, also delved into *hayʾa*. Manuscript records, including ownership statements and colophons that offer information about who copied or owned texts, affirm that specific textbooks on *hayʾa*, such as Jaghmini's *Epitome*, were widely used and commented upon by madrasa-affiliated individuals. Robert Morrison's research on Nizam al-Din Nisaburi (d. 1328/9), a fourteenth-century polymath associated with Tusi's student Qutb al-Din Shirazi, vividly illustrates the place of the astral sciences within an established madrasa scholar's broader intellectual quests, which also involved philosophical theology and Qur'anic exegesis.[103] Similarly, Sonja Brentjes, Sally Ragep, İhsan Fazlıoğlu, and the late Cevat İzgi have documented, using manuscript evidence, that specific books on *ilm al-hayʾa* and other facets of the mathematical sciences gradually became part of the canon taught in certain madrasas.[104] For Ragep, Fazlıoğlu, and İzgi, the proliferation of surviving copies of titles such as Jaghmini's *Epitome* and Qadizada al-Rumi's later commentary on Jaghmini's text, some of which were copied by madrasa instructors and students across diverse regions and periods, signifies their widespread transmission within the formal madrasa canon.[105] Beyond such manuscript evidence, various other sources from the period also confirm that works on *hayʾa* were studied within madrasas. In addition to the letters of Jamshid al-Kashi, extant teaching certificates from the fifteenth century, and some catalogs of sixteenth-century Ottoman madrasa libraries, there are certain autobiographical accounts documenting the study of *hayʾa*. Taşköprülüzade, one of the pivotal scholar-bureaucrats of the mid-sixteenth-century Ottoman intellectual realm, recounts in an autobiographical passage that while he was a madrasa student in the early sixteenth century, he studied ʿAli Qushji's treatise on *ʿilm al-hayʾa* under Mirim Çelebi.[106] Taşköprülüzade does not specifically mention where this study took place, but his remarks, together with other supporting evidence mentioned above, demonstrate that mathematical and theoretical astral knowledge, which had not been a foundational focus of madrasas in earlier centuries, began to percolate into some of the established institutions from the thirteenth century onward.

When it came to the more practical, astrologically more relevant facets of *munajjims*' expertise, including proficiency in using *zijes* and astronomical instruments, crafting annual almanacs with astrological prognostications, and rendering astrological forecasts in the form of birth horoscopes or elections, madrasas were of limited utility. This was in contrast to late medieval and early modern Europe, where astrology was part of the curriculum at several universities and there were even special chairs for the teaching of astrology.[107] Certainly, there were madrasa instructors who, at various stages in their careers, compiled annual almanacs, held copies of *zijes* and treatises on astronomical instruments, copied manuals on different astrological techniques, or taught these techniques to their madrasa students.[108] Yet, unlike *hay'a* texts, evidence for the systematic penetration of astrologically more pertinent knowledge within madrasas is scant, making it difficult to argue that madrasas operated as a site for producing and disseminating the specialized astrological knowledge aspiring *munajjims* required. Moreover, reservations regarding the epistemological foundations and principles of *ahkam* were frequently harbored by madrasa-affiliated individuals, which made the systematic teaching of this science in madrasas less likely. Additionally, endowment deeds (*waqfiyya*) that offer insight into the structure and content of instruction in endowed madrasas, along with extant inventories of madrasa libraries, present little to no evidence for the teaching of strictly astrological knowledge in madrasas. For example, multiple versions of the *waqfiyya* for the Fatih Mosque complex in Istanbul, produced between the late fifteenth and early sixteenth centuries, do not describe or imply any astrological content within the required knowledge of the instructors to be appointed.[109] Similarly, the *waqfiyya* prescribing the functions and functionaries of the Süleymaniye complex in the mid-sixteenth century, including the instructors teaching at the Süleymaniye madrasas, does not specify the exact scope of the expected erudition of instructors other than prescribing that they should be learned in both the transmitted and rational sciences.[110] Specifically, in the *waqfiyya* of the Süleymaniye complex, which also housed a medical school, there are references to proficiency in the medical sciences for potential instructors at the medical school. Despite the occasional use of astrological knowledge in medical practice, no specific mention of it is found in the document. Finally, the two detailed inventories produced in the sixteenth century, listing the library holdings of the Fatih madrasas and registering nearly two thousand titles available in the most prestigious madrasas of the Ottoman capital, reveal that the total number of titles strictly related to astrology is marginal (less than 0.5%).[111]

In contrast to madrasas, grand imperial mosque complexes that often employed personnel for timekeeping tasks—known as *muwaqqits* (also referred to as *sa'ati* or *miqati*), from the Arabic word *waqt* meaning time—would seem to be a more viable institutional framework for aspiring *munajjims* to hone their expertise. Modern historians of science tend to draw a stark distinction between *muwaqqits* and *munajjims*. While *munajjims* are commonly conceived of as simple fortune-tellers involved in the practice of astrology, *muwaqqits* are thought to be esteemed scholars in the revered discipline of timekeeping (*'ilm al-miqat*), focusing solely on calculating the five daily prayer times and the direction of *qibla* and ostensibly steering clear of astrology.[112] George Saliba and François Charette have argued that because of their affiliation with religious institutions, *muwaqqits* refrained from astrological practices.[113] Sonja Brentjes, however, has provided evidence, drawing from the Mamluk biographical accounts, that some *muwaqqits* in the Mamluk realm did indeed engage in astrological pursuits.[114] More evidence for this argument has also been documented by David King and Fien de Block in their research into the extant manuscripts of Mamluk-era *muwaqqits*.[115]

The widespread categorical demarcation of *muwaqqits* from *munajjims* stands in contradiction to historical evidence, particularly that originating from the Ottoman period. These sources highlight instances where *muwaqqits*, thanks to their all-encompassing expertise in the astral sciences, either smoothly transitioned from their timekeeping roles in mosque complexes to serve as court *munajjims* or performed both roles simultaneously. For example, Necmeddin b. Seyyid Muhammed (d. after 1553), who served as the timekeeper at the Old Mosque in Edirne in the 1520s and later at the Fatih Mosque complex in Istanbul in the 1540s and 1550s, was also busy crafting almanacs with detailed astrological predictions, five of which have survived.[116] During his tenure as the *muwaqqit* of the Old Mosque in Edirne, archival records list Necmeddin as one of the court *munajjims* receiving a regular stipend.[117] Later decades and centuries offer more examples of *muwaqqits* from various mosque complexes being recruited into the courtly cadre of *munajjims*. A case in point is an imperial decree from April 1574, which records the (dual) appointment of Ahmed, the *muwaqqit* of the Sultan Selim Mosque in Istanbul, as a student-assistant to the court *munajjims*, with a daily remuneration of two *akçes*.[118]

The overlapping roles of *muwaqqit* and *munajjim*, as vividly portrayed in Ottoman sources, nicely capture the fact that these two professions fundamentally drew upon the same scientific expertise. They engaged with similar texts—some of which unambiguously fell within the realm of *ahkam* (astrology)—and provided analogous services grounded in the expertise of measuring and interpreting time in its various forms. A further illustrative case is the extant oeuvre of Mustafa (d. 1571), the timekeeper of the Sultan Selim Mosque in Istanbul by the mid-sixteenth century, who reportedly held the position of chief court astrologer in the 1560s.[119] Even within his numerous extant treatises, which focused primarily on astronomical instruments or timekeeping, there are extensive sections dedicated to astrological principles.[120] Manuscript copies of various astrological treatises or archival registers also abound with traces of *muwaqqits* as authors, readers, copyists, or owners of these texts. For instance, registers of gifts and allowances from different periods in the sixteenth century list several timekeepers receiving payments in return for the almanacs with prognostications that they presented.[121] The extant birth horoscope of Prince Ahmed (d. 1513), the favorite son of Bayezid II, was penned by a *muwaqqit* named Muhammad.[122] One of the extant copies of ʿAli-Shah Bukhari's aforementioned astrological textbook, *Athmar va ashjar*, belonged to a *muwaqqit* from Edirne in the later sixteenth century.[123] Similarly, the Turkish translation of Qasrani's authoritative textbook in horary astrology was completed in the late seventeenth century by Kasımpaşalı ʿOsman b. Muhammed (d. after 1691), who was the *muwaqqit* of the Fatih Mosque complex at the time.[124] Delving further into extant copies from the *zij* tradition would only uncover more examples underscoring this connection.

A question of profound significance for our exploration here, as it relates to the locales of *munajjims*' learning, is whether any designated physical space was allocated to *muwaqqits* within mosque complexes. The extant endowment deeds from the late fifteenth- and sixteenth-century imperial mosque complexes, including those of the Fatih, Bayezid, and Süleymaniye complexes, often contain clear provisions outlining the required qualifications and daily remuneration of the individual to be appointed as a *muwaqqit*. The Süleymaniye *waqfiyya*, for instance, stipulates that the *muwaqqit* ought to possess "an in-depth comprehension of the overarching theories and specific practicalities of the science of the stars," and that in return for his services he would receive ten *akçe*s daily.[125] However, what is notably missing from these endowment deeds and contemporary narrative sources is any reference to a particular space or chamber within the mosque complex assigned as the *muwaqqit*'s office. It is mostly in written sources and surviving architectural remnants from later centuries that we discern a distinct spatial unit called "the timekeeper's chamber" (Tur. *muvakkithane*), strategically situated on the exterior of the complex or sometimes erected as a separate, individual building, facing the bustling thoroughfares of the street.[126] The majority of these structures, typically comprising a small, single-story building equipped with mechanical clocks of varying sizes and astronomical instruments, were constructed after the eighteenth century, particularly during the times of Sultan Mahmud II (r. 1808–39) and ʿAbdülmecid I, as part of the surging public demand for precise timekeeping for the regulation of modern urban life.[127] Today, some of the extant *muwaqqit* houses in Turkey are used as coffee shops and kiosks, with most of their customers and passersby unaware of their past. Based on the architectural evidence originating from later centuries, modern scholars studying the institution and profession of *muwaqqit*s have surmised that these spaces must have been prevalent, too, in earlier centuries and that they operated almost like a small observatory and a site dedicated to the study of the astral sciences.[128] Our understanding of the architectural presence and functions of it in earlier centuries, however, remains shrouded in uncertainty owing to scant references to a dedicated space, or even a room, for *muwaqqits*.[129]

Court and Patronage

In the relative absence of an educational institution that would have provided space for aspiring *munajjim*s and enabled the transmission of their astrological expertise, one might wonder—echoing Sabra's line of inquiry—how this scientific tradition maintained itself for centuries. The answer, in many respects, lies at the court. Predominantly the princely household and, in some cases, the residence of ruling elites and wealthy urbanites, the court served as the primary institutional platform where *munajjims* emerged and thrived.[130] More than just providing a livelihood for practicing *munajjims*, this setting fostered private instruction and master-disciple relationships, ensuring the transmission of their expertise. As will be detailed in subsequent chapters, many actors who later rose to prominence as *munajjims* in the Ottoman and contemporary or earlier Persianate realms had learned and apprenticed their art at the hands of an erudite court *munajjim*. Although the details about the relationships between master and apprentice *munajjims* do not shed light on the exact locations where instruction took place, the master *munajjim*'s presence at the royal court no doubt functioned as a pull attracting novices and facilitating the transmission of knowledge thanks to their easier access to tools and texts kept at royal repositories.

Perhaps few disciplines leaned so heavily upon royal patronage as theirs, a dependency that had both material and sociocultural ramifications. Compared to knowledge in many other disciplines that relied simply on texts, the *munajjims*' art possessed an additional material layer, requiring access to specialized, precision instruments, such as astrolabes, celestial globes, quadrants, and even mechanical clocks. Unlike texts and books, these instruments were not always readily available or affordable, a complaint occasionally voiced by *munajjims*. Thus, the princely courts, with their rich collections of astronomical instruments in addition to books, provided *munajjims* with the materiel they needed. Moreover, beyond the material concerns of access to tools and texts, *munajjims* encountered even more intricate challenges in the sociocultural context within which they had to operate. While their expertise was deemed useful—in the end, the royal and public appeal to *munajjims* continued for many centuries—the scholarly prestige enjoyed by *munajjims* was arguably lower than that of madrasa scholars.[131] Moreover, their specialty was continually subjected to skepticism, from both an epistemological and a religious standpoint. Such suspicions were perhaps the primary factor keeping aspiring *munajjims* out of madrasas, which were populated by those devoted to the study of law and other scholastic disciplines linked to judicial training. Large mosque complexes did employ *muwaqqits*, whose expertise closely mirrored that of the *munajjims*, but the compensation they offered was notably meager. Furthermore, unlike their European peers, who had the luxury to pub-

FIGURE 5. An Ottoman astrologer observing the 1577 comet. From Mustafa ʿAli, *Nusretname*. Topkapı Palace Museum Library, Ms. H 1365, 5b. Reproduced by permission from the Directorate of National Palaces.

lish, sell, and widely circulate their works, such as annual almanacs and prognostications, *munajjims* in the Ottoman and broader Islamicate world were mainly bereft of the benefits of a profitable book market.[132] Adding to these structural voids was the absence of a professional corporation for *munajjims*, akin to the guild structures that safeguarded members' benefits in other business sectors of the medieval and early modern Islamicate realm.[133] Confronted with these constraints, *munajjims* had little choice but to seek the support of wealthy benefactors, preferably ruling sovereigns, to ensure their livelihood and sustain their craft.

In the patronage of sciences and arts in any court culture, it is essential to distinguish between support for individual experts in a specific craft and systematic institutional backing for their domain of expertise. As Sonja Brentjes astutely notes in her analysis of courtly patronage of the mathematical sciences in the pre-modern Islamicate world, a central tenet of the patronage culture, especially of royal patronage, was that "[r]ulers supported scholars, not disciplines."[134] While royal backing for individual experts did at the same time encourage the growth of that particular expertise, at least during the benefactor's lifetime, Brentjes's observations serve as a reminder of the transient and personalized nature of scientific patronage in pre-modern cultures, which was largely shaped by individual patrons' personal intellectual appetites and ideological aspirations. Thus, in the pre-eighteenth-century Ottoman and Islamicate milieu, where it was difficult to find a deliberate and systematic state policy promoting science and technology, the whims and wishes of individual patrons played a pivotal role for experts in those sciences that lacked other institutional means to fulfill their intellectual pursuits. Yet as the tastes and pressing concerns of these patrons or princes evolved, so did their preferences and priorities. Hence, *munajjims*, like certain artists and experts in various domains, often found themselves grappling with the capricious nature of this intricate patronage system.

A notable feature of the early modern Ottoman court's patronage of *munajjims* was the emergence of a special position, even a quasi-bureaucratic office for these experts, by the late fifteenth century. Known as the office of the chief *munajjim*, this position remained intact until the empire's official end in the twentieth century. Over its more than four-centuries-long life, commissioning services that ranged from submitting annual almanacs to reporting auspicious times for imperial actions, the position housed around forty chief court *munajjims*, along with many others as their associates and assistants.[135] The Ottoman case thus exemplifies a remarkable institutional framework that facilitated the transmission of expertise and the training of future generations of *munajjims*. Yet even within such a structured system, the affinities and predilections of individual patrons played a significant role. The number of *munajjims* recruited to the office at a given time together with the overall vogue for their expertise, traceable in their remuneration and the amount and frequency of astrological texts produced during that period, fluctuated considerably across different eras, reflecting the varying personal tastes and attitudes of the sovereign patrons.

Just as *munajjims* depended on the backing of their patrons, those patrons in turn had distinct interests in supporting the *munajjims*. The ruling elites' appeal to astral experts is a globally common theme in court life throughout the ancient, medieval, early modern, and even modern eras. From the viewpoint of these ruler-patrons, having a *munajjim* or a legion of them had several obvious practical benefits. With their proficiency in mathematics and geometry, their knowledge of different calendar systems, and their astrological know-how, *munajjims* could perform myriad computations and predictions vital for imperial decision-making. In theory, *munajjims*' expertise in the mathematical and astral sciences could have been utilized in a wide range of areas, from calendar conversions and intercalation to navigation, land surveys, and, naturally, prognostications about the future. Whether or not this expertise was harnessed to its fullest potential remains a matter of debate. Evidence culled from Ottoman documents suggests that *munajjims*' expertise was sought exclusively for astrological purposes, at least during the period in question. A case in point concerns the periodic fiscal complications arising from the discrepancy between solar and lunar calendars and the role (or lack thereof) of *munajjims* therein. While the collection of agricultural revenues was based on the solar calendar, salaries were disbursed according to lunar months. Given the eleven-day disparity between a lunar and a solar year, the Islamic lunar calendar that the imperial administration utilized effectively lost a year's worth of revenue every thirty-three years (called a *sıvış yılı* in Turkish; "absconded year").[136] Such occasional financial challenges caused by calendrical discrepancies could, in theory, have been elucidated and even averted by *munajjims*. However, by the mid-sixteenth century, when the financial impact of the calendar controversy became pronounced, maybe for the first time, Ottoman finance officials and ruling elites seem not to have considered seeking the *munajjims*' expert insights.[137]

The enduring appeal of the astrological expertise of *munajjims* for ruling elites from antiquity even down to modern times, across both Eastern and Western cultures, is vividly captured by a late fifteenth-century encyclopedist in his treatise on the taxonomy of sciences presented to the Ottoman sultan Bayezid II. In this account, the encyclopedist Shirvani (fl. late 15th century) underscores the importance of the astral sciences for rulers and sultans. He posits that natural disasters and societal upheavals—be they earthquakes, floods, wars, famines, or plagues—occasionally arise in the sublunary world as the result of the influence of planetary conjunctions, eclipses, and various celestial configurations. One can hope to avoid misfortune and harm, says Shirvani, by mastering this science and closely tracking these celestial phenomena.[138] A similar rationale was echoed by Tusi when advocating for Hülegü Khan's financial support for his new observational enterprise in Maragha.[139] For Tusi and many like-minded astral experts, *munajjims* possessed the distinctive ability to discern the *kairos* (the right, opportune moment) amid the enigmatic flow of *chronos* (time), thereby providing some semblance of order and predictability to the inherent chaos and uncertainty of the world.[140] The proposition that *munajjims* bring some form of order to this inherent chaos constitutes their cosmological function, especially when the cosmos is considered, in Federico Campagna's terms, "the product of an act of ordering chaos."[141] For ruling elites, this presumed authority of *munajjims* in tracking and interpreting the movements of celestial bodies, hence, providing a measure of order, was perhaps the most enticing facet of their expertise, resonating deeply with the perennial aims of medieval and early modern rulers to establish order and maintain hierarchy as heavenly ordained rulers, mirroring the cosmic order of the heavens.

In addition to relying on the interpretations and predictions of *munajjims* for imperial initiatives, the royal patronage of these experts also served as a potential political tool and even a formidable medium of propaganda.[142] Its utility functioned on two levels: First, by supporting a body of scientific experts, the sovereign or other ruling elites could project an image as a generous patron of men of knowledge—even when the expertise of the endorsed group was viewed skeptically by their contemporaries. If the royal patron was also personally invested in studying the science they supported, this would further burnish his reputation, elevating him in the eyes of at least some contemporaries to the status of a learned ruler or even an idealized "philosopher-king."[143] Second, and more specifically related to the patronage of astral experts, the *munajjims*' interpretation of worldly events based on heavenly phenomena—coupled with their often sycophantic praise for the reigning sovereign, adorned with celestial metaphors and jargon—imbued the ruler's reign with an aura of heavenly sanction and cosmic importance. This served to legitimize and enhance the sovereign's standing, casting it as preordained by celestial signs.

Conclusion

Despite—or perhaps because of—the consistent presence of *munajjims* at the courts of medieval and early modern Muslim rulers and their close ties to ruling elites, the practice of *munajjims* regularly sparked controversy within society. Previous scholarship on the perception of *munajjims* in medieval Muslim society has explored a vast body of polemical literature, very much parallel to relevant writings from medieval and Renaissance Europe, where individuals from diverse intellectual backgrounds, including philosophers, theologians, jurists, historians, imams, Sufi shaykhs, and others, vehemently criticized *munajjims*.[144] When we look at the writings of some of the staunchest opponents of astrology in medieval Islamicate culture, including the physician Ibn Sina, the Asharite theologian-turned-mystic al-Ghazali (d. 1111), the Hanbali jurists Ibn Taymiyya (d. 1328) and his student Ibn Qayyim al-Jawziyya (d. 1350), and the polymath historian Ibn Khaldun (d. 1406), it appears clear that although each relied upon and sprang from a distinct intellectual tradition, none of them in fact denied the unbroken chain of causation between the heavenly and terrestrial realms. Ibn Sina, as a good Aristotelian, for example, conceded the premise that terrestrial events are influenced by celestial causes. However, he disputed *munajjims*' assertions of fully understanding the dynamics of this causality and of predicting future events.[145] To Ibn Sina, as indeed to many practicing *munajjims* as explored above, humans were innately incapable of comprehending such an extensive cascade of causes.

Critics of *munajjims* in medieval Islamic society put forth a diverse range of arguments, touching on the epistemological and methodological limitations inherent in their science, as well as potential societal issues stemming from the belief in *munajjims*' false prognostications. These criticisms frequently highlighted instances of failed predictions, miscalculations, or blatant errors made by *munajjims*, without specifying any particular practitioner. Such instances were shared as proof that the *munajjims*' knowledge lacked credibility and therefore should be dismissed. Central to these critiques is associating *munajjims* with astral determinism. Within this discourse, *munajjims* were often relegated to the status of magicians, sorcerers, and street diviners, who were thought to blasphemously ascribe independent creative powers to celestial objects. Moreover, certain sayings attributed to the Prophet Muhammad, such as "All *munajjims* are liars" and "Whoever believes in the stars commits heresy," were repeatedly invoked in these discussions to further discredit *munajjims* from a strictly religious standpoint.

The intensity, frequency, and fervor of these critiques in the medieval and early modern Islamicate world have been interpreted by new generations of scholars of the occult sciences as evidence that *munajjims*, as occultists, had formidable influence within society. While such an interpretation is not incorrect—after all, who would have bothered assailing a group with only marginal impact?—it is insufficient to take these polemics solely as an indication of the *munajjims*' prominence and popularity in the medieval and early modern Islamicate world. What is often missing here, amid all the criticism, is the perspective of the *munajjims* themselves regarding their expertise. The widespread skepticism toward *munajjims* and their expertise, per se, indicates how arduous a business it was to be a *munajjim*. Beyond the challenges they regularly encountered in their education and practice—from mastering the necessary technical knowledge and accessing the required instruments to finding the right people to sustain their scientific endeavors—they also had to withstand persistent attempts to undermine their credibility and authority. Under these manifold challenges, some *munajjims* chose to leave their specialization and transitioned into other professions, while others persevered, continuing their work despite the obstacles and financial concerns. A closer look into the personal stories and social biographies of *munajjims* from the Ottoman domain, as gleaned from available autobiographical and biographical fragments, will inform our discussion in the following chapter.

TWO

PERSIANATE FOUNDATIONS

Having provided a general outline of *munajjims* and the scope of their expertise, we narrow our focus in this chapter to those practitioners of the astral sciences operating within the Ottoman Empire, specifically during the late fifteenth and early sixteenth centuries—a pivotal period for the institutionalization of court patronage for such experts. The Ottoman practice of employing *munajjims* and utilizing their expertise stands as a riveting, yet largely unexplored, chapter in the broader history of the sciences in Ottoman, Islamicate, and world history. While the Ottomans were neither the first nor the only pre-modern polity that harnessed the expertise of astrologers, their creation of a position—almost an office—within the court bureaucracy for *munajjims*, which ensured a constant presence of astral experts in the courtly milieu for over four centuries, is particularly striking. This institution brought forth a vast paper trail, distinguishing the Ottoman experience of recruiting astral experts from other Islamicate and even European societies. From the mid-fifteenth to the early twentieth century, nearly forty *munajjims* served the Ottoman ruling elites in the capacity of chief court *munajjim*, and this does not include their deputies and others in court service.[1] The scope of court *munajjims*' service is well documented in the rich archival materials preserved by the bureaucratically minded Ottomans. These materials include not just astrological content—such as annual almanacs and brief reports of auspicious moments routinely submitted by court *munajjims*—but also correspondence and petitions written either by or about *munajjims* when issues and concerns arose about their appointments and services. Additionally, detailed payment registers bearing information on the appointment and compensation of various court officials provide insights into the identities and remunerations of court *munajjims*.

Each set of these extensive sources offers peculiar insights while also presenting certain limitations. The technical writings of Ottoman court *munajjims*, particularly the almanacs, birth horoscopes, and brief reports of their calculations for auspicious moments, illuminate the scope of their expertise in ways that are hardly paralleled across other periods and regions of Islamic history. While the scholarship on astronomy and astrology in the Islamicate world is undeniably rich, especially when it comes to the publication of foundational texts and treatment of key figures, the current literature does exhibit certain gaps and biases.[2] A prominent concern with existing works is their pronounced emphasis on Arabic materials from the early centuries of Islam, reinforcing—maybe without deliberate intent—the misleading Golden Age narrative, which, since its genesis in the nineteenth century, has shaped and still guides both scholarly and popular discourse on the histories of science in the Islamicate world. Notwithstanding the foundational contributions of modern scholars regarding early Islamic luminaries in the astral sciences, including Abu Maʿshar, Mashaʾallah, and Biruni, and the transmission of astrological and astronomical knowledge from the Indian, Sassanian, and Hellenistic traditions to the Islamic world, an excessive focus on earlier periods and a heavy reliance on Arabic sources have inadvertently perpetuated these widespread misconceptions. For that reason, the rich technical corpus of Ottoman *munajjims*, written predominantly in Ottoman Turkish and Persian, along with Arabic, provides invaluable perspectives on the practice of the astral sciences during the "postclassical" Islamic era. Moreover, compared to Arabic textbooks and manuals of earlier periods, which by nature offer scant insight into how the rather theoretical and technical knowledge of the celestial spheres was put into concrete practice, the textual output of Ottoman *munajjims* elucidates how multiple strains of celestial knowledge were applied in specific historical contexts to meet varying needs under different circumstances.

Technically less relevant archival materials produced by and about *munajjims*, which include petitions, letters, and voluminous bureaucratic ledgers related to court *munajjims*, further reveal the distinctive nature of the Ottoman experience. Unlike previous studies on the social and cultural history of *munajjims* in the medieval and early modern Islamicate world, which rely almost exclusively on anecdotal evidence derived from narrative sources and accounts penned by figures other than *munajjims*, the study of *munajjims* within the Ottoman realm offers a chance to reconstruct their courtly presence and service based on hard, state archival evidence. Two significant limitations of these archival documents, however, should be noted. First, they contain information only about the practitioners connected with the dynastic court, thus overlooking those outside its patronage network. It goes without saying that there were *munajjims* operating independently in the provinces and core territories of the empire without any ties to the dynastic court. However, the shortage of documentation about these practitioners precludes their integration into the narrative of *Forgotten Experts*. Second, the details found in archival documents about court *munajjims*, especially in registers of payments, are typically restricted to individuals' names and salary brackets, with little else to glean about their life stories, educational background, social relationships, or daily interactions.

Beyond the technical sources composed by *munajjims* and the corresponding archival documents about them, historical narratives and contemporary biographical accounts sporadically provide useful anecdotes. Nonetheless, such anecdotal tidbits can rarely be verified or supplemented by archival or autobiographical data. Only a select few *munajjims*, whose names occasionally appear in payment ledgers or are briefly mentioned in narrative sources, afford us a window onto reconstructing and recounting their lives in all their complexities. These life stories, however, are not always readily presented in comprehensive autobiographical accounts. They need to be pieced together from scattered fragments of *munajjims*' writings (many of which remain in hitherto unexamined, even uncataloged, manuscripts), traces of their annotations in books they once possessed, and brief testimonials of their contemporaries. Although we may never fully uncover, at least for the earlier centuries under examination here, the intricate details of Ottoman *munajjims*' daily interactions with clients, the specific methods of their education, favored techniques in the exercise of their craft, or the entirety of their career trajectories, the available details still offer a colorful perspective on the scope of *munajjims*' expertise and practice, their scholarly influences and affiliations, relationships with their patrons and peers, and the challenges they faced when aspirations went unrealized.

Building on the previous chapter, which broadly covered *munajjims'* craft in the pre-modern Islamicate world, this chapter will examine the social and intellectual lives of two *munajjims* from the late fifteenth- and early sixteenth-century Ottoman realm: ʿAbdurrahman (d. after 1510), curiously referenced in some early modern Ottoman sources as the Ottomans' first chief court *munajjim*, and Khitabi (d. between 1512 and 1515), who was also known as the *sayyid munajjim*, a title indicating both his professional prominence among his peers (*sayyid* literally means chief or prominent) and his self-proclaimed lineage from the Prophet Muhammad (*sayyid* also refers to those descended from the Prophet). The two men's stories exhibit both shared traits and idiosyncratic distinctions reflecting the practice of the astral sciences amid the Ottoman sociopolitical and imperial order in the making. Flourishing in the late fifteenth and early sixteenth centuries, their careers coincided with a surge in courtly interest in the astral sciences, owing largely to the genuine curiosity of Sultan Bayezid II and his inner circle about this field of expertise, a topic that will be explored in greater depth in the next chapter. Unlike later court *munajjims* who acquired their expertise in the Ottoman domains and wrote primarily in Ottoman Turkish, these two, along with several others whose lives cannot be fully reconstructed except for their evident Persianate connections, were trained in Iranian lands and authored their works in Persian. This heritage allows us to trace Persianate influence on the formation of indigenous Ottoman cadres and customs of the astral sciences.

Although not all their compositions have survived, the shared diversity of these two practitioners' compositions—ranging from astrological almanacs and horoscopes to medical compendia—as well as the particular circumstances under which these texts were composed shed light on their common strategies when jockeying for royal support and rebranding their expertise whenever needed. Both also sought to advance their scientific endeavors, refining the technology they relied upon—specifically, astronomical handbooks with tables and astronomical instruments—and undertaking scholarly journeys for the sake of expanding their specialized knowledge. Moreover, the former served in provincial settings in the service of a contending prince, while the latter progressed from a provincial prince's service to a central role for and direct acquaintance with the regnant sultan. Notably, neither seems to have honed his expertise in the astral sciences within the confines of the madrasa, which supports our earlier discussion about the limited role of astrological expertise within these primary institutes of learning in the Islamicate and Ottoman worlds.

Given the fact that these individuals were scarcely remembered in their own times, rarely enjoyed an afterlife, and have remained virtually unknown to modern scholars and readers—despite their self-proclaimed and peer-recognized credentials, numerous extant compositions across a wide array of astral and other sciences, and long tenures at princely courts as experts in the celestial arts—this chapter also serves as a tribute to these forgotten scientists of a bygone age or even as the fulfillment of a humanistic duty to recognize their presence. By weaving together the fragmented autobiographies of these two distinctive *munajjims* drawn from traces found in manuscripts, an archetypal "Ottoman *munajjim*" comes to life, a figure navigating a complex terrain of challenges and opportunities, from his initial training to offering his expertise to patrons.

A Disillusioned *Munajjim* in the Provinces: ʿAbdurrahman

Throughout much of the twentieth century, a prevailing sentiment among scholars of Islamic history held that the voices of early modern individuals expressing their experiences in their own words were unlikely to be found. Adherents of this absence thesis believed that the dearth of diaries, memoirs, and personal letters was an inescapable byproduct of a society that had yet to embrace a robust sense of individuality in western European fashion. However, since at least the 1980s, a fresh wave of research exploring "ego-documents" and other autobiographical texts from the early modern Ottoman world has revealed the fallacies, as well as implicit or sometimes overt Orientalist biases, that underpinned the assumptions of previous generations of scholars. Thanks to at least four decades of scholarly labor, it has become increasingly clear that a rich body of first-person narratives, in diverse forms and styles, did indeed exist among early modern Ottomans. The tradition of penning autobiographical pieces was particularly evident among but not entirely exclusive to those who enjoyed greater access to the tools and means of writing—scholars (*ʿulamaʾ*), members of various Sufi orders, and bureaucrats, to name but a few.[3]

However, when it comes to what we may characterize as "scientific experts"—namely, those practitioners of different forms of technical knowledge and crafts, including physicians, astrologers, alchemists, architect-engineers, herbalists, mariners, miners, and any kindred group of skilled individuals whose epistemic foundations and practical pursuits were not always congruent with those of the ʿulamaʾ—there remains a noticeable void in both the discovery of sources and studies examining them.[4] This absence is, without a doubt, regrettable, for the self-reflections of these individuals concerning their lives, the ambivalent and occasionally contentious nature of their expertise, their social status and relationships, and the peculiar sociopolitical dynamics of their times hold immense value. Whether expressed in stand-alone autobiographical accounts or deduced from first-person narrative fragments hidden within their works, these insights are of paramount importance for those who imagine the history of science and intellectual history not merely as the transmission of texts, tools, and ideas from one cultural zone to another but as the broader sociopolitical, cultural, financial, and emotional history of sciences and their practitioners, regardless of the scope of those individuals' intellectual influence and legacy.

Not a single stand-alone autobiography or memoir has been found among the long list of textual artifacts left by *munajjims* active in the Ottoman world before the eighteenth century.[5] This lack should not deter us, however, from exploring the self-portrayals inconspicuously embedded within the texts they authored to lay out the scientific intricacies of their pursuits in various disciplines. An excellent case in point is an Arabic medical treatise written by the late fifteenth- and early sixteenth-century *munajjim* named ʿAbdurrahman.[6] Composed in February 1502 and presented shortly thereafter in Amasya to his chief patron, Prince Ahmed—the eldest living son of Bayezid II and favorite contender for the throne at the time—this medical treatise is prefaced by an expansive introduction that chronicles its author's travels and travails on his odyssey of becoming and unbecoming a *munajjim*. As such, the text offers a vivid portrayal of the ebb and flow of a court *munajjim*'s career as it evolved in tandem with the changing tastes and priorities of his patron.

Much like many other unsung Ottoman scientific experts of the early modern era, ʿAbdurrahman remains shrouded in obscurity, his identity and scientific output having received no attention from modern scholars. Leaving open the possibility of unearthing additional works in the future, only three of his accounts have been identified and cataloged thus far;[7] none of these texts, however, has been studied, let alone edited and published. The latest of these is a *taqwim* that ʿAbdurrahman crafted in March 1510 and presented to Prince Ahmed in Amasya.[8] At the time, Prince Ahmed was already locked in a fierce struggle for the throne with two of his brothers, Selim I (r. 1512–20) and Korkud (d. 1513) as the demise of their father, the reigning Sultan Bayezid II, loomed.[9] As a *munajjim* in service at Prince Ahmed's gubernatorial seat in Amasya, ʿAbdurrahman likely produced and presented almanacs on an annual basis. Unfortunately, only this singular exemplar of his *taqwims* remains, leaving us to imagine the existence of numerous others that may have been lost to time. The earliest surviving work of ʿAbdurrahman is the aforementioned medical treatise, which he also presented to Sultan Bayezid II in 1503, just a year after submitting the first version to Prince Ahmed, by altering the dedication passages and omitting the autobiographical excerpt.[10] The reasons behind including his personal grievances in the earlier version presented to the prince and later excising them in the subsequent rendition submitted to the sultan becomes apparent upon closer examination of his relationship with Prince Ahmed, as elucidated in the autobiographical passage that reflects ʿAbdurrahman's growing frustration: after being encouraged and supported by Prince Ahmed to travel to Iranian lands to "perfect his expertise in the science of the stars," ʿAbdurrahman had returned Amasya only to find himself overshadowed by other scientific experts.[11] Before we investigate the later stages of his relationship with the prince, let us retrace the steps of ʿAbdurrahman's formative years.

His Master: Yazdanbakhsh

In the autobiographical preface to his medical treatise, ʿAbdurrahman omits specifics about his birthplace and date but mentions that he began his studies of the astral sciences under a *munajjim* named Yazdanbakhsh, also known as Mawlana Kuchak.[12] Regrettably, ʿAbdurrahman offers no further clue about Yazdanbakhsh, but fortuitously, we find the latter's name in two separate contemporary historical records. The first is an archival register of payments compiled in 1478 during Mehmed II's reign that enumerates courtly stipendiaries and their corresponding remunerations. One of the oldest surviving examples of such registers, this ledger lists only one *munajjim*, named Kuchak, who is receiving ten *akçes* per diem, a sum comparable to a messenger's wage but notably less than that of a falconer or storyteller.[13] The register suggests that by the time Yazdanbakhsh, aka Kuchak, appeared in imperial court records, he likely resided in the capital. A second historical document, however, establishes that in 1477, Yazdanbakhsh presented his Arabic astronomical handbook with tables, *al-Zij al-mujmal* (The synoptic *zij*), to the future Sultan Bayezid II, then a prince and regional governor in Amasya.[14] A note on the flyleaf attaches the epithet "Amasi" (someone from Amasya) to "Yazdanbakhsh." The strong Persianate flavor of his name (lit. "gift of God" in Persian) suggests that he might have been the descendant of a Persian émigré–learned man who ventured into Ottoman lands and settled in the city of Amasya.[15] Nevertheless, we lack definitive information about both his background and his later life, other than his being the son of a certain Pir ʿAli. It therefore remains unclear whether ʿAbdurrahman studied with Yazdanbakhsh in Istanbul or in Amasya; however, given that ʿAbdurrahman spent the greater part of his career in Amasya in the service of Prince Ahmed, the training between the two probably took place in that province sometime in the later 1470s or early 1480s.

ʿAbdurrahman remains tight-lipped about the texts and authoritative sources that guided his initial training in the astral sciences at the feet of Yazdanbakhsh. Nor does he refer to any indication of where this instruction took place, whether at the princely court, Yazdanbakhsh's private abode, or within the hallowed halls of a madrasa. Given the absence of references to Yazdanbakhsh in the contemporary or near-contemporary biographical compendiums of madrasa scholars, it seems improbable that he served as a madrasa instructor. Yazdanbakhsh's legacy is limited to his *al-Zij al-mujmal*; thus, any discussion on his scientific inspirations and models, which he likely passed on to his student ʿAbdurrahman, must remain in the realm of conjecture. Yet his *al-Zij al-mujmal* stands as a noteworthy example of *zij* production and circulation in the fifteenth-century Ottoman landscape.

As delineated in greater detail in the previous chapter, *zijes* were a necessary component of any *munajjim*'s toolkit.[16] These astronomical handbooks typically comprise a set of tables detailing various astronomical functions and data about celestial movements, accompanied by instructions and parameters for utilizing the tabulated data for crafting annual almanacs and casting horoscopes. Not unlike modern digital tools and software that assist practitioners of the astral sciences in making their calculations, *zijes*, particularly those containing detailed planetary-motion tables, facilitated manual calculations of projected positions for celestial objects. Modern scholars posit that from the ninth through at least the seventeenth century, over two hundred *zijes* were produced in the Islamicate world, although not all the *zijes* cited have survived and been identified,[17] and to date, very few of the extant examples have been edited or studied in detail.[18] While many *zijes* served as commentaries or expositions of earlier compilations, others emerged from systematic celestial observations carried out either individually or as a collective endeavor. The two most renowned *zijes* produced in the Islamicate East that were widely used in the Ottoman world—the *Zij-i Ilkhani* of Tusi and his associates from the thirteenth century and the *Zij-i Sultani* of Ulugh Beg and his colleagues from the fifteenth century—were compiled through the collaborative research of numerous astral experts who convened at the Maragha and Samarkand observatories, respectively.

Yazdanbakhsh's *al-Zij al-Mujmal* emerged not as the product of recurrent celestial observations by a group of astral experts but rather as its author's modest attempt to compare his individual data at hand with those found in the *zij* of the Abbasid-era astral expert Abu'l-Wafaʾ al-Buzjani (d. 988)—a source Yazdanbakhsh held in high esteem.[19] His choice of an Abbasid-era *zij* to rely upon when compiling his own one is particularly intriguing. By the time Yazdanbakhsh composed his work, various astronomical handbooks were in circulation throughout Anatolia. The most prevalent among them was the *Zij-i Ilkhani* corpus that included the original *Zij-i Ilkhani*, completed at the Maragha observatory around 1270, together with the subsequent editions and revisions, crafted by late thirteenth- and fourteenth-century astral experts of the Islamicate East who sought to correct the original's imprecision.[20] Ottoman *munajjims* and almanac-makers of the later fifteenth century were also familiar with Ulugh Beg's state-of-the-art *Zij-i Sultani*, originally compiled in Samarkand in the 1440s. This familiarity largely stemmed from ʿAli Qushji's relocation from Herat to the new Ottoman capital in the early 1470s, together with many family members, students, and associates. As the leading light of the Samarkand mathematical-astronomical school[21] and one of the creators of the *Zij-i Sultani*, Qushji's arrival in the Ottoman world was occasioned by a personal invitation from Mehmed the Conqueror, who, as some near-contemporary sources state, promised to support Qushji's observational endeavors by having an observatory constructed.[22] Qushji lived for only two years after his arrival in Istanbul, but his presence likely facilitated, if not single-handedly effected, the transmission of the *Zij-i Sultani*, along with an array of other works of the eastern Islamicate tradition in the astral and mathematical sciences, into the Ottoman world.[23] An examination of extant Ottoman *taqwims* from the mid-fifteenth to the late sixteenth century reveals that while the *Zij-i Ilkhani* corpus was the chief source for fifteenth-century *munajjims* and almanac-makers in the Ottoman world, there was a gradual and almost definitive shift toward the use of *Zij-i Sultani* starting in the early sixteenth century.[24] In that regard, Yazdanbakhsh's reliance in the late 1470s on al-Buzjani's rather dated *zij* from the Abbasid era, instead of more recent and up-to-date ones, is confounding, showing the multilayered and sometimes unexpected pathways of scientific influence and the enduring presence of earlier Islamic astral traditions in Anatolia during the fifteenth century.

Yazdanbakhsh was not the only scientific expert with whom ʿAbdurrahman studied. Once he grasped the basics of the astral sciences under Yazdanbakhsh's tutelage, ʿAbdurrahman embarked on learning a sister discipline in the natural sciences, studying medicine at the feet of a certain Hekim İshak, whom he hails in his autobiographical preface as the Hippocrates of the day, the Socrates of the time, the Galen of his age, and the Avicenna of his era.[25] Biographical accounts from the sixteenth century, such as Taşköprülüzade's *Shaqaʾiq*, make mention of a physician named İshak, who, originally a Christian by confession, was active in the last quarter of the fifteenth century as a highly skilled member of his profession, fully deserving of ʿAbdurrahman's encomia.[26] The same source also notes that İshak studied logic, philosophy, and religious sciences with Molla Lutfi (d. 1495), an utterly eccentric professor teaching at one of the eight newly constructed Fatih madrasas—known as *Sahn-ı Seman*—in Istanbul. Molla Lutfi was executed in 1495 on the charge of heresy, though the real reasons for his execution were likely related to a deadly intra-ʿulamaʾ feud.[27] During his studies with Molla Lutfi, İshak apparently converted to Islam and eventually renounced both medicine and philosophy. If the details of Taşköprülüzade's portrayal of İshak are accurate, then ʿAbdurrahman's study with him must have transpired in the 1480s, prior to İshak's ultimate conversion and withdrawal from the medical profession.

ʿAbdurrahman and His Patron Prince Ahmed

Sometime in the 1480s, ʿAbdurrahman established a relationship with Prince Ahmed, who took notice of him, probably thanks to ʿAbdurrahman's thorough education in the astral and medical sciences under the guidance of two renowned practitioners of the time. Prince Ahmed had resided in Amasya as its governor since his father's enthronement in the capital in 1481. Regrettably, ʿAbdurrahman's brief autobiographical sketch does not provide any clues as to how and when he was introduced to the prince. Intriguingly, he writes that soon after their acquaintance, Prince Ahmed advised him to venture to the "Iranian provinces" if ʿAbdurrahman had any desire to further his knowledge of the astral sciences.[28] What we have here is a curious recollection about a provincial patron of skilled individuals, not only eager to financially support aspiring experts to travel elsewhere to develop their expertise but, more important, well-informed concerning ideal destinations for his servants to cultivate their skills in their respective fields. The Persianate Islamicate East was indeed the hotbed of practical astral science at the time despite the ongoing political turmoil and frequent changes of ruling houses in the wake of the dissolution of the Timurid Empire.[29] Prince Ahmed's reported recommendation also testifies to a self-awareness regarding the relatively inferior level of training in the astral sciences available in Ottoman lands at the time.

Heeding his patron's astute advice, ʿAbdurrahman journeyed to Iranian lands—province unspecified—and studied under a certain Sayyid Niʿmatullah Shirazi, whom he extols, echoing his lofty characterization of İshak the physician, as the paragon of philosophers, the "chief of the *munajjims*" (*sayyid al-munajjimin*), the Nasir al-Din al-Tusi of the time, and the Ptolemy of the age.[30] Despite the formulaic nature of these remarks, which sometimes owe their origin to the concern for rhyme in the rhymed-prose style of Arabic, Persian, or Ottoman Turkish, these stock wordings are, in truth, invaluable for tracing how one's expertise in a given field was articulated by people in that same profession and who the authorities were that served as benchmarks against which one's erudition was measured in the field. In the imagination of Ottoman *munajjims*, Ptolemy and Tusi never ceased to be regarded as authoritative figures, although, as we will see later, other names were also frequently featured, such as Jamshid al-Kashi, a key figure from the Samarkand observatory; Ibn al-Shatir (d. 1375), the famed Mamluk astronomer; and Abu Maʿshar, the Abbasid-era astrologer who was arguably the most renowned of all time in the Islamic history of astrology.

Unfortunately, ʿAbdurrahman's autobiography omits precise details about who Sayyid Niʿmatullah Shirazi was, where and how long ʿAbdurrahman studied with him, and what specific branches and genres of the astral sciences he pursued to deepen his expertise.[31] Upon returning to Prince Ahmed's court in Amasya, probably in the mid-1490s, ʿAbdurrahman began serving in the capacity of a *munajjim* in the lower echelons of Ahmed's provincial court bureaucracy. However, for five restless years, he says, he longed for a promotion and the better allowances that he believed his glowing credentials and demonstrated merit deserved.[32] In the meantime, he realized that Prince Ahmed's interests had shifted from the astral sciences toward medicine, as the prince was lavishing much more attention and generosity on physicians regardless of their proficiency.[33] This realization prompted ʿAbdurrahman, as he says, to compose his medical treatise in February 1502, a synthesis of and extracts from two widely circulating medical books at the time, Ibn Sina's *al-Qanun fi al-Tibb* (The canon of medicine) and Ibn al-Nafis's (d. 1288) *Muʾjaz al-Qanun* (The epitome of the canon, an abridgement of Ibn Sina).[34] In doing so, he sought to remind the prince of his expertise in the broader field of the natural sciences and his worthiness of a higher status in the courtly hierarchy.

Did ʿAbdurrahman attain his aspirations by presenting his medical treatise to Prince Ahmed? It is highly unlikely, at least within the immediate context of the year 1502, as we find ʿAbdurrahman presenting the same text, with the dedication altered and the autobiographical introduction removed, to Sultan Bayezid II in Istanbul a mere year later. ʿAbdurrahman probably wished to try his luck on the grander royal stage. For the ensuing few years, ʿAbdurrahman's whereabouts remain a mystery until he appears in a brief note found in a comprehensive archival register documenting all the gifts, allowances, and payments dispensed to court officials and others during the final decade of Bayezid II's reign from 1503 to 1512. According to the record, on August 12, 1509, ʿAbdurrahman, identified in the relevant entry as "the *munajjim* of Sultan Ahmed," received 1,500 *akçes* for an unspecified reason.[35] A few months later, ʿAbdurrahman presented Prince Ahmed with his almanac for the new solar year spanning March 1510 to March 1511, beginning on 30 Dhu al-Qa'dah 915 AH (11 March 1510 CE).[36] These two pieces of evidence suggest ʿAbdurrahman's presence at Prince

Ahmed's court in Amasya as his royal *munajjim* as late as spring 1510. However, ʿAbdurrahman's fate before and after Prince Ahmed lost the battle for the throne to his brother Selim I and his ultimate murder by Selim's agents in 1513 remains obscure.

While no records explicitly place ʿAbdurrahman in the Ottoman capital of Istanbul—except for the copy of the medical treatise he presented to Sultan Bayezid II in 1503—some subsequent sources descending from the community of early modern Ottoman *munajjims* refer to him as the first Ottoman chief court *munajjim*. For example, the marginalia of an astronomical treatise written by an eighteenth-century court *munajjim*, who seems curious about the professional lineage of his role, includes a short list of chief court *munajjims* that spans from the early to the late sixteenth century. Interestingly, the list starts with ʿAbdurrahman and sequentially features such names as Yusuf b. ʿÖmer es-Saʿati, Muvakkit Mustafa, and Taqi al-Din. Although the veracity of the entire list is subject to debate when cross-referenced with contemporary archival documents and registers, it is notable that ʿAbdurrahman's name found recognition among later generations of Ottoman *munajjims*. This enduring recognition likely arises from his surviving *taqwims*, which remained in circulation and were probably viewed by remote colleagues even two centuries after his passing.

ʿAbdurrahman's Taqwim

Only a single copy of ʿAbdurrahman's *taqwims* has survived to the present day. Given that almanacs, as ephemeral objects, typically lost their relevance at the end of each solar year and that texts were copied manually up until the nineteenth century, when the movable-type printing press finally began to be widely adopted, it is perhaps unsurprising that just one manuscript copy of what might have been many remains. Almanacs' ephemeral nature and content may have rendered their long-term preservation unlikely. More intriguing, if perhaps unanswerable, questions are why the surviving copies of *taqwims* were preserved at all and how ʿAbdurrahman's sole extant almanac, presented to Prince Ahmed in Amasya in spring 1510, found its way to the library of a learned nineteenth-century Ottoman gentleman in Istanbul.[37] While scattered archival references and anecdotal evidence indicate that multiple manuscript copies of annual almanacs were indeed produced for distribution to various ruling elites, oftentimes only the presentation copies submitted to the palace were safeguarded in palace repositories.[38] The remaining copies likely met their end by being discarded or by wear or other natural occurrences. As to ʿAbdurrahman's 1510 *taqwim*, the extant volume, a meticulously crafted forty-six-folio codex intended for Prince Ahmed in Amasya, does not offer any clue as to why or how it was preserved. What happened to the manuscript between its production in Amasya in the early sixteenth century and its arrival at Esʿad Efendi's library in Istanbul in the mid-nineteenth century remains a mystery.

ʿAbdurrahman's extant *taqwim*, which he described as the "complete almanac" (*al-taqwim al-tamm*), offers an illuminating opportunity to delve deeper into the *taqwim* genre, one of the most neglected scientific corpora of the medieval Islamic world. The previous chapter presented the general characteristics of the genre, which wove together astronomical data, calendrical knowledge, and astrological forecasts and guidance on a yearly basis. It should be noted, however, that despite the gradual standardization of the content of *taqwims* from the early sixteenth century—a development reflecting the growing professionalization and specialization at the Ottoman court—extant almanacs from the fifteenth century exhibit notable variations, encompassing various strains of useful knowledge. Among these diverse elements are chronology tables dating in brief entries some key political, religious, or natural events from Creation onward. Several other tables, dedicated to different divinatory techniques, were also frequently inserted into earlier almanacs—from dream interpretation and spasmatomancy (a type of divination based on interpreting involuntary bodily convulsions and twitches) to a method attributed to Aristotle for predicting the victorious side in battle. While the undeniable utility of such a diverse body of knowledge for readers is evident, the variety of topics also allowed authors to showcase the breadth and versatility of their expertise. By composing a "complete" *taqwim* that included not only the standard components related to calendrics and astrological forecasts but also those additional sections, such as chronology tables, figures for different divinatory tools, and detailed ephemeris charts showing the exact degrees of planetary positions, ʿAbdurrahman likely attempted to display his wide range of knowledge and skill to Prince Ahmed.[39]

The first four folios of his almanac are devoted to universal chronology, where ʿAbdurrahman provides dates for some epochal events in a universal-historical fashion that culminates in "the history of the Ottoman house.'"[40] He starts this section with the Creation of the universe, which he calculates to have taken place 18,614 (lunar) years prior to the year his almanac was compiled.[41] As in many other *taqwims* from the fifteenth century that include chronology sections, the events are not dated by the Hijri year in which they occurred but are instead identified by the number of years that have elapsed between their occurrence and the year the almanac was compiled. In chronology sections, *munajjims*, almost as a rule, put the entries in the following format: "It has been {X} years since . . ." (Tur . . . *dan berü {X} yıldır*; Per. *az gah-i . . . {X} sal ast*). These entries are grouped under various headings that reflect the historical understanding of their compilers. Accordingly, the first unit of the chronologies in almanacs was the canonical history of pre-Islamic prophets from Adam onward, which was then followed by the life of the Prophet Muhammad and early Islamic history through the end of the Abbasid dynasty. The Umayyads were typically omitted, with the entry on Karbala directly followed by the rise of Abu Muslim, the founder of the Abbasids. Next would come the most exhaustive section, detailing the political and natural history of the Islamicate East during the Seljuq, Mongol, and post-Mongol periods. The history of the Ottoman dynasty was usually appended here, tactfully situating the Ottomans within the political genealogy of the Turko-Persian-Mongolian Islamicate East. The entries on Ottoman history typically began with the rise of ʿOsman and concluded with the most recent event deemed significant near the date of the almanac's compilation.[42]

Although these almanacs with chronology tables generally cover a consistent set of historical events, they are not entirely identical. Most discrepancies occur in the count of elapsed years, particularly regarding pre-Islamic prophetic history, owing to the difficulty of establishing accurate pre-Islamic chronologies, a fact acknowledged by several medieval experts in the astral sciences, including Biruni.[43] The scope and range of events included from more recent Islamic history also vary among almanacs. For biblical and prophetic history, the variety is rather limited: almanacs basically replicate the same list of events, sometimes phrased differently, from the Creation and birth of Adam to the life of the Prophet Muhammad. The sections on early Islamic history also typically list similar events, from the birth, prophecy, and death of Muhammad to the era of the four rightly guided caliphs and the eventual rise of the Abbasids. Discrepancies abound, however, when charting more recent Anatolian and

Ottoman history. For example, the plague outbreak in the Karasi region of northwestern Anatolia in 1348–49 is mentioned solely in the surviving 1421 almanac.[44] Similarly, the full solar eclipse of 1388 and the appearance of a celestial novelty the same year are recorded only in a *taqwim* for the year 1432.[45] These examples could easily be multiplied; as such, they reflect *munajjims*' individual judgments about what was worthy of memorializing and what was not. Their choices may have also been influenced by their geographical affiliations and professional connections, especially if they were associated with different courts in fifteenth-century Anatolia. Such affiliations may have endowed them with local information on the regions from which they emerged or in which they were active. Such a parochial bias might explain why events like the death of İlyas Beg of the Menteşe principality in western Anatolia, the Samsun fire of 1421, and the passing of Yaʿkub Çelebi b. Germiyan are mentioned only in specific almanacs.[46]

Modern scholars often assume that chronology sections were present in all Ottoman *taqwims*, but it is essential to emphasize that they were primarily featured in extant fifteenth-century almanacs and began fading by the early sixteenth century. Although the inclusion of chronological information in almanacs has deep roots in Islamic history, no early Islamic *taqwims* with historical chronology sections have survived for comparison. Writing in the early eleventh century, Biruni aptly noted that chronology sections were added to almanacs for "the intellectual pleasure found in such things," suggesting that their audience, primarily ruler-patrons, enjoyed reading them for their instructive value.[47] Given *munajjims*' longstanding expertise in computing time by tracking celestial configurations, it is hardly surprising to find them engaged in recording dates of epochal events. What is perhaps more intriguing is why chronology tables began to disappear from the Ottoman almanacs by the early sixteenth century.

The full truth behind this shift is unrecoverable, thanks to a lack of definitive evidence. One possible explanation is the concurrent rise of new forms, actors, and styles of historical writing championed at the time by the Ottoman court, which demonstrated, in Baki Tezcan's words, the "attempt to directly control the development of historiographical expression."[48] Unlike the rather dry and brief chronological data of *munajjims*' almanacs, which portrayed the Ottomans as just another, unexceptional political entity in the Islamicate East and Muslim Anatolia, new dynastic histories and regnal and campaign narratives of the Ottoman house, highlighting the Ottomans' triumphs and written in a more embellished language—either in Persian or Ottoman Turkish—and style, might have better aligned with the greater ideological stakes at the Ottoman court from the early sixteenth century on. This shift in taste likely rendered older chronologies inadequate and obsolete. The chronology tables' falling out of fashion by the early sixteenth century might also be related to the ongoing bureaucratization, professionalization, and routinization of services at the imperial court. The more refined and specified roles such as chief *munajjim* and chief physician, with clearer delineations of duties and services, may have prompted the gradual removal of diverse and vague contents from *munajjims*' almanacs, pushing them to focus solely on astrological counsel and calendrical information. Regardless of the precise reasons underlying this gradual shift, my examination of nearly two hundred surviving Ottoman almanacs between the early fifteenth and early nineteenth centuries reveals that chronology sections disappeared from the early sixteenth century. I located only three *taqwims* as exceptions, including ʿAbdurrahman's extant almanac from 1510. Notably, these exceptions were not products of the court *munajjims* in the capital but rather the work of provincial practitioners and outsiders, possibly eager to demonstrate the breadth of their expertise in order to (re)gain the court's favor or perhaps unaware of the changing fashions in almanac composition and consumption at the imperial center.[49]

In his universal chronology, ʿAbdurrahman jumps immediately from prophetic history to the history of the Ottoman dynasty, bypassing the rather extensive Islamic history section typically found in fifteenth-century almanacs. Linking the history of Osman's house straightforwardly to the prophetic past, this preference accentuates the Ottomans' distinct place in the broader sweep of history even more strongly. Like other almanacs with Ottoman chronologies, he commences with the rise of ʿOsman b. Ertugrul and the conquests of Bilecik and Yarhisar 216 (lunar) years ago. Compiling his almanac in 915 AH, ʿAbdurrahman thus dates, conventionally, the dawn of the Ottoman enterprise to 699 AH/1299 CE. He then chronicles key dates, including conquests and the births and deaths of Ottoman rulers without adding any value judgments or personal commentary. One exception is the information he gives about the defeat of the Ottomans by Timur 110 (lunar) years earlier (corresponding to 805 AH/1402 CE), which he asserts occurred on account of the "betrayal" of Tatar soldiers.[50] Not surprisingly, he underscores the births of Bayezid II and Prince Ahmed, with the latter even highlighted in red describing both as the "glory of the Ottoman house."[51] His chronology concludes with details about Prince Ahmed's family and his military ventures, including the births of his sons (Murad, ʿAlaʾaddin, Süleyman, and ʿOsman), his triumph over the Karaman principality, and his two separate endeavors to overthrow the "heretic Qizilbash," the first of which took place eight years ago in 1501 and the more recent three years back (912 AH/1506 CE).[52] These brief notes in ʿAbdurrahman's chronology offer valuable insights into the early phases of the Ottoman-Safavid rivalry, shedding light on the initiatives undertaken by one member of the Ottoman dynasty against the growing threat of Shah Ismaʿil and the Ottomans' labeling of his supporters as heretics and infidels from the beginning.

Following the chronology section come the annual astrological predictions, grounded in precise calculations of the celestial configurations at the moment the Sun entered the sign of Aries, 8 hours and 33 minutes into Monday on 11 March 1510. He explicitly states that these conclusions, based on the latitude of Amasya, were made by drawing on what he names as the *Zij-i Muhaqqaq-i Ilkhani*.[53] This must be Wabkanawi's *Zij* unless it is another revision of the Ilkhanid tables produced after the thirteenth century. As per tradition, his astrological predictions commence with the dedicatee of the almanac, Prince Ahmed. Lauding him with lofty titles, such as the Most Exalted Sultan and the Most Judicious Khaqan, and further, as master over all rulers in the lands of the Rum/Romans, Turks, Arabs, and Persians, ʿAbdurrahman generally paints a favorable year for the prince. This portrait is, of course, within the usual bounds of *munajjims*' conventions and limitations. Still, he alludes to potential disputes, animosities, and court intrigues involving various rulers without ever specifically naming them. The predictions regarding Prince Ahmed and other rulers are then followed, just as in any other *taqwim*, by similar remarks on the fortunes of various societal groups—viziers, ʿulamaʾ, men-at-arms, mystics, physicians, scribes, merchants, women, spies, and more—and predictions concerning earthly affairs such as wars, diseases, harvests, and prices. Each of these clusters is linked to a specific planet and interpreted on the basis of expected positions of the planet within that year. While these forecasts, bound by the conventions and limitations of *munajjims*' craft, are expressed in an imprecise manner, ʿAbdurrahman occasionally offers predictions that might warrant closer scrutiny—for example, his anticipation of "the rise of someone staking a claim to sovereignty," or the imminent demise of a ruler residing in the fifth clime, which encompasses cities such as Erzurum, Shirvan, and Tabriz.[54] Notably, in his detailed astrological predictions with respect to the seven climes, which in classical geography divided the inhabited world into seven zones, each with an array of cities, ʿAbdurrahman refers to different conventions adopted by "Persian and Greek(-speaking) natural philosophers."[55] While occasionally he tries to reflect both traditions, he predominantly shows a preference for Persian natural philosophers over the Greeks, revealing his deeper familiarity with the Persianate astral tradition.[56]

A Persian Émigré *Munajjim* at the Sublime Porte: Khitabi

Slightly before ʿAbdurrahman left Anatolia and arrived in the Iranian lands to deepen his knowledge of the astral sciences per his patron Prince Ahmed's advice, an émigré practitioner in the same field, originating from the Gilan district of northern Iran, arrived in Amasya. This practitioner, Husam b. Shams al-Din al-Lahijani al-Gilani, known to have signed his works as Khitabi *munajjim* al-Husayni, found a way to connect with Prince Bayezid, who was governing the region at the time. Khitabi, like ʿAbdurrahman, has escaped the attention of modern scholars despite the array of texts on astral and medical sciences he left behind. The sparse references to his life in modern bio-bibliographical sources and studies are markedly discordant, sometimes confusing him with other contemporary scholars.[57] Unlike ʿAbdurrahman, who provides us with useful autobiographical clues for piecing together at least some portions of his life, Khitabi remains almost silent about his family, teachers, and peers. One of the few autobiographical vignettes he offers, in his treatise on a novel astronomical instrument, *Risala tashrih al-alat* (Treatise on the explication of instruments) that he dedicated to Bayezid II in 1483, is a mention of a certain Rukn al-Amuli as his former master.[58] Rukn al-Amuli is a more familiar figure to modern-day scholars. His astronomical handbooks with tables, composed in the late 1430s, have at least been cited, if not extensively studied, by modern historians of astronomy. Moreover, as some of his extant texts contain curious autobiographical details that help us reconstruct the social and intellectual lives of *munajjims* in action, it is appropriate at this point to consider Rukn al-Amuli's story before returning to that of his student, Khitabi.[59]

His Master: Rukn al-Amuli

Rukn al-Amuli, or Rukn b. Sharaf al-Din al-Husayni al-Amuli, as he signed his extant works, was a contemporary of the renowned scientists of stars gathered in Samarkand under the patronage of Timurid prince Ulugh Beg. It appears that he pursued his celestial interests in other regions of Timurid Iran and Central Asia, often facing challenges and inconsistent royal support. It is unclear whether Rukn al-Amuli was ever invited to Samarkand when the new observatory of Ulugh Beg and its affiliated madrasas were welcoming contemporary aspirants and experts in these sciences from around the broader Persianate world.[60] There is no mention in his extant writings of his presence in the Samarkand observatory, either because he never received an invitation or because he chose not to participate. Nevertheless, during the 1430s, Rukn al-Amuli was busy compiling an astronomical handbook with tables that paralleled the *zij* of Ulugh Beg and his associates in Samarkand. In addition to his Persian *Zij-i jamiʿ-i [A]bu Saʿidi*, composed in Shiraz and dedicated to Timurid prince Abu Saʿid Gurakan (r. 1450–69), he also seems to have crafted another *zij* in Herat (present-day Afghanistan).[61] In March 1456, Rukn al-Amuli wrote a comprehensive Persian manual on astrolabes, *Panjah bab-i sultani dar ʿilm-i usturlab* (Fifty sultanic chapters on the science of the astrolabe), composed of fifty chapters, each exploring a distinct application of the instrument, including astrological operations. Unlike his *zij*es that have not survived in large numbers, his how-to manual on astrolabes enjoyed an afterlife, with at least four copies in Istanbul and several others located elsewhere.[62]

Rukn al-Amuli's *Fifty Sultanic Chapters* is a gem for its technical details on the myriad uses of the astrolabe and especially its autobiographical preface, which reveals *munajjims*' ideas about the contents and conditions of their expertise. Rukn al-Amuli begins by recounting that he has devoted most of his life to the study of the "philosophical sciences," particularly the mathematical sciences that encompassed the three "H"s: *hayʾa* (astronomical theory), *handasa* (geometry), and *hisab* (mathematical calculation).[63] To Amuli, the ultimate goal, or as he puts it, the "fruit" (*thamara*), of learning these three sciences is the ability to make astrological judgments and accurately measure time, in addition to discerning God's omnipotence and acknowledging the grandiosity of his cosmic design.[64] *Munajjims*' main business, in Rukn al-Amuli's phrasing, is computing time and drawing astrological forecasts, which requires a precise understanding of planetary movements, the ability to calculate their mean motions, and the skill to designate the ascendant of any given moment. Rukn al-Amuli contends that the astrolabe is the best instrument for executing these astronomical operations, which are essential for casting horoscopes and practicing electional astrology. He nonetheless posits that this craft, along with the mathematical sciences more broadly, is a type of intellectual endeavor that cannot flourish without "the support of rulers and statesmen."[65] For the past twenty years—since he completed his *zij* and another treatise on the celestial globe that has yet to be discovered—Rukn al-Amuli laments that he has been bereft of royal support. Notwithstanding possible exaggeration as a plea to his new patron, Abu'l-Qasim Babur Mirza (d. 1457), the Timurid ruler in Khurasan, to whom he dedicates his treatise on the astrolabe, al-Amuli's life over the past two decades seems to have devolved into an unmitigated disaster. He details a series of afflictions that beset him, not the least of which was his prolonged separation from loved ones and constant relocations to distant places. More recently, his odyssey took him to India and Kerman, during which he was plagued by political turmoil, massacres, and famine.[66]

Neither in Khitabi's writings nor in those of his master Rukn al-Amuli can one unearth any specifics pertaining to when and where they first met, how long they collaborated, or which texts and methods featured in Khitabi's training. Still, Rukn al-Amuli's oeuvre undoubtedly left an indelible mark on Khitabi. For instance, in his *Risala tashrih al-alat*, presented to the Ottoman sultan Bayezid II, Khitabi extols his master's work as one of the three paramount *zij*es of the period, alongside the Ilkhanid and Ulugh Beg tables.[67] Moreover, in Khitabi's sole surviving almanac, compiled in March 1489, he states that he utilized Rukn al-Amuli's *Zij-i Jami*ʿ*-i [Abu]Sa*ʿ*idi* when calculating the horoscope of the year-transfer and determining all the other planetary positions for the upcoming year.[68] This is particularly striking, in view of the fact that there is no mention whatsoever of Rukn al-Amuli's *zij* or his astrolabe treatise in the long list of extant *taqwim*s and other textual productions by other Ottoman *munajjim*s from the fifteenth and sixteenth centuries.[69] It is entirely plausible that some of the four copies of Rukn al-Amuli's astrolabe treatise, now housed in the collections of Istanbul's manuscript libraries, were brought to the Ottoman domains by Khitabi himself, even though no ownership statement or other manuscript evidence exists to substantiate this conclusively.

Khitabi's Early Career

Khitabi arrived in the Ottoman realm in the late 1470s. Like many of his contemporaries who traveled from Iranian lands to Anatolia and eventually to the Ottoman capital of Istanbul, his first major stop was the province of Amasya, where Prince Bayezid (the future Sultan Bayezid II) had been holding court since 1454 as the provincial governor.[70] The earliest extant work by Khitabi is a treatise that he claims to have hastily composed in 1479 in Tokat—located just fifty miles from Amasya—and presented to Prince Bayezid.[71] The content and structure of this treatise, which was dedicated to the merits of natural philosophy over religious science and the supremacy of (natural) philosophers over religious scholars, are quite telling as to Khitabi's aspirations.[72] By titling his text *Jamiʿ al-qismayn* (The gatherer of two parts) and dividing it into two main chapters, one devoted to the astral sciences and the other to the science of medicine, Khitabi cast himself as a versatile expert equally proficient in both fields. In doing so, he likely hoped to improve his chances of securing princely favor—a motive he candidly mentions as the primary impetus for writing the treatise in such haste in the first place. Khitabi begins his text by expounding on the astral sciences and embarks on an exclusively astrological discussion, laying out the qualities and indications of the twelve astrological houses. For Khitabi, the astral sciences, which has a more eminent status than medicine because of its subject matter (i.e., celestial bodies), is a useful, divinely ordained field of knowledge that enables people to understand the influence of the motions of celestial objects upon the sublunary world, shield themselves from harm and ruin (as mandated by the Qur'an), and gain insight into the divine order as it pertains to their personal lives.[73] He then moves on to medicine, cataloging maladies and prescribing treatments.

Later that year, in December 1479, Khitabi traveled to Istanbul and sought to impress the reigning sultan, Mehmed II, at court. He presented the sultan with an extensive commentary on Tusi's *Si fasl* (Treatise of thirty chapters), one of the most popular introductory textbooks on the astrological significance of planets, zodiac signs, and planetary aspects, as mentioned in the previous chapter.[74] Although Khitabi claims in his preface that his primary goal in writing an exhaustive commentary on Tusi's *Si fasl* is to render the concepts and terminology used in the text more accessible to novices in the field, Tusi's original treatise—a rather straightforward

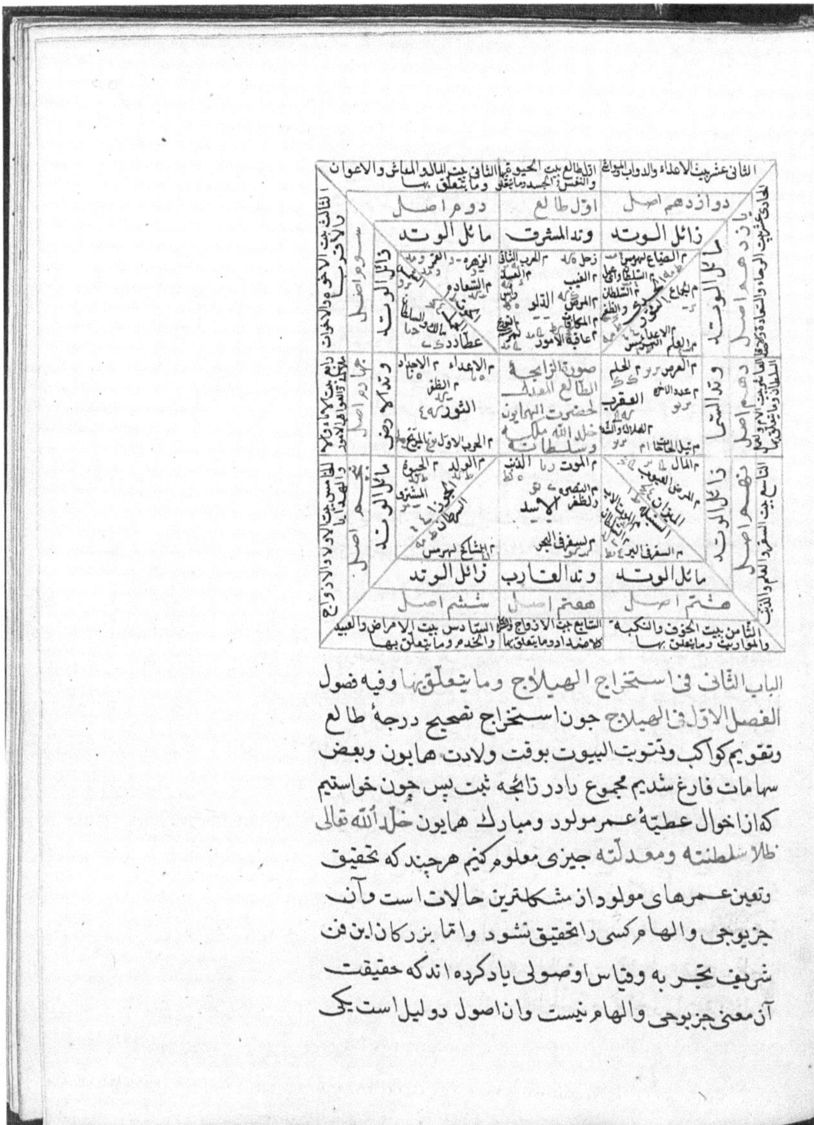

FIGURE 6. The birth horoscope of Sultan Mehmed II. From Khitabi, *Taliʿ-i Mavlud*. Süleymaniye Library, Esad Efendi Ms. 1997, 28b–29a. Reproduced by permission from the Directorate of the Turkish Institution for Manuscripts.

textbook written in plain Persian—was already well known among aspiring *munajjims* and almanac-makers in Ottoman Anatolia and Rumelia at the time.[75] Numerous copies of the text, along with commentaries, had been widely circulated in Ottoman territories since the latter half of the fourteenth century. Moreover, Turkish translations of the text began to appear as early as the late fourteenth century, indicating wider recognition and increased demand among a Turkish-speaking audience.[76] In that regard, Khitabi's pronounced objective of penning an expository commentary in Persian and reaching a broader audience relates more to his loftier aim of showcasing to the reigning sultan his profound knowledge and expertise in celestial matters. The fact that the work has survived in only a single copy and been preserved in the palace library further suggests that, contrary to Khitabi's expectations, the text did not receive a widespread readership among his contemporaries or subsequent generations.

Khitabi's Masterpiece: The Horoscope of Mehmed the Conqueror

Despite its limited circulation in its time and afterward, Khitabi's lengthy commentary on Tusi's *Si fasl* seems to have achieved his desired outcome and secured him a position in the service of Sultan Mehmed II toward the end of the ruler's life, as is corroborated by the exquisite birth horoscope Khitabi crafted upon the order of the sultan to compute and interpret the Conqueror's natal chart .[77] More than one copy of this text survives—the earliest dated June 1480—and a flamboyant royal copy was soon commissioned to the chief imperial calligrapher and bookbinder at the time, Ghiyath al-Din *al-mujallid* al-Isfahani.[78] Although this royal copy has no colophon or explicit date of completion, dates in the text establish that it was likely completed close to the death of Mehmed II in 1481. As an invaluable example of the ornate and extensive natal horoscopes compiled for reigning sultans or contending princes—surviving examples of which have come down to us especially from the Timurid realm, including the more famous horoscope of Iskandar Sultan, compiled by *munajjim* Kashi, or the lesser-known one produced by Yahya b. ʿImad b. Yahya *al-munajjim* al-Kashani for another Timurid prince and brother of Iskandar Sultan, Mirza Rustam b. ʿUmar-Shaykh (d. 1424/5)—Khitabi's voluminous work, spanning over 260 folios, or 520 pages, stands as a perfect testament to our *munajjim*'s engagement with Babylonian, ancient Greek, Abbasid, and medieval Islamicate and Persianate schools of the astral sciences and their patrimonies.[79]

Khitabi's lengthy horoscope exhibits the main characteristics of other surviving horoscopes from the Timurid realm, skillfully blending complex astronomical computations and explanations—which often comprised the first two-thirds of the volumes—with vague astrological predictions regarding the destiny of the dedicatee presented in the final third of the book. After setting the precise date (26 Rajab 835 AH / 29 March 1432 CE) and location (Edirne) of Mehmed II's birth, based on the available information he says that he found, Khitabi meticulously elucidates and illustrates the specific techniques an accomplished *munajjim* should employ when preparing an individual's nativity.[80] The methods he cites include the *namudar*s of Ptolemy, Hermes, Abu Maʿshar, and Zarathustra, a technique used to calculate the ascendant (*taliʿ*) at the moment of a nativity, as well as other specific astrological techniques deployed to establish the life expectancy of the newborn—*haylajat*, *kadkhudah*s, and *firdarat*.[81] At times, Khitabi turns his work into a comprehensive textbook for aspirants in this branch of knowledge, addressing essential information on the dispositions of stars and planets, the indications of each astrological house, and the signs of planetary aspects. These intricate explanations of astronomical, mathematical, and astrological technicalities spanning dozens of folios—which likely had little or no direct bearing on the sultan-patron—were intimately bound up with his authorial aspirations as well as the conventions of the genre. As their expertise required royal support in the absence of alternative funding structures, *munajjim*s seized every opportunity and textual platform to display their proficiency in the fundamentals of the astral sciences. Thus, Khitabi's daunting references to major authorities in the astral sciences from the ancient and medieval world (or name-droppings, one might say) may be taken as reflecting his desire to portray himself as a well-read and erudite astral expert seeking a secure position within the court hierarchy. A less skeptical reading of Khitabi's interests might also suggest that these horoscopes, with their extensive technical explanations, served a pedagogical purpose for aspiring astral scientists, as was the case in the early modern European context.[82] However, given the scarcity of manuscript evidence or anecdotal references concerning the use of these nativity books by students of the astral sciences, it is hard to confirm a pedagogical function for the existing horoscopes in the Ottoman realm.

*Khitabi's Career under Bayezid II as the Chief Astral Expert (*sayyid munajjim*)*

Khitabi continued to hold a position, and even thrived, at the Ottoman court in the capital after the death of Mehmed II in 1481. As the institutionalization of the palace corps for scientific experts, including physicians and *munajjims*, was still in its infancy, the tenure of these professionals might not have been renewed upon the ascension of a new sultan, owing to potential risks associated with the type of intimate knowledge they might have possessed. Khitabi, however, not only managed to remain in the palace corps serving his dynastic patrons for the next three decades, until the mid-1510s, but also attained a higher status, as reflected in his adoption of the moniker "sayyid munajjim" (lit. the prominent *munajjim*, as well as a possible claim to descent from the Prophet Muhammad's family). Khitabi's earlier acquaintance with Mehmed's son and successor, Bayezid II, evident in the abovementioned dual treatise on natural philosophy that he presented in Amasya in 1479, must have smoothed the transition around the succession. Still, in January 1483, Khitabi presented another text, along with an instrument of celestial observation, to the new sultan, whom he lauded, among other praises, as "wise and knowledgeable in both the 'high' (i.e., heavenly/celestial) and 'low' (i.e., worldly/terrestrial) sciences."[83] This text, titled *Risala tashrih al-alat* (Treatise on the explication of instruments), manifests Khitabi's objective of testing the accuracy of the three most favored *zijes* of the time: Tusi's, Ulugh Beg's, and his master Rukn al-Amuli's. In order to do so, Khitabi conducted a solar observation based on Istanbul's latitude and predicted a solar eclipse to occur in October 1482.[84] He anticipated two conjunctions and two full lunar eclipses as taking place in 1483: the first conjunction involving Mars and Jupiter, and the second Saturn and Mars. However, as he says, the calculations derived from the tables of Rukn al-Amuli and Ulugh Beg differed significantly from those of the Ilkhanid tables.[85] While Khitabi's solar observation in Istanbul by the early 1480s is noteworthy in itself, his endeavor to ensure the accuracy of astronomical data is even more striking, showing how essential it was for *munajjims* to have precise astronomical data for the more accurate practice of their business.

In *Risala tashrih al-alat*, Khitabi does not explicitly state a preference between the Ulugh Beg and Rukn al-Amuli tables. However, while drafting the almanac for the solar year spanning from March 1489 to March 1490, he relied on the work of his master. Only one almanac has survived that was indisputably written by Khitabi, as it bears on its flyleaf his autograph.[86] In this almanac, too, Khitabi praises Bayezid II on the occasion of the new year and conveys his best wishes to the sultan, whom he refers to as the "caliph of the All-Merciful, shadow of divine affection, strengthener of the world and religion, supporter of Islam and all Muslims, and the glory of kings and sultans, who triumphs over his enemies with the assistance of the Benevolent King."[87] He then proceeds to list important astrological variables and presents his annual predictions regarding the fortunes of the sultan. Adhering to the genre conventions of almanacs, he couches his predictions in markedly imprecise language. According to his predictions, the sultan's glory and majesty will remain unblemished, and his health and temperament will stay balanced. He will show sympathy toward his subjects and establish complete control over civil order, but occasionally, and particularly during winter, his anxiety will heighten owing to his enemies and adversaries. Khitabi continues by detailing the fortunes of people from various sectors, including viziers, scholars, mystics, and many others. His general prognostications for the year conclude with predictions about diseases, meteorological conditions, crops and prices, and wars and battles. He then draws two tables for the horoscopes of the upcoming year, one based on the Chinese Uighur animal calendar and the other derived from his master Rukn al-Amuli's astronomical tables. The section that follows comprises a month-by-month elaboration of calendrical information and accompanying astrological judgments. It is worth noting that the predictions Khitabi conveys in the monthly sections of his only extant almanac exhibit a discernible emphasis on potential conflicts and battles involving Turks, Arabs, and Kurds.[88] This focus may lead one to consider the possibility that these predictions were related to the repercussions of the then-ongoing Ottoman-Mamluk confrontation in southeastern Anatolia.[89]

Khitabi's signature in his sole surviving almanac for the year 1489–90 does not include any mention of the honorific *sayyid munajjim*. However, several archival documents from the period indicate that he embraced this title as early as the late 1480s. For instance, an endowment record from 1488–89, which lists estates in the Eyüb district of Istanbul, records him under this title as the proprietor of the endowed estate.[90] Another contemporary register, listing all individuals who received stipends from the endowment of the Muradiye Mosque complex in Edirne from April 1488 to May 1489, includes him as the recipient of 15 *akçes* per diem in supplementary income.[91] There is also an undated, single-page Persian document in the collection of administrative archival documents housed in the Topkapı Palace, containing astrological predictions for a certain month from an unspecified year. This precious document provides a glimpse into impromptu astrological forecasts that *munajjims* occasionally offered to the court, in addition to their more standardized duties of submitting annual almanacs and reporting auspicious moments for imperial ventures, for which we have much more substantial documentation. Like the intentionally vague language employed in almanacs' annual predictions, this stand-alone document avoids direct references to any specific individuals or historical events. Instead, it predicts, for example, that between the fifth of Shawwal and the fifth of Dhu al-Qa'dah of that year, celestial signs indicate the strength of governors and Turks, suspicions surrounding the ʿulamaʾ, tribulations for pregnant women, the fragility of natural philosophers and poets, soldiers mobilizing from the northern and western quarters, and the passing of an individual among the learned.[92] The real significance of the document, particularly for our purposes here, lies in the signature at its conclusion: "the servant Khitabi *sayyid munajjim*."

The title *sayyid munajjim*, beyond marking the lineage tracing back to the Prophet's family, constituted a recognition or even certification of certain astral experts, acknowledging their self-proclaimed or peer-recognized prominence within their field of expertise. Spanning a period of over a century, from the early fifteenth to the mid-sixteenth century, there were at least four *munajjims* in the Persianate and Ottoman worlds who either used this title in their works or were referred to as such by their peers or students. As mentioned earlier in this chapter, ʿAbdurrahman praised his master in Iranian lands, Niʿmatullah Shirazi, as the *sayyid* of *munajjims*. Additionally, an almanac written in plain Turkish for the year 1531–32 and dedicated to the Ottoman sultan Süleyman was signed by another self-proclaimed *sayyid munajjim* from the provincial town of Tokat.[93] A more renowned *sayyid munajjim* was Muhammad al-Husayn from the early fifteenth-century Timurid realm, who authored at least two astronomical and astrological works. One of these works, *Lataʾif al-kalam fi ahkam al-aʿwam* (Delicacies of words on the decrees of years), gained popularity among late medieval and early modern enthusiasts of the astral sciences. Based on his extant writings, which he autographed as "Muhammad al-Husayn known as *sayyid munajjim*," he was active as early as 1400–1401, when he personally observed a comet appearing in the direction of Anatolia that he believed foreshadowed Timur's victory over the Ottomans at the Battle of Ankara in 1402.[94] He later became affiliated with the Samarkand observatory and presented his treatise on astronomy and geometry to Ulugh Beg in April 1434.[95] Owing to the relative fame of this Timurid *sayyid munajjim*, some modern scholars have confused him with our Khitabi, whose 1483 treatise on a new astronomical instrument was incorrectly attributed to the earlier and more famous *sayyid munajjim* from Timurid Iran. Aside from the significant gap in time between the two individuals' lives and careers, their real names, as revealed in their extant works, provide conclusive evidence that they are not the same person.

Our Ottoman *sayyid munajjim*, Khitabi, seems to have enjoyed a prosperous life at the court of Bayezid II throughout his reign. In the comprehensive expense register that meticulously recorded the names of individuals who received gifts and allowances from the palace between 1502 and 1512, Khitabi is listed—under his sobriquet *sayyid munajjim* rather than his name—as the beneficiary of sultanic favor on at least thirty separate occasions.[96] According to the detailed records in this register, Khitabi seems to have held a status superior to the other monthly-salaried court *munajjims* (their numbers fluctuating between four and six at any one time) and the unaffiliated enthusiasts who presented almanacs to the court at different times. Unlike many of his peers, he was not only compensated when he presented an almanac but also received payments and/or robes of honor on four or five additional occasions in certain years, the specifics of which are unfortunately not provided.[97] The Topkapı Palace archives contain a single-page panegyric in Persian that Khitabi presented to Bayezid II, but since it is not dated, we cannot link it to one of the occasions when he was rewarded.[98] On several unspecified occasions, he was even paid a sum of 7,000 *akçes*, an amount equal to the pay rate of high-ranking statesmen and top scholars at the grade of professors at the eight royal *Sahn-ı Seman* colleges. The amounts he received for presenting an almanac also consistently exceeded those given to other *munajjims* and almanac-makers: he received 1,500 *akçes* for each annual almanac submission, while other *munajjims* were paid between 500 and 1,000 *akçes*, likely depending on their rank or seniority. Furthermore, in October 1503 he was given a garment following the loss of his son, suggesting that he had a close relationship with the sultan, for it was only Bayezid II's intimate companions who received gifts upon such occasions of death or marriage.[99]

Khitabi passed away sometime between March 1512 and February 1515. Documentary evidence exists for a few of his students, including a certain Receb, his son İbrahim, and Yusuf b. ʿÖmer es-Saʿati—the last two of whom served as chief court *munajjim* for the Ottoman dynasty at various points between the 1520s and 1560s.[100] Despite this, Khitabi's legacy did not enjoy an afterlife. The reasons for his lack of recognition among his successors, in spite of his elevated status during his lifetime and the availability of his textbook-like accounts written in Persian, are difficult to determine. One possible factor for the scant appeal of his works to subsequent generations might be his exclusive use of Persian, for the Ottoman learned elites and knowledge seekers from the sixteenth to the nineteenth centuries who were interested in the astral sciences showed a more marked preference for materials written in Ottoman Turkish.[101] Yet Khitabi's descent into oblivion is in fact typical and emblematic of many other Ottoman court *munajjims* of the early modern era, who faded into obscurity soon after—and sometimes even before—their passing. Khitabi's extant writings, with the exception of Mehmed II's horoscope, exist only in a single copy and were stored in the palace treasury with little to no demand from internal or external readers.

Considering explicit references in Khitabi's works to authoritative texts in ancient and medieval astral sciences, he must have possessed a substantial collection of books on various aspects of celestial knowledge. To date, however, I have come across his name only in a couple of manuscripts, either in the marginalia or on a flyleaf. The first is a copy of Wabkanawi's *al-Zij al-muhaqqaq* that contains several marginal notes by a "Sayyid Munajjim," some of which are related to intended computations for the year 1483–84 CE.[102] The other extant manuscript, which was once possessed by Khitabi, is a copy of Kushyar's *al-Zij al-Jamiʿ*, now located at Leiden University Libraries. This copy bears an ownership statement on its flyleaf dated 1514 or 1515 by Yusuf b. ʿÖmer es-Saʿati, who says he was a *munajjim* at the Ottoman court by then. More intriguingly, for our purposes, Yusuf's note registers that the copy came into his possession following the death of his teacher, *Sayyid Munajjim*.[103]

As will be discussed in more detail in Chapter 4, Yusuf b. ʿÖmer es-Saʿati remained on the palace payroll as one of the court *munajjims* for five de-

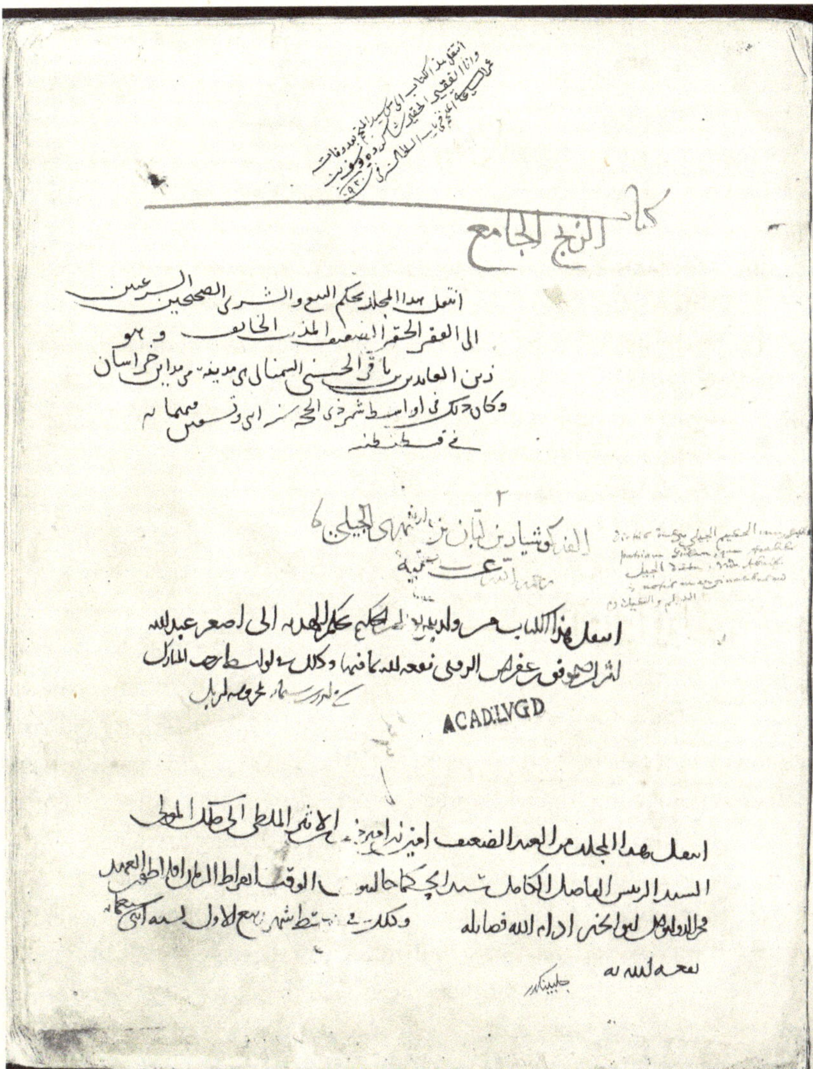

FIGURE 7. Ownership records by Khitabi and his student Yusuf b. ʿÖmer on a copy of Kushyar's *al-Zij al-Jamiʿ*. Leiden University Library, Or. 8, 1a. Courtesy of the Leiden University Library, Netherlands.

cades, leaving behind at least nine extant almanacs, the latest of which is from March 1560.[104] The oldest of these surviving almanacs was compiled in 1511.[105] In this earliest-extant *taqwim*, Yusuf attests to his reliance on the *Zij-i muhaqqaq-i Ilkhani* for calculating the horoscope at the precise instant when the Sun would transit into the sign of Aries, marking the onset of the new (solar) year.[106] This reliance suggests that Wabkanawi's revised astronomical tables, *al-Zij al-muhaqqaq*, a copy of which was in Khitabi's possession, likely played a role in their master-student interactions. While Khitabi himself did not particularly favor using Ulugh Beg tables—a fact documented by his aforementioned astronomical treatise and single extant almanac—his student Yusuf eventually came to prefer it. In the majority of his extant *taqwims* from 1514 through 1560, Yusuf executed his horoscope computations based on the Ulugh Beg tables.[107] These subtle yet telling details regarding both the continuity and change in the authoritative sources utilized are indicative of the strong impact of post-Mongolian Persianate traditions upon the formation and development of Ottoman astral practice.

None of Yusuf's extant almanacs contain even a cursory mention of his master, as one might expect, nor do they shed any light on where their paths may have intersected or where they may have pursued their endeavors. Based on the absence of any contemporary or subsequent reference to Khitabi or Yusuf b. ʿÖmer in historical narratives and biographical dictionaries pertaining to the Ottoman ʿulamaʾ, it can be inferred that their connection was formed beyond the confines of the madrasa. This supposition is further supported by the widespread reluctance among madrasa scholars to formally study the practical astral sciences, including astrology, within these traditional institutes of higher learning—an observation elaborated upon in the previous chapter. Therefore, the training Khitabi provided to Yusuf b. ʿÖmer should be contextualized within the specific position created at the imperial court for *munajjims*. This is not to suggest, however, that the training took place in a designated physical site or spatial "office" reserved for court *munajjims*. No such allocated space within the palace has been identified as a base of operations for the *munajjims*. When used in this context, the term "office" refers not to a spatial unit but to an organizational structure by way of which the chief experts and their students congregated, likely in private residences and other venues where less formal scholarly gatherings (*majlis*) were convened.[108]

Conclusion

Khitabi and ʿAbdurrahman were, needless to say, not the only astral experts within the late fifteenth- and early sixteenth-century Ottoman milieu. More than twenty-five extant *taqwims* from the period between 1421 and 1500—70 percent of which were compiled in Persian—as well as several birth horoscopes, original treatises on astrological techniques, and copies of authoritative texts on the astronomical and astrological study of the heavens, provide clear evidence that many more practicing and aspiring *munajjims* were active in the Ottoman territories.[109] However, due mostly to the shortage of substantial biographical and autobiographical information about these figures, it is difficult to thoroughly reconstruct the social and intellectual lives of these Ottoman-court astral experts. ʿAbdurrahman and Khitabi stand out as representative cases; the details derived from their extant works, albeit incomplete and fragmented, offer us a valuable glimpse into their world, revealing not only their scientific inspirations and preferred techniques but also the challenges and opportunities they encountered.

Both ʿAbdurrahman and Khitabi lived and flourished at a time when the Ottoman enterprise was slowly evolving into a major imperial power in Eurasia. This transformation was catalyzed by the epochal conquest of Constantinople in 1453 and the ensuing imperial, ideological, cultural, and architectural initiatives undertaken by Mehmed II and his son, Bayezid II, to transform their domain, especially their newly acquired capital, into an enviable hub of arts, sciences, learning, and commerce. This crucial period of imperial formation witnessed an interplay of various ideological and intellectual currents; some seamlessly blended while others occasionally in conflict. These ranged from Byzantine administrative practices and ancient monotheistic and philosophical traditions to more recent Persianate achievements in science, literature, and poetry. In particular, in the latter half of the fifteenth century, the decline of the Timurid Empire in western Asia, which was not immediately replaced by a single successor dynasty with unquestioned political and intellectual authority, made the emerging Ottoman empire—open to diverse traditions and people—an attractive destination for men and women of knowledge from former Timurid territories.[110] In fact, the courts of Mehmed II and Bayezid II were so populated with Persian-speaking litterateurs, bureaucrats, and scientific experts that

Turkish-speaking indigenous Ottomans, seeking a place for themselves in the emerging bureaucracy or the intellectual circles of the expanding capital, began to voice concerns over the pro-Persianate cultural dynamics and orientations at the time.[111]

As a scientific tradition with deep roots in the late medieval Persianate landscape, *munajjims*' expertise gained special recognition during this period when Persianate influences were particularly pronounced and visible in the Ottoman capital. The next chapter will further explore the court culture of Ottoman rulers from the second half of the fifteenth century, placing particular emphasis on the reign of Bayezid II, a time when the *munajjims* and their expertise enjoyed their halcyon days.

THREE

HEAVENLY PATRONAGE

The stories of *munajjims* detailed in the previous two chapters make it evident that few sciences are as desperately reliant on royal favor as theirs. The norm in the broad sweep of Islamic history was that the primary endowed institutions of higher learning, the madrasas, did not serve as sites where prospective *munajjims* acquired their proficiency in astrology and eventually earned a living while practicing their trade. Observatories, as grandiose imperial projects, were constructed only once in a long while, with the only pre-nineteenth-century observatory in the Ottoman world having been established by the late sixteenth century and operating for only a few years.[1] Large, endowed mosques often created a position, and hence some steady income, for the adepts of the astral sciences in the role of *muwaqqit* (timekeeper), but the salaries they offered were meager. Furthermore, the critical absence of the printing press deprived early modern Ottoman astral experts and almanac-makers of a profitable book market, while their European counterparts thrived on the sale of their books and highly sought-after annual prognostications.[2] Along with these structural absences, Ottoman *munajjims* lacked a professional corporation like the guilds that offered support and protection to members of other trades. Although certain Ottoman archival and narrative sources, as well as accounts from European observers, mention fortune-tellers' shops within bustling commercial districts of major cities, no documentary evidence has been unearthed that shows an Ottoman court *munajjim* conducting business in the marketplace.[3] Faced with these constraints, *munajjims* had little choice but to seek the patronage of a wealthy benefactor, ideally the sovereign himself, as a means of securing their livelihood and sustain their craft.

This familiar tale of *munajjims* relying on the patronage of rulers and affluent urbanites took on a somewhat different character and even an institutional veneer in the early modern Ottoman courtly context. By the late fifteenth century, more specifically during the reign of Bayezid II, a discernible administrative position for *munajjims*, almost an office, surfaced within the nascent Ottoman court bureaucracy and remained intact for over four centuries until the empire's official demise after World War I. This was an office without walls; it occupied no designated space in the palace or elsewhere. Rather, it played an income-generating and status- and expertise-certifying role, keeping certain practitioners of the astral sciences on the palace payroll in return for their routine services. These services included submitting annual almanacs, reporting auspicious hours, and providing celestial guidance as needed. Such a broad remit required officeholders, particularly the chief court *munajjim*, to commit to it as a full-time job, preferably in close proximity to the sultan and ruling elites.

The number of *munajjims* staffing the unit varied depending on the esteem in which their expertise was held at a given time. While some periods proved more favorable for *munajjims* than others, there was always at least one individual listed as the chief practitioner among the monthly-salaried palace personnel from the late fifteenth through the early twentieth century. No specific procedural path or written code is known to have governed entrance to and promotion within the corps of court *munajjims*. Decisions regarding individual merits and qualifications were subject to personal evaluation, periodic reassessment, and constant negotiation. Theoretically, the position was open to any individual capable of demonstrating and sustaining the superiority of their expertise; however, failures in the accuracy of computations and predictions were not without consequences.[4] Despite this implied meritocratic ethos, the inner workings of the unit, like many other early modern Ottoman institutions, hinged heavily on personal acquaintances, family connections, and trusted views of trustworthy parties.

The emergence of a distinct corps of *munajjims* in Bayezid II's reign paralleled the development of similar cadres designed to harness specialized skills and technical knowledge for more effective imperial governance.[5] The corps of royal architect-engineers (Tur. *miʿmaran-ı hassa*), for instance, emerged during this very period as a discernible unit, differing from the earlier, somewhat amorphous recruitment of architect-engineers nestled among the vaguely defined ranks of distinguished royal servants (Tur. *müteferrika*).[6] This unit of royal architect-engineers was not only tasked with executing and maintaining architectural projects financed by the royal family but also with overseeing active architect-engineers, even at the provincial level. Similarly, the body of royal physicians (Tur. *etıbba-yı hassa*) and the corps of artillerymen (Tur. *topçıyan*), each helmed by a practitioner appointed as their chief (e.g., the *hekimbaşı* or *ser-etibba*, the chief physician), began to leave their bureaucratic footprints during this period.[7] In fact, within the particular context of the formation of Ottoman bureaucratic practice, Bayezid II's era left a lasting legacy. Some of the precursors of certain Ottoman bureaucratic registers, which would take on their definitive forms only by the mid-sixteenth century, such as the registers of important affairs (*mühimme defterleri*), made their first appearance at this time.[8]

It may seem to be an oversimplification and a monocausal explanation to attribute these developments solely to Bayezid II and his determination to build an empire with a robust bureaucratic structure. After all, Ottoman archival practices and record-keeping capabilities had a longer history, and the drive toward creating predictable, well-ordered, and rationalized governance with clearly defined administrative hierarchies and ranks was already a priority in earlier decades.[9] The most evident example of this drive was Mehmed II's "constitutional" law code (Tur. *kanunname*), which was designed, among other things, to establish rules and protocols for recruitment and promotion in various administrative positions.[10] However, as Derin Terzioğlu also notes, "It fell to Mehmed's much less charismatic son and successor Bayezid II . . . to sort out the institutional arrangements that would sustain this newly forged empire."[11] Considering the standard underappreciation of his long tenure in modern Ottoman historiography compared to the more celebrated reigns of his father Mehmed II, his son Selim I, and his grandson Süleyman, it is of extreme importance to recognize Bayezid II's era as a pivotal moment in the arc of Ottoman imperial formation in the long sixteenth century, during which time the foundations of what would eventually be considered "classical" Ottoman institutions, conventions, bureaucratic culture, and language were laid down.[12]

Bayezid II also deserves special attention as an individual deeply interested in learning the mathematical, astral, and philosophical sciences, as well as for his unparalleled patronage of *munajjims* and other men of knowledge, including certain Jewish natural philosophers. At first glance this may appear contradictory vis-à-vis his prevailing image in modern scholarly literature and lay discourse as a saintly and pious figure, even zealously religious.[13] His piety has been held responsible, perhaps more often in vulgar publications, for causing an unforgivable glitch in the upward trajectory of the Ottoman Empire toward becoming a modern, capitalist, and science-minded world power. In the counterfactual scenarios and speculative musings of certain modern interpreters, the Ottoman Empire's ultimate fate would have been markedly different if Mehmed II's favored son, Cem Sultan (d. 1495), had succeeded his enlightened father instead of his sanctimonious brother Bayezid.[14] In truth, however, it was Bayezid II who proved to be the true champion of sciences and institutor of a well-functioning imperial structure. His three-decade rule was the seventh heaven for *munajjims*, and this openness and open-handedness toward a variety of learned people and scientific experts was not incompatible with his renowned piety. On the contrary, his commitment to leading a pious life, pursuing astral scientific interests, surrounding himself with a large group of astral experts and other learned individuals, and striving to create a rationalized imperial administration were all interrelated facets of his complex profile that seemed to make him, in some contemporary observers' estimation, something of an idealized philosopher-king.

In this chapter, our gaze shifts from *munajjims* to their patrons, exploring the extraordinary scope and far-reaching repercussions of the celestial fascination that gripped the court of Bayezid II. His incumbency was certainly neither the first nor the last time in Ottoman history when the expertise of *munajjims* exerted a profound attraction upon the sovereign. Moreover, Bayezid II was not the sole driving force behind the marked fascination with the astral sciences during his tenure. Several other key figures from the sultan's inner circle, such as Müʾeyyedzade ʿAbdurrahman and Mirim Çelebi, also deserve special recognition for their contributions to the creation of an exceptional intellectual milieu that facilitated an influx of astral experts and their knowledge into the Ottoman capital. These scholars came in particular numbers from the post-Timurid Persianate realm,

FIGURE 8. Sultan Bayezid II with a book in his hand. BnF Ms. Turc 140, 7b. Courtesy of the Bibliothèque national de France, Paris.

as well as the Mediterranean world, in the wake of the 1492 Jewish expulsion. Although enthusiasm for the *munajjims*' expertise was not always maintained to the same degree by Bayezid's successors and courtly interest in their knowledge diminished markedly, though did not vanish entirely, by the reign of Süleyman, the formation during Bayezid's rule of a formal administrative framework for recruiting astral experts and routinizing their service shaped the manner in which the institutional deployment of *munajjims*' expertise was preserved in the Ottoman realm for centuries thereafter.

Ad Hoc Royal Patronage before Bayezid II

Before the ascent of Bayezid II, the Ottoman court, like many other medieval Islamic dynasties, was no stranger to the presence of *munajjims*. Yet, precise details of their identities and origins are elusive. The earliest verifiable evidence of a *munajjim* serving an Ottoman ruler dates back to the late 1430s. Given the frequent references to *munajjims* in historical narratives about the Seljuqid, Ilkhanid, and various other Muslim and non-Muslim polities in Anatolia throughout the thirteenth and fourteenth centuries, coupled with the surviving copies of almanacs produced at the courts of the principalities of Eretna and Karaman, as well as Pontus Trebizond, and other scientific treatises from the period, it is entirely plausible that astral experts had been within the orbit of the Ottoman ruling circle from its nascent days.[15] While astrological almanacs have come down to us from the time of Mehmed I (r. 1413–21), it is not until the reign of Murad II (r. 1421–44; 1446–51) that we encounter a *taqwim* bearing a *munajjim*'s signature. In 1439, a certain *munajjim* by the name of İbrahim crafted and presented to Murad II a Persian almanac brimming with his astrological prognostications for the year's impending fortunes and tribulations.[16] The exact nature of İbrahim's status, however, remains enigmatic, leaving us uncertain as to whether he served as a court *munajjim* or was simply a freelance savant lending his expertise to the Ottoman ruler. Regardless, the manuscript's very existence attests to the proximity of *munajjims* to the seat of Ottoman power in the first half of the fifteenth century. Such manuscript evidence is further bolstered by contemporary historical narratives. In *Gazaname-i Rum* (The book of campaign in the Rum), written by a certain Kaşifi around 1456, the author recounts that, upon the birth of his son Mehmed, Murad II summoned his *munajjims* and had them interpret his son's birth horoscope, which foretold that Mehmed would rule as a world conqueror and master of auspicious conjunction (*sahib-qiran*).[17]

No fewer than four *taqwims* and two separate birth horoscopes in multiple copies have survived from the more than thirty-year rule of Mehmed II. The latter of these horoscopes, compiled by Khitabi toward the end of Mehmed II's reign, was introduced in the previous chapter. The earlier horoscope, which contains both the Persian original and its Turkish translation bound together in the same volume, was initiated in May 1451 and completed in October 1451, according to its dual colophon.[18] This earlier work suggests that a *munajjim* well-versed in Persian was close to Mehmed II after his second and ultimate ascension to the throne in Edirne in February 1451. Unfortunately, the copy does not offer any clues to the identity of its compiler. The same is also true for the surviving almanacs from the time of Mehmed II, which lack autographs that would reveal the identity of the *munajjims* at his court. Nevertheless, several contemporary narrative accounts allude to a cohort of *munajjims* from whom Mehmed II sought counsel, seeking their guidance in designating auspicious times before military campaigns or the construction of imperial edifices. One such primary source is the account of Cardinal Isidore (d. 1463), the Greek metropolitan of Kyiv. As a firsthand witness to the siege of Constantinople in 1453, Isidore mentioned in a letter to Pope Nicholas V (d. 1455) that Mehmed II solicited the advice of his "Persian *munajjims*" (*astrologi persiani*) to ascertain the most favorable time for the siege.[19] Similar anecdotes appear in contemporary Ottoman sources as well. For instance, Tursun Beg (d. circa 1491), an Ottoman bureaucrat and eyewitness author of the *Taʾrih-i Ebu'l-Feth* (History of Mehmed the Conqueror), briefly notes in his chronicle that prior to the siege of Constantinople, "skilled geometers and expert *munajjims*" calculated a favorable moment for constructing the fortress of Rumeli Hisarı on the Bosporus.[20] In a similar vein, a high-ranking bureaucrat during the latter years of Mehmed II's reign, Tacizade Caʿfer Çelebi (d. 1515), refers to the role of "experts in the sciences of the stars" in calculating the opportune hour to commence the attack on Constantinople.[21] Another contemporary historian, Kıvami, narrates in his *Fethname-i Sultan Mehmed* (History of Sultan Mehmed's conquest) that when the sultan ascended the throne for the second time in 1451, he asked *munajjims* to interpret the horoscope for the moment of his enthronement.[22] Despite such valuable contemporary anecdotes, these Ottoman sources fall short of specifying the names or ethnogeographic affiliations of the *munajjims* in question.

The presence of *munajjims* in close proximity to Mehmed II is hardly surprising. Indeed, one might expect to find more traces of *munajjims* and their scientific writings under the aegis of such a venerated patron and possessor of knowledge. The portrait of Mehmed II as an avid patron of learning and an enlightened ruler—a man with a remarkable breadth of intellectual interests, from Ptolemy's geography and ancient Greek history to pre-Islamic monotheistic traditions and Islamic philosophical theology—has become so deeply, and of course justifiably, rooted in modern Ottoman historiography that it is difficult for students of Ottoman history to envision another figure in the Ottoman past with an equal thirst for knowledge and patronage of the sciences.[23] Even when they conjure a commensurate figure, Bayezid II rarely comes to mind.[24] In fact, Bayezid is often considered and presented as the antithesis of his learned father. For instance, Bayezid's sale of paintings and Christian relics from his father's collection to European agents is frequently cited as a sign of Bayezid's religious bigotry. However, this interpretation rarely acknowledges the possible role of these sales in the more than decade-long cold war between Bayezid and his brother, Cem Sultan, which took on an international dimension after Cem sought refuge in France and the Vatican.[25] Likewise, the execution in 1495 of Molla Lutfi, an eccentric Ottoman scholar who railed against the self-appointed scholarly authorities of his time, is often seen as emblematic of the decline in free thought and intellectual spirit under Bayezid II's rule.[26] Yet such an interpretation fails to consider the agency of those very scholarly authorities who made the decision to execute Molla Lutfi despite Bayezid's efforts to protect the scholar.

The stark contrast often drawn between Bayezid II and his father, Mehmed II, serves as a convenient narrative in modern Ottoman historiography, which often seeks to interpret the Ottoman past through such binary lenses as West (Europe) versus East (Islam), science versus religion, or progress versus decline.[27] Within this rather anachronistic framework, Mehmed II embodies all that is associated with the imagined West: commissioning portrait paintings, lavishing funds on scientific experts, and displaying a willingness to adopt new technology, whereas Bayezid II is disparaged for his presumed reversion to Islam and the East. This contrast is occasionally fueled by various decontextualized or unfounded allegations leveled against Bayezid II. For example, Leonardo da Vinci's unfulfilled proposal to Bayezid II to construct a bridge over the Golden Horn is presented as an indication of the sultan's reluctance to support European scientists and experts.[28] Yet what made Bayezid II such an attractive candidate for renowned Renaissance artists and polymaths to approach has seldom been addressed. Likewise, the likely apocryphal account of Bayezid's rejecting Christopher Columbus when the Genoese navigator sought him as a potential patron before embarking upon his costly expeditions is often portrayed as another inexcusable mistake by the sultan, which, according to modern interpreters, resulted in the Ottomans' missing out on the entire age of exploration and subsequent development of mercantile capitalism.[29] However, little effort has been expended on examining why Columbus might have considered approaching Bayezid II, if he truly did so, and whether Bayezid II's deliberate attempts to implement a naval policy, which was somehow linked to the promotion of the celestial sciences, played any role in it.[30] Finally, an undocumented imperial decree banning the printing press is erroneously attributed to Bayezid II.[31] Not only is there no contemporary evidence or reference to support such an unwarranted claim; it was in fact during his reign that the first printing presses ever established in Ottoman lands were founded by the empire's Jewish subjects.[32]

Lest I be misunderstood, it is essential to state that there were noteworthy differences between Mehmed II and Bayezid II, as construed by their contemporaries. However, these distinctions pertain more to their divergent approaches, styles, and bases of rule than to their intellectual and scientific interests. Mehmed II, for instance, was perceived by some of his contemporaries—especially those from the ghazi milieux or individuals with ties to particular Sufi networks—as an oppressive ruler who unjustly confiscated their endowed lands.[33] In contrast, Bayezid II received praise from these very same circles for simply his restoring those confiscated lands to their rightful owners upon his ascension. Even the attribution to Bayezid of the epithet *veli* (Ar. *wali*; lit. the "friend" of God [and perhaps also the Sufi circles]) might have resulted from his showing favor to the Sufi networks that had been disenfranchised by his father's fiscal policies, though we cannot easily trace when exactly and among which circles this epithet came into the fore.[34]

On the intellectual front, the two sultans shared more common ground than one might initially expect. For example, both were passionate about convening scholars with philosophical leanings in their presence for vigorous intellectual debates, including discussions on mathematical problems.[35] They each endeavored to transform their empire—and specifically, their capital—into a sublime arena for learning and sciences. To achieve this grand vision, they invested in building expansive educational complexes and acquiring thousands of books for their palace library, culminating in the colossal book collection for which an exquisite catalog was compiled between 1502/3 and 1503/4.[36] The two sultans also welcomed learned individuals from outside their domains, from both the so-called East and West, whose reputations and expertise had reached their ears. Mehmed II's invitation to ʿAli Qushji, a luminary of the Samarkand observatory, among his other scholarly accomplishments, is well known. Less known, however, is Mehmed's pledge to create an ideal environment for Qushji to conduct his observational program.[37] Alas, ʿAli Qushji's arrival in Istanbul in 1472—together with a sizable entourage of his family and students—was soon followed by his untimely death in 1474, leaving Mehmed's promise unfulfilled, even if it had been sincere. Nevertheless, the tradition of welcoming practitioners of the astral sciences to the Ottoman realm continued at an even greater pace under the reign of Bayezid II, who had already displayed strong signs of support during his tenure as a princely governor in Amasya.

At the time when ʿAli Qushji and his retinue arrived in the Ottoman world, the study of celestial sciences in the Ottoman realm was sporadic, if not entirely in its infancy.[38] As mentioned in the previous chapter, several treatises by Tusi had been in circulation throughout the lands of Rum in the fourteenth and first half of the fifteenth century, including his introductory textbook (*Si fasl*), his astronomical handbook with tables (*Zij-i Ilkhani*), and his memoir on astronomy (*Tadhkira*). The Fenari circle in western Anatolia featured such members as the stellar scholar Molla Fenari (d. 1431) and ʿAbdulwajid b. Muhammad (d. 1435), who were well-versed in the scientific output of the Maragha tradition or the corpus of Jaghmini's *Epitome*.[39] Yet anecdotal and archival records about scientific pursuits in the Ottoman realm up until the late fifteenth century imply that Ottoman territory was far from the ideal place for an aspirant to delve deeper into the astral sciences. For instance, Qadizada al-Rumi, who eventually became one of the intellectual cornerstones of the Samarkand school, was born and raised in the lands of Rum (more specifically, in Bursa), as his appellation suggests, and received his first education within the Fenari circle. Nevertheless, his master Muhammed Şah Fenari, the son of the renowned Molla Fenari, felt compelled to suggest that he should journey to Iran and Central Asia to expand his knowledge of the astral sciences.[40] A similar experience was shared by an unnamed scholar, likely from the early sixteenth century, who in his petition to the palace asks for financial support to travel from Anatolia to Azerbaijan and Central Asia to advance his knowledge in the mathematical sciences.[41] We should also recall ʿAbdurrahman *munajjim* from the first chapter, who was urged by his patron to venture to the Iranian lands to hone his skills in the astral sciences. These details make it clear, besides testifying to the inadequacy of the expertise and opportunities available to aspirants for excelling in the astral sciences in Anatolia, that in the eyes of fifteenth-century scholars native to Ottoman lands, the Persianate domains were the ultimate point of reference and cynosure for cultivating expertise in the astral sciences.

The relatively nascent state of the astral sciences at the time is also apparent in the limited number of *munajjims* who found courtly recognition and remuneration prior to the reign of Bayezid II. Despite the vague references in contemporary narrative sources to a number of (Persian) *munajjims* at the court of Mehmed II, the archival records from the period tell a somewhat different story. It is true that compared to later centuries, not many archival documents have survived from his time; thus, the matter might simply be an absence of evidence rather than the evidence of absence. Nevertheless, we do possess a relatively comprehensive payroll book from 1478, compiled near the end of Mehmed II's reign, which enumerates the palace personnel, including *munajjims*. Contrary to the image portrayed by the literary sources of the era, the register reveals only one *munajjim*, Yazdanbakhsh, whom we have previously met as one of the masters of ʿAbdurrahman. [42] Aside from the noteworthy detail of Yazdanbakhsh's receiving ten *akçes* per diem for his services, the true significance of the record lies in its reference to his status: he is listed in the register under the loosely defined corps of *müteferrika*, which corresponded to a varied group of individuals with distinguished skills and/or lineage. This suggests that, at the time, the Ottoman ruling elites had yet to establish a specialized court unit for *munajjims* within their emerging bureaucracy, making it challenging to determine the boundaries between servitors receiving patronage ad hoc and formal administrative units with specific tasks.[43]

While there was only one *munajjim* listed in the sole surviving register of payments from Mehmed II's time, by the latter years of Bayezid II's reign, the number, status, and salaries of *munajjims* had undergone a remarkable transformation. Another payroll book, which must have been drafted sometime between 1490 and 1500, according to internal evidence, lists six *munajjims* receiving a combined total of 6,068 *akçes* monthly, yielding an average daily salary of 33.7 *akçes* per court *munajjim*.[44] Regrettably, the list does not specify the names of these *munajjms* but groups them, for the first time, as an individual unit under the rubric of the monthly salaried palace personnel (Tur. *müşaherehoran*), as opposed to the earlier, more loosely defined status of *müteferrika*.[45] The better-known register of gifts and allowances, which thoroughly covers the last decade of the reign of Bayezid II from 1503 to 1512, also supports the information provided in the older payroll. In this voluminous register, over twenty different men are recorded as compensated *munajjims*, timekeepers in sultanic mosque complexes, or individuals presenting the court with an annual almanac.[46] Within these names, at least five of them—Khitabi aka Sayyid Munajjim, Yusuf b. ʿÖmer es-Saʿati, İshak, Salman-i ʿAjam, and Erdeşir—are consistently listed with reference to their status as monthly salaried palace personnel, suggesting that by that time they constituted the corps of court *munajjims*.

When it comes to the number of *munajjims* securing stable employment at the Ottoman court, Bayezid II's reign not only outshines that of his enlightened father but also surpasses his successors. For example, in a register compiled circa 1514, during Selim I's reign, the court's corps of *munajjims* included four practitioners, although details regarding their remuneration are absent.[47] Two different pay registers from the dawn of Süleyman's reign also list four *munajjims* under the same rubric of the unit of court *munajjims*: Seyyid İbrahim (the son of Khitabi), İshak, Salman-i ʿAjam, and Yusuf b. ʿÖmer, receiving 15, 14, 13, and 10 *akçes* per diem, respectively.[48] In another register prepared slightly later, three of the same *munajjims* are listed, with the omission of Salman.[49] It is worth emphasizing that all of these four *munajjims* started their careers under Bayezid II, illustrating how the corps of court *munajjims*, which began to take on an institutional and impersonal character during his reign, shaped the future of the unit in the coming decades. Additional archival documents from later in the reign of Süleyman confirm that after a brief period of stability, the number of court *munajjims* had dwindled to two by the 1540s, accompanied by a concomitant decrease in their compensation.[50] For instance, Yusuf b. ʿÖmer, apparently the chief court *munajjim* in the 1550s, received 12 *akçe*s per diem for his monthly regular service and 500 *akçe*s for submitting his annual almanac.[51] Forty years prior, when the young Yusuf was serving merely as one of the junior *munajjims* in the corps, he was making more or less the same amount. The chief *munajjim* back then enjoyed far more generous remuneration than Yusuf: 1,500 *akçe*s for the submission of the annual almanac and an estimated 40 to 50 *akçe*s per diem for regular services. Taking the eroding effects of inflation over those four decades into consideration, the discrepancy in court *munajjims*' remuneration between the reigns of Bayezid II and Süleyman proves all the more striking.[52]

Intellectual Pursuits of Bayezid II as a Philosopher-King

The flourishing of the astral sciences during the reign of Bayezid II was unparalleled in Ottoman history, a fact discernible just from the sheer number of *munajjims* employed at his court and the comparatively generous funds allocated to them. Yet Bayezid II had much more to offer as a learned patron of knowledge, showing unwavering personal dedication to and fascination with the area of expertise *munajjims* possessed. In fact, the unprecedented scope of interest in the Ottoman court in the study of the heavens and related fields of expertise during his rule was inextricably linked to Bayezid II's personal intellectual inclinations. From his twenty-seven-year governorship in Amasya to his even lengthier sultanate in Istanbul, Bayezid II ardently sought expertise in the astral sciences, surrounded himself with a veritable cadre of astral scientists and natural philosophers, commissioned various treatises on different branches and genres of the astral sciences, and dedicated his own leisure time to the study of the science itself under the guidance of a distinguished astral expert. It was during his reign, for instance, that Istanbul became the site of modest celestial observations and other scientific experiments involving alchemical operations, conducted by émigré experts gathered at his court.[53] At least two large astrolabes were cast for the sultan: the first, crafted in 1504–5, likely by a Persian émigré scholar named Shirvani, and the second, an elaborate brass instrument recently auctioned off for one million pounds, manufactured the following year by an expert apparently from Anatolia, Nücumi el-Rumi (d. after 1505/6).[54] The catalog of the palace library, compiled under the direction of

Bayezid's chief librarian Hayreddin ʿAtufi, abounds with textbooks on astrological principles, treatises on astronomical instruments, detailed horoscopes, and astronomical handbooks with tables, copied and dedicated to Bayezid II.[55] Some of these items, as their colophons and dedication remarks reveal, were even composed at the sultan's personal request and initiative, such as an Arabic treatise on a novel astronomical instrument, which was apparently translated from Hebrew by a convert named ʿAbdüsselam at the behest of the sultan (*bi-talqin al-sultan*).[56]

Bayezid II's profound interest in the astral sciences and related areas of knowledge was so widely acknowledged during his lifetime that both contemporary learned elites and foreign observers with direct access to court circles explicitly noted the sultan's penchant for these sciences. Andrea Gritti (d. 1538), the renowned Venetian merchant and statesman who spent a significant portion of his early life in Istanbul and maintained close connections with high-ranking members of the Ottoman court, wrote in one of the reports he sent to the Venetian senate that the sultan was esteemed as a learned individual in both *astrologia* and *theologia*, and that he ardently pursued these disciplines while also taking delight in the arts of mechanics and alchemy.[57] Another contemporary testimony to Bayezid's learned interests comes from Ibn al-ʿUlayf (d. 1520), a noted poet residing in Mecca who had visited Istanbul and presented the sultan with a panegyric well before composing an Arabic chronicle in Mecca that extolled the virtues of the Ottoman dynasty and Bayezid II. Within this chronicle, Ibn al-ʿUlayf describes Bayezid's scholarly character and identifies various fields of knowledge the sultan endeavored to master. According to Ibn al-ʿUlayf, in addition to certain branches of the religious sciences, including hadith and jurisprudence, Bayezid II distinguished himself in the "study of the celestial spheres" (*ʿilm al-falak*) and the "science of the stars" (*ʿilm al-nujum*).[58] Similar observations concerning Bayezid's erudition and scholarly interests, particularly his passion for natural philosophy and astral science, also feature in the collections of contemporary poets. For instance, in one of his Persian *qasidas* presented to the court, Afsahi (d. after 1512) praises Bayezid II's learnedness, asserting that it surpassed even the likes of Ibn Sina and Tusi in the realms of "wisdom" (*hikmat*) and "astronomy" (*hayʾa*).[59]

Another indication of Bayezid II's peculiar intellectual interests is the fact that several scholars and literati chose to compose and present works related to celestial knowledge as initial attempts to seek Bayezid II's patronage. Naturally, this was unsurprising for self-proclaimed *munajjims* and natural philosophers, such as the previously mentioned Khitabi, who penned a compendium on the sciences of the stars and medicine as his first treatise dedicated to Bayezid when he was still a prince in Amasya, or another *munajjim* by the name of Bali, whose initial gift to the sultan in 1481–82 was a book on astronomical instruments.[60] However, that several other learned individuals and literati not otherwise known as specialists in the knowledge of the heavens composed texts about the astral sciences and presented them to the sultan in their pursuit of his support is highly telling of contemporary assessments of Bayezid II's interests. For instance, a prominent jurist and religious scholar by the name of Kıvamuddin Yusuf, better known as "the judge of Baghdad" (Qadi-yi Baghdad) for his longtime presence in the city before the rise of the Safavids, arrived in the Ottoman territories after he was forced to flee in the face of the Safavid Shah Ismaʿil's advance toward the region and persecution of Sunni scholars. Upon his arrival, Qadi-yi Baghdad decided to compose a Persian almanac with prognostications as his first gift to the sultan, whom he praises as, among other lofty cognomens, someone who "brought together philosophical knowledge with Solomonic rule."[61] For this debut, on 25 April 1508, Qadi-yi Baghdad received a handsome amount of 1,500 *akçes* and was later appointed to teach at one of the *Sahn* schools, mainly in jurisprudence and other madrasa sciences.[62] No further evidence is available regarding Qadi-yi Baghdad's subsequent engagement with the astral sciences after his initial approach to the sultan with an astrological almanac. In a similar vein, a better-known figure from Bayezid II's time, Firdevsi-i Tavil (d. after 1512), made his debut in 1487 by presenting the sultan with an encyclopedic work on celestial magic and talismans.[63] This seems to have been successful in garnering the sultan's favor, at least for some time, before this curious character ultimately failed to continue enjoying that royal support for reasons that are mostly unknown.

The most intriguing example among contemporary testimonies regarding Bayezid II's interest in studying the mathematical, astral, and philosophical sciences is an anonymous letter addressing the sultan. Adopting the idiom of a self-assured and assertive Sufi shaykh steeped in alchemical terminology, the author refers to Bayezid's close engagement with learning "divine art" (*sanʿat-ı ilahiyye*)—or astronomy (*sanʿat el-heyʾe*), depending on the reading of the word[64]—along with another formidable but unspecified branch of natural philosophy. In the author's opinion, Bayezid lacked intuitive erudition; his endeavors were based solely on experience. In penning the letter, the author intends to remind the sultan, whom he characterizes as a fervent servant on the path of Islam, of the transitory nature of life and the inconsequence of worldly possessions.[65] He then reveals that, in accordance with the portents from one of his recent dreams, he has decided to send a disciple to impart to the sultan his true essence. The author of the letter contends that the training should continue until Bayezid attains the spiritual stage that his disciple has already reached under his guidance. Once Bayezid achieves this stage, then the author pledges to write a talismanic note for the sultan to assist him in unlocking hidden secrets.[66]

In addition to offering insights into how relationships between sultans and Sufi shaykhs were initiated and negotiated, as well as alluding to contemporary debates on competing epistemologies and authorities of knowledge, this letter clearly reveals that Bayezid II's engagement with various branches of learning, including the astral sciences, was a subject of discussion and even contention in his own day. The particular manner in which the anonymous author of the letter employs the concept of experience (Tur. *tecrübe*, Ar. *tajriba*) merits further consideration. While it carries diverse connotations in medieval Islamicate intellectual discourse, this concept often refers to judgments based on repeated testimonies, observed regularities, or recurrent memories of phenomena over an extended period. In Islamicate taxonomies of knowledge, *tajriba* is generally considered a lower epistemological form, unable to produce "certain knowledge'" (*yaqin*) because it is concerned with particular, contingent, and changing facts rather than necessary, exact, immutable truth.[67] Yet in the writings of some mystically inclined learned individuals, *tajriba* enjoyed a more elevated status. For example, al-Ghazali, in his autobiographical *Deliverance from Error* (*al-Munqidh min al-dalal*), states that experiencing and tasting in one's soul the positive effects of a prophet constitutes "necessary knowledge" (*ʿilm daruri*) and provides certainty about his prophecy (*yaqin bi'l-nubuwwa*).[68] The anonymous author of the letter, however, despite speaking in strong mystical undertones, belittles *tajriba*, which he had heard was Bayezid II's customary approach to studying these sciences. Thus, this author's intended meaning of *tajriba* seems to indicate a type of knowledge acquired not from spiritual tasting, as it was understood among Sufi mystics, but rather from reading texts as well as conducting firsthand observations and perhaps some empirical study.

Concerning such a scientific culture at Bayezid II's court, prizing book learning, experiential knowledge, and observation, additional details of interest are provided by Jewish natural philosophers with an extensive scholarly network across the Mediterranean who contributed to the intellectual milieu surrounding the sultan. Among these erudite individuals was Moses Galeano (d. after 1542), known as Musa Jalinus in his extant Arabic treatises, who traveled frequently between Venice and Istanbul from 1497 to 1502. A frequenter of the scholarly salons held at the sultan's abode, as well as at the house of Müʾeyyedzade ʿAbdurrahman, the chief military judge and the sultan's closest intellectual companion, Galeano is credited with engineering a spherical astrolabe, composing a medical treatise at the request of Bayezid II's chief physician, Ahi Çelebi (d. 1524), and translating—from Latin into Arabic—Abraham Zacuto's *Almanach Perpetuum*, one of the principal sources of fifteenth- and sixteenth-century European astronomical and astrological practice.[69] In 1500, Galeano completed his compendium of knowledge in Hebrew, *Taʿalumot Hohmah* (Puzzles of wisdom), in which he surveyed numerous errors and fallacies in a wide array of scientific disciplines, such as medicine, astronomy, and mechanics, while also recounting firsthand anecdotes about the scholarly atmosphere at Bayezid's court. One such anecdote tells of an alchemical operation performed in the presence of Bayezid II and his companion Müʾeyyedzade. Apparently, an alchemist had captured the sultan's interest by promising to turn lead into gold. As the alchemist conducted the experiment, Samuel Abulafia, a rabbi and a Jewish refugee from Spain at Bayezid's court, asked Galeano to deliver a note to the sultan exposing the alchemist's performance as a fraud. According to Galeano, Bayezid II recognized the deception upon reading the rabbi's note, and the disgraced alchemist subsequently took his own life by consuming a lethal poison in the bathhouse.[70]

Regardless of questions about the authorial intentions behind this story—whether it served as a moral reminder to avoid alchemy or as an example of rabbinic authority's superiority over other sorts—it opens a gate to further exploring Bayezid II's possible interests in alchemy, among other subjects. The sultan was presented with a few surviving alchemical treatises on the properties of the seven metals. More intriguingly, his name is specifically celebrated in the alchemical corpus of subsequent periods as an idealized patron engaged in alchemical pursuits. In one seventeenth-century text, Bayezid II is even claimed to have mastered alchemy.[71] Tuna

Artun, who has unearthed several such texts from the seventeenth century that feature Bayezid II as a sultan initiated into alchemical learning and practice, rightly points out that these ascriptions may be apocryphal, but the fact that Bayezid II, rather than any other Ottoman sultan, was singled out as a model ruler-patron suggests that the sultan's interests in this field may really have been witnessed and transmitted within the circles of early modern Ottoman alchemists.[72]

While contemporary evidence documenting Bayezid II's actual involvement in the study of alchemy is scarce, we possess a substantial amount of proof regarding his engagement with the mathematical and astral sciences. In fact, Bayezid II is the only Ottoman sultan documented to have systematically studied the mathematical and astral sciences under the guidance of a distinguished expert in the astral sciences. By the second half of the 1490s, the sultan had summoned Mirim Çelebi, a prominent expert in the astral sciences with an illustrious lineage connecting him to the leading members of the Samarkand school, to tutor him in these disciplines. This association was not a typical relationship between an ascending sultan and his official tutor, where the latter, in his official capacity and status as a princely tutor (*hace-i padişahi*), would instruct the sultan primarily in the fields of the religious and literary sciences, as well as ethics and politics.[73] Instead, Mirim Çelebi's teaching had a specialist focus and involved only the astral and mathematical sciences. Furthermore, there is no bureaucratic paper trail in the available documents to suggest that Mirim Çelebi was ever granted the official status of princely tutor.

As the paternal grandson of Qadizada al-Rumi and the maternal grandson of ʿAli Qushji, two luminaries of the Samarkand school, Mirim Çelebi was indisputably the foremost scientist-qua-scholar of celestial knowledge in the Ottoman lands at the turn of the sixteenth century.[74] Unlike the *munajjims* whose lives and textual output we discussed in the previous chapter, Mirim Çelebi, following in the footsteps of his illustrious grandfathers, was deeply involved in the production of madrasa scholarship, in addition to his activities related to the practice of astronomy and astrology. In the absence of any autobiographical writings by Mirim Çelebi, we must have recourse to Taşköprülüzade, who became Mirim Çelebi's student in the 1510s and read at his feet ʿAli Qushji's treatise on astronomical theory, when Mirim was writing a gloss of it, which he eventually submitted to Sultan Selim I.[75] In his concise account of Mirim's education, Taşköprülüzade notes that Mirim studied under eminent scholars of the time, including Hocazade (d. 1489) and Sinan Pasha (d. 1486).[76] In his 1499 commentary on the Ulugh Beg tables, Mirim also mentions his collaboration with a certain ʿAtaʾullah (d. after 1481), another Persian émigré scholar engaged in the astral sciences in the Ottoman capital during the late fifteenth century.[77] Although the precise date and place of his birth remain uncertain, Mirim Çelebi appears in historical records from the late 1480s onward. In the earliest extant document featuring his name, Mirim is listed as receiving 15 *akçes* per day from the extra income of the Ayasofya endowments between June 1488 and February 1489.[78] The record's explicit reference to Mirim as the "kinsman" (*hiş*) of ʿAli Qushji suggests that his prestigious lineage was the reason for these gratuities. The same record also adds that his gratuities ceased on 28 February 1489, likely because Mirim had secured a teaching appointment at a madrasa by then. During the 1490s, Mirim taught at several madrasas in Bursa and Edirne before being called upon by Bayezid II to serve as his private tutor in the mathematical sciences.[79] Mirim's exceptional family lineage and notable expertise in the mathematical and astral sciences was undoubtedly the reason behind his appointment as the royal tutor for providing instruction in these disciplines.

During his service at Bayezid II's court, Mirim Çelebi not only instructed the sultan in the mathematical sciences but also prepared almanacs with prognostications, authored manuals on astronomical instruments, probably provided on-the-spot astrological advice, and contributed to the training of a new generation of *munajjims*.[80] Two of the almanacs he produced have survived to our time—one of them seemingly for the year 1495 and the other for 1499—as documented in the catalog of a modern private auction company, but regrettably, they are now in private collections inaccessible to researchers.[81] The fact that Mirim Çelebi was composing almanac prognostications at the time is further corroborated by a minor payment register from the early sixteenth century, which records Mirim Çelebi as the recipient of 1,000 *akçes* for an almanac he presented.[82] In 1499, Bayezid II personally requested that Mirim Çelebi compose a commentary on the Ulugh Beg tables to elucidate their more impenetrable points. In the dedicatory remarks of the text, which was more an expositional work on the concepts and parameters mentioned in the original Ulugh Beg tables than a revision of celestial data based upon fresh observations, Mirim praises Bayezid as, among other appellations, "the most perfect and enlightened of the Caesars of the world," as powerful as Alexander, who orders the affairs of the world in accordance with "Platonic wisdom," and as "the messianic savior of the end times."[83] Mirim repeats similar remarks in the epilogue, where he identifies Bayezid as the sovereign of the entire inhabited world and the prophesied world ruler (*sahib-qiran*).[84] In his works on astronomical instruments, composed during the first decade of the sixteenth century, Mirim employs similar titulature. For instance, in his Persian manual on the use of the sine quadrant that he dedicates to the sultan Mirim expresses his gratitude toward Bayezid II, whom he characterizes as "the vehicle of eschatological consummation" and "the messianic savior of mercy in the end times."[85] Although Mirim does not elaborate on the celestial foundations of the titles he ascribes to Bayezid II—and such sycophantic remarks should always be treated cautiously—his employment of these expressions, as a close companion of the ruler and a prominent expert in the astral sciences, is likely not arbitrary or without intellectual endorsement. In any case, whatever his thoughts, the titles reveal the broad extent of the discussion on the vocabulary of sovereignty within the late fifteenth- and early sixteenth-century Ottoman realm.[86]

Mirim is never referred to as a court *munajjim* in the extant bureaucratic registers from the period, though he is consistently mentioned among the monthly-salaried palace personnel without a specific professional designation.[87] The entries in the voluminous gift register, spanning the years from 1503 to 1512, demonstrate that he was an esteemed figure in the sultan's inner circle. In addition to numerous annual and notably generous payments made to Mirim Çelebi, the sultan even sponsored his wedding in November 1505 and later presented Mirim's wife with several items of clothing as a gift in February 1510, indicative of the sultan's profound respect and care for him and his family.[88] Idris Bidlisi, another important contemporary figure and an émigré Persianate literatus close to the courtly environment of Bayezid II, also acknowledges the high esteem in which the sultan held Mirim. As Bidlisi says, Mirim received special treatment at the sultan's routine scholarly gatherings, distinct from all other learned people.[89] Further contemporary testimony to Mirim Çelebi's proximity to Bayezid II is that during the pro-Selim rebellions of the Janissaries in the capital in late 1511, Mirim was among those high-ranking individuals specifically targeted—alongside Müʾeyyedzade ʿAbdurrahman, the chancellor Tacizade

Caᶜfer Çelebi, and the chief physician Ahi Çelebi—on the grounds that they supported Bayezid's favored son Ahmed against Selim.[90]

In addition to the annual almanacs, the commentary on the Ulugh Beg tables, and the works on astronomical instruments, Mirim Çelebi composed at least two treatises on specific branches of astrology—namely, elections and interrogations.[91] Although the colophons in the extant copies of these works do not provide us with the exact date of their composition, it is likely that Mirim compiled them after the death of Bayezid II, as they lack dedicatory passages to the sultan. In fact, Mirim's astrological works appear to be geared more toward practicing *munajjims* who needed to upgrade their skills in the relevant astrological techniques. In his work on interrogations, *Masaʾil dar Taʾsirat* (Inquiries on the influences [of the stars], for instance, Mirim addresses virtually every possible question a client might pose to a practicing *munajjim*, ranging from queries related to travel by sea to the fate of purchased slaves.[92] As such, these prescriptions constitute a valuable gallery of scenes depicting social occasions and events significant in the daily lives of Ottoman subjects at the time. The text also showcases Mirim's vast knowledge of the subject, displaying his mastery over the teachings of both ancient and more recent authorities, including Vettius Valens, Hermes Trismegistus, and Mashaʾallah. Furthermore, he occasionally invites his readers, presumably his students, to consult authoritative texts in this discipline, such as al-Qasrani's *Kitab al-masaʾil* (Book of interrogations).[93] Mirim also highlights astrolabes and quadrants as essential tools in a practicing *munajjim*'s paraphernalia and contends that one of the primary reasons for inaccurate predictions made by *munajjims* is the presence of defects in these instruments that compromise their accuracy.[94]

As suggested by his remarks, particularly in his extant texts on electional and interrogative astrology, Mirim Çelebi trained several *munajjims*. One of these students was ᶜAli, who, according to the corresponding entry in the extensive gift register, presented his first almanac in March 1507.[95] Regrettably, the gift register does not provide any further details besides his name and affiliation with Mirim Çelebi.[96] Another student of Mirim, based on the testimony of sixteenth-century biographers, was ᶜÖzri Çelebi, who is said to have excelled as an expert in casting birth horoscopes. A near-contemporary biographer of poets, ᶜAşık Çelebi (d. 1571), states, without mentioning where, that ᶜÖzri learned the intricacies of astrological practice from Mirim Çelebi but ultimately lost his life after a series of events stemming from Sultan Süleyman's mounting wrath at ʿÖzri's displeasing astrological predictions.[97] As ʿAşık Çelebi narrates, ʿÖzri composed his first and only almanac before the siege of Rhodes in 1522. In his *taqwim*, which is unfortunately not extant, ʿÖzri allegedly predicted that the island of Rhodes would ultimately be conquered but the siege would be prolonged and result in the loss of numerous soldiers. Infuriated by the predictions in the almanac, Süleyman ordered his retainers to bring the author into his presence. Upon hearing the sultan's order, ʿÖzri visited one of his close friends, Mustafa the geomancer, who, according to ʿAşık Çelebi, was a fellow student of Mirim Çelebi.[98] Together, they cast and interpreted ʿÖzri's horoscope and reached the conclusion that ʿÖzri would perish during or after his visit to the palace.[99] While there are legitimate reasons to question the veracity of this anecdote, it unequivocally demonstrates, by highlighting Mirim Çelebi as the master of both ʿÖzri and his friend Mustafa the geomancer, that in the eyes of the later sixteenth-century Ottoman learned class, Mirim Çelebi was the paramount expert in the astral sciences.[100]

Another student of Mirim Çelebi was, of course, Sultan Bayezid II. In the absence of contemporary records and manuscript traces or marginalia, it is difficult to establish the exact duration of their studies and the specific books involved in the sultan's education. One indication of the books Bayezid II likely studied with Mirim Çelebi lies in a collection of surviving manuscripts containing unique inscriptions that document the volumes' being personal copies of Bayezid II ("Its owner is the sultan, Bayezid son of Muhammad Khan" [*sahibuhu al-sultan Bayezid b. Muhammad Khan*], or sometimes "*Ex libris sultani* Bayezid son of Muhammad Khan" [*min kutub al-sultan Bayezid b. Muhammad Khan*]), together with the standard oval seal of the sultan found in over five thousand volumes registered in the palace library during his time. While there are a handful of works in logic, philosophy, and jurisprudence that attest to being Bayezid II's personal copy, works in the astral sciences constitute the prevailing majority of surviving manuscripts that document the sultan's own use. Those books Bayezid II possessed and possibly studied, either under Mirim Çelebi's supervision or independently, include:

(1) Tusi's textbook on almanac-making[101]

(2) Tusi's *Astronomical Handbook with Tables*[102]

(3) Tusi's *Commentary on Ptolemy's Almagest*[103]

(4) Tusi's *Memoir on Astronomy*[104]

(5) Jurjani's *Commentary on Jaghmini's "Epitome"*[105]

(6) Abu Maʿshar's abridged *Introduction to Astrology*, both in the Arabic original and a Persian rendition[106]

(7) The *Epistle of the Ikhwan al-Safaʾ* on the astral sciences[107]

(8) An anonymous treatise on the uses of the astrolabe[108]

(9) An introductory work in verse on the astronomical and astrological calculation of time[109]

(10) Marrakushi's (fl. second half of the 13th century) *summa* on astronomical instrumentation and the computation of time[110]

(11) A collection containing Qusta ibn Luqa al-Baʿlabakki's (d. circa 912–13) treatise on the use of the celestial globe, as well as (Pseudo-)Aristotle's short treatise on a prognosticative technique employed to predict the outcome of a battle, a method quite popular among almanac writers[111]

Bayezid's reading tastes and interests, as revealed through abundant manuscript evidence, archival records, and the testimony of his contemporaries, appear to have encompassed both theoretical and practical iterations of celestial knowledge, including the strictly astrological aspects. Apart from the books in his own possession that primarily address astrological principles, a few surviving archival documents attest to Bayezid's keenness to hear the astrological predictions of his *munajjims*. For example, the anonymous petition submitted to Bayezid II, which was introduced in Chapter 1 in the context of self-doubts and epistemological anxieties of practicing *munajjims*, provides clear evidence of the sultan's eagerness to heed astrological advice. In this petition, the anonymous *munajjim*, writing in Persian, seeks the sultan's sympathy and forgiveness for his recent failure to present almanacs.[112] He implores the sultan to reinstate him in his service, as it seems he lost his position owing to his recent negligence in preparing almanacs with prognostications.[113] The fact that a *munajjim*'s

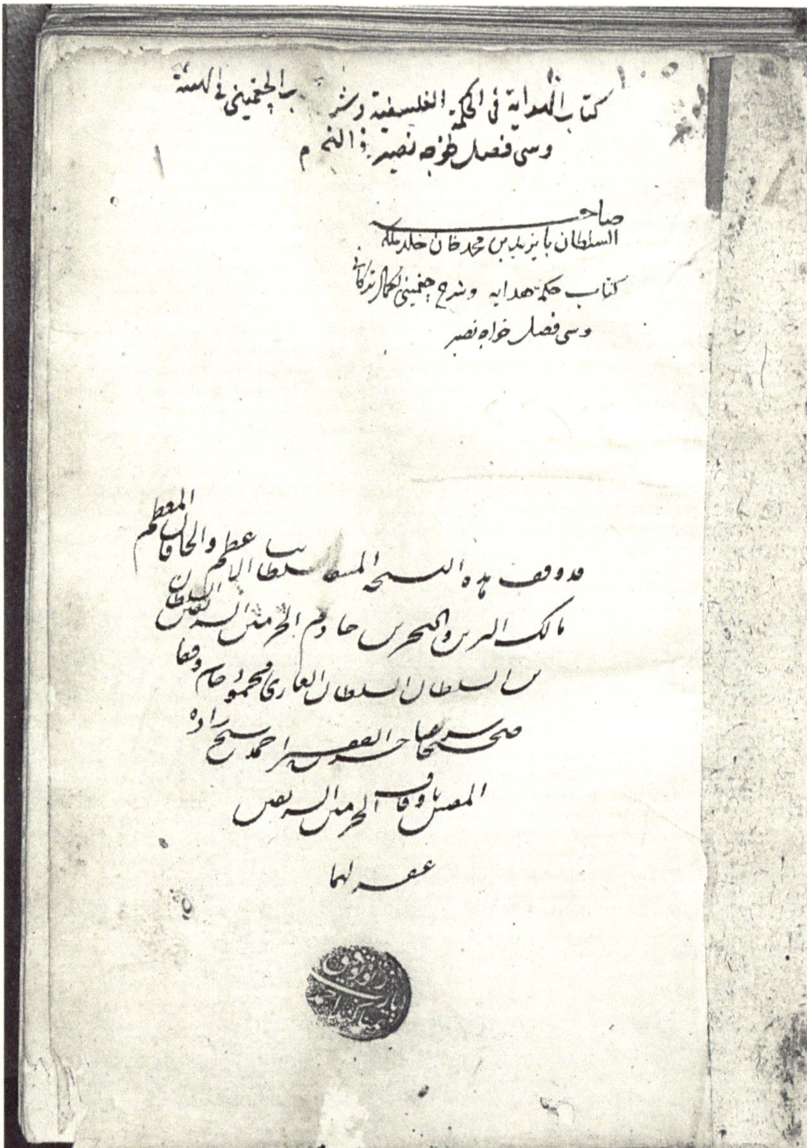

FIGURE 9. Sultan Bayezid II's personal copy of a *majmua*, containing Tusi's *Si Fasl* and a commentary on Jaghmini's *Mulakhkhas*. Süleymaniye Library, Ayasofya Ms. 2474, 1a. Reproduced by permission from the Directorate of the Turkish Institution for Manuscripts.

failure to produce annual astrological predictions in a timely manner could potentially result in the loss of his position indicates the importance Bayezid II placed on the astrological advice found in *taqwims*.

Intellectual Companions of Bayezid II as a Philosopher-King

The sources presented thus far on astral-scientific inquiries at the court of Bayezid II indicate that the peculiar cultivation of the astral sciences and scientists during his reign was closely connected to Bayezid II's personal intellectual inclinations. Two significant questions, however, remain unaddressed. First, what factors and motivations underlay Bayezid II's genuine and documented interest in not just supporting but also becoming learned in this body of knowledge? Was he alone on his journey—apart from the astral experts he collaborated with—or were there other individuals within his inner circle who shared his dedication and possibly exerted a lasting influence on his intellectual formation and scientific priorities? Second, what were the ramifications and immediate or long-term consequences of his deliberate efforts, perhaps even his "policy," to support a particular type of science and a group of scientific experts that otherwise would have been met with some degree of suspicion?

In addressing the first question, an easy answer would be to consider the long-standing reliance of rulers and decision-makers on astrologers and other experts of divinatory arts, from ancient through modern times, in both the East and West. From kings of ancient Babylonia and Chinese emperors of different dynasties to Abbasid caliphs and some twentieth-century presidents, such as Ronald Reagan and François Mitterrand, sovereigns have often sought the expertise of astrologers for practical benefits.[114] These sought-after benefits encompassed, among other things, these experts' judgments on political and military decisions using astrological reasoning and the provision of meaning for otherwise enigmatic celestial phenomena. The royal support extended to astral experts in particular and scientists in general also held propaganda value that operated on two levels. First, the support provided to scientific experts enabled the sovereign to disseminate an image as a generous patron of knowledge and learning. If the royal patron was also personally engaged in studying the science they supported, then they merit further acclaim on the part of contemporaries as a learned ruler, even an idealized philosopher-king. Second, and more specifically concerning the patronage of astral experts, their interpretation of worldly events with reference to celestial phenomena and the sycophantic praise they have for the reigning sovereign—replete with heavenly metaphors and vocabulary touting the rule of their patron as supreme, certified by celestial portents—endowed the monarch with irrefutable divine significance and recognition. Against this broader background, then, there is indeed nothing extraordinary about Bayezid II's support of *munajjims*.

Bayezid II's devotion to the study of the astral sciences and his diligent efforts to learn its various aspects were also not entirely unheard of, even in the relatively narrower Islamicate context. For instance, the thirteenth-century Rasulid ruler of Yemen, al-Ashraf ʿUmar II (r. 1295–96), not only patronized *munajjims* but also authored at least two treatises on the subject—one on the general principles of the astral sciences, *Al-Tabsira fi ʿilm al-nujum* (A beacon for knowledge of the stars) and another on the use of astrolabes. The latter text was, in fact, written as a manual for an actual astrolabe of al-Ashraf ʿUmar II's own construction, for crafting which the Rasulid ruler received a certificate from his teachers.[115] Apart from the Rasulid ruler, perhaps the most renowned of all rulers in Islamic history to display a marked interest in learning and teaching the astral sciences is the Timurid sultan Ulugh Beg. Having gathered in Samarkand numerous experts from nearby and distant regions, including Jamshid al-Kashi, Qadizada al-Rumi, and ʿAli Qushji, and having utilized their efforts in the newly established Samarkand observatory and the accompanying madrasas, Ulugh Beg is mentioned in contemporary Timurid sources both as a patron of and an active participant in this ambitious scientific endeavor.[116] In the letters of Jamshid al-Kashi to his father and the teaching certificate given to Fathallah Shirwani (d. 1486) by his teacher Qadizada al-Rumi, Ulugh Beg is often identified as an enthusiastic participant in classes and other gatherings held for debating astral and mathematical subjects.[117] The latter source even extols Ulugh Beg literally as the "philosopher-king" (*al-sultan al-faylasuf*) of the age.[118]

Did Bayezid II ever envision himself as a philosopher-king, striving to establish a court resembling Ulugh Beg's in Samarkand, where experts in the astral sciences and other natural philosophers from various regions would converge? Without any first-person narratives by the sultan that might elucidate his self-fashioning and inspirations, we can assert little with certainty. His sole textual composition, *Divan*—a compilation of his poetry written under his pseudonym ʿAdli (the Just)—is also not very helpful for tracing Bayezid II's exact intentions through his own eyes and words.[119] Nevertheless, the sultan's close collaboration with Mirim Çelebi, a direct descendant of some key names from the Samarkand observatory, might have imbued Bayezid II with a keener understanding of and profound appreciation for Ulugh Beg as a model ruler. Though not direct evidence, substantial indications pointing to Persianate, and more specifically Timurid, influences on the artistic and cultural orientations at the fifteenth-century Ottoman court, already pointed out by several historians of art and architecture, hint at Bayezid's likely admiration for Ulugh Beg's philosopher-kingship.[120]

With specific regard to Bayezid's potential engagement with the Platonic ideal of the philosopher-king, it is worth noting that his library housed an Arabic translation of Plato's *Timaeus*, a book on the mathematical order of the divine craftsman's creation, bound together with another short piece by the philosopher on politics.[121] The extant copy, now housed in the Süleymaniye Library, does not contain any traces or marginalia documenting Bayezid II's reading it. İhsan Fazlıoğlu has claimed to find marginalia by Mehmed II and Bayezid II in the single surviving copy of *al-Muthul al-aflatuniyya* (Platonic ideas) that was also listed in the inventory of the palace library at the turn of the sixteenth century.[122] This latter copy does contain extensive marginalia, though I found no explicit reference in these notes to Bayezid II or Mehmed II. Still, considering the availability of such key Platonic texts in his immediate reach and contemporary references to his scope of learning, Bayezid II, who adopted for himself the moniker ʿAdli (the Just) while composing poetry, may well have been familiar, perhaps deeply so, with the Platonic ideal of philosopher-kingship, which in its original formulation by Plato required the sovereign to individually possess an advanced understanding of the mathematical and astral sciences in order for him to institute and maintain a just, orderly rule.[123]

To further trace Bayezid II's intellectual endeavors, we must journey back to his formative years in Amasya, where he forged lifelong relationships with some curious personalities and polymaths, such as Müʾeyyedzade ʿAbdurrahman, who may have left an indelible imprint on the sultan's cultivation of interest in the mathematical and astral sciences. Bayezid was only around six years old when his father, Mehmed II, dispatched him to Amasya in 1454 to govern the region.[124] Situated along the northern corridor of the Silk Road in Anatolia, Amasya was a vibrant gateway connecting the lands of Rum to the Persianate realm. Its strategic location rendered the city and its surrounding areas a well-trodden path for scholarly, mystical, and commercial elites, as well as for commoners, journeying between the Persianate and Ottoman worlds.[125] Before the administrative marginalization of the city in the sixteenth century—resulting largely from Selim I's efforts to subordinate it to the newly established province of Sivas amid the rising Ottoman-Safavid rivalry—Amasya flourished as an important hub of cultural and intellectual life in the fifteenth century. Bayezid II's twenty-seven-year rule in the city, followed by another long-term governorship, that of his favored son Ahmed, invigorated the region, paving the way for the rise of numerous outstanding artists, poets, scholars, and physicians, Muslims and non-Muslims alike.[126] There is a veritable wealth of eminent Amasyans to highlight from the fifteenth century, including such luminaries as Amirdovlat Amasiatsi (d. circa 1495), an Armenian natural historian and physician who later entered the service of Sultan Mehmed II;[127] Mihri Hatun (d. 1514), the esteemed female poet from Bayezid's inner circle;[128] Sabuncuoğlu Şerefeddin (d. 1468), the author of an illustrated treatise on surgical and medical practices;[129] Şeyh Hamdullah (d. 1520), the calligrapher and inventor of one of the canonical styles of Ottoman calligraphy;[130] and Müʾeyyedzade, a bibliophile polymath who merits closer scrutiny for his potential influence on Bayezid II's interests in the astral and mathematical sciences.

Müʾeyyedzade ʿAbdurrahman was born in Amasya in 1456 into a distinguished family of shaykhs and scholarly elites whose lineage traced back to the Iran of centuries past.[131] His father, ʿAli Çelebi, served as the chief secretary of Bayezid's princely court from 1461, tasked with drafting the prince's correspondences. Thanks to his father's immediate presence in the prince's service, young Müʾeyyedzade found himself in close proximity to Bayezid, who was eight years his senior. The two boys literally grew up together from the 1460s, forging a friendship that would remain unbroken until Bayezid's death in 1512. Their bond was not merely rooted in their shared passion for science, learning, and poetry but also in the ample time they spent together in their youth, sometimes indulging in various intoxicants as well as hashish. Müʾeyyedzade's pernicious influence on Bayezid, drawing the prince into drug-fueled parties, so alarmed and angered Sultan Mehmed II that by the latter half of the 1470s he dispatched inspectors to Amasya to apprehend and execute all those deemed "mischief-makers," responsible for leading his son astray.[132] Having been forewarned of the impending danger, Bayezid ensured the safe departure of his beloved friend from the city and penned a response letter to his father in Istanbul, stating that he had significantly reduced his consumption of narcotic drugs and stimulants, which he had previously taken only for the sake of losing weight.[133]

It was likely after 1477 that Müʾeyyedzade had to leave Anatolia, first visiting Aleppo for a brief period of time and eventually settling in Shiraz, where he would spend several years studying the rational and traditional sciences with such stellar scholars of the city as Jalal al-Din Davani (d. 1502), Sadr al-Din Dashtaki (d. 1498), the young Ghiyath al-Din Mansur Dashtaki (d. 1542), and Mir Husayn Maybudi (d. 1504), all of whom were known for their prowess in the mathematical and philosophical sciences.[134] During his sojourn in Shiraz, Müʾeyyedzade seized the opportunity to deepen his learning in these disciplines, specifically in the astral sciences, which he had already begun cultivating in Amasya. Two particular manuscripts bear witness to Müʾeyyedzade's early interest in the astral sciences while he was still in Amasya in his early twenties. One of these is the aforementioned Yazdanbakhsh's *Zij*, which was initially dedicated to Prince Bayezid in Amasya but subsequently gifted to Müʾeyyedzade in January 1477, as documented by a note on its flyleaf,[135] which suggests that Bayezid and Müʾeyyedzade also shared books during their time together. The copy of Yazdanbakhsh's *Zij* contains on pages 1b–2a several fragments of poetry attributed to ʿAli b. Abi Talib and some instructions on the computation of horoscopes, arguably written by Müʾeyyedzade. The second manuscript is a copy of al-Wabkanawi's *Zij*, which, as revealed by an ownership statement on the first page, was purchased by Müʾeyyedzade in Ladik, a town situated just twenty-five miles from the center of Amasya, in June 1477.[136]

Müʾeyyedzade resided in Shiraz at least until June 1483, as evinced by the teaching certificate he received from the illustrious Jalal al-Din Davani.[137] This certificate delineates the books and disciplines Müʾeyyedzade studied under Davani's tutelage and the permission granted to him to transmit this knowledge to future generations of students. According to the certificate, Müʾeyyedzade read four works in varying measures, two of which were in the fields of theology and philosophy, namely Jurjani's *Sharh al-mawaqif* (Commentary on [al-Iji's] *Stations*) and Tusi's *Sharh al-isharat wa'l-tanbihat* (Commentary on [Ibn Sina's] *Remarks and Admonitions*). The other two books Müʾeyyedzade studied with Davani pertained to the mathematical sciences, specifically astronomical theory and geometry: Qadizada al-Rumi's commentary on Jaghmini's *Epitome* and Tusi's exposition on Euclid's geometry (*Tahrir Uqlidis*). Beyond these four books, Müʾeyyedzade was exposed to a dazzling array of other scholarly works circulated within the intellectual circles of Aleppo and Shiraz, copies of which he assiduously sought, acquired, and ultimately brought back to the Ottoman lands after Bayezid II ascended the throne and invited his dear friend to return. While we may not be able to trace all the manuscripts once in Müʾeyyedzade's possession, we are fortunate to have access to a detailed inventory compiled shortly after his death in 1516, listing the remaining books in his estate and amounting to 2,112 volumes.[138]

By the global standards of its time, Müʾeyyedzade's private book collection rendered him an impressive bibliophile, on par with or even surpassing better-known Renaissance polymaths and humanists. For comparison, Erasmus's (d. 1536) personal library is estimated to have comprised a little more than four hundred volumes, while Galileo Galilei's (d. 1642) private collection is considered to have contained between five and six hundred volumes.[139] In reality, Müʾeyyedzade's collection was substantially larger than 2,112 volumes, as the inventory was compiled by Selim I's order after it was heard that Müʾeyyedzade's surviving sons had sold or lost numerous items from their father's estate, necessitating intervention to curtail their negligence.[140] Although the exact number of books lost or sold remains unknown, reliable sixteenth-century sources with an insightful view of contemporary scholarly life, such as Taşköprülüzade's biographical compendium of scholars, assert that Müʾeyyedzade's original book collection comprised around seven thousand volumes.[141]

When we scrutinize the inventory of books that remained from Müʾeyyedzade's collection, his keen interest in astral and mathematical sciences comes into clear view. In her study of Müʾeyyedzade's book collection, Judith Pfeiffer has determined that the astral and mathematical sciences constitute 10–15 percent of his collection, encompassing nearly all the essential texts a practicing *munajjim* would have possessed in various branches and genres of celestial knowledge, both theoretical and practical.[142] For instance, the inventory lists over fifteen copies of *zij*es, including all the major astronomical handbooks produced in the post-thirteenth-century Islamicate East, such as the Ilkhanid and Ulugh Beg tables, as well as their revisions and commentaries on them.[143] The inventory does not even list some *zij*es that we know were commissioned and possessed by Müʾeyyedzade when he was occupying the highest-ranking position in the scholarly establishment as the chief military judge of Rumelia. In 1505, during a scholarly gathering held at Müʾeyyedzade's mansion, the aforementioned Moses Galeano was specifically asked by Müʾeyyedzade to translate the *zij* of Abraham Zacuto. Within a year, Galeano completed the translation from a European language into Arabic and presented it to Müʾeyyedzade. The wording of Galeano for its title (*The Book of Astronomical Tables Translated into Arabic from the Frankish Tongue*) suggests that he used the 1502 print edition of Zacuto's *Almanach Perpetuum* published in Latin in Venice, not the original Hebrew version.[144] In any case, the Arabic translation of Zacuto's astronomical tables popular across Europe, completed by a Jewish polymath at the request of a high-ranking Ottoman scholar-bureaucrat during a scholarly gathering in the bureaucrat's private mansion, nicely captures the extent of the transcultural intellectual milieu around Bayezid II and his scholarly interlocutors who showed a keen interest in state-of-the-art astral knowledge.

The extant book inventory also documents that, in addition to the *zij* corpus, Müʾeyyedzade was seriously interested in acquiring the Arabic and Persian manuals and textbooks mentioned in Chapter 1 for learning the basic scientific premises and vocabulary of *munajjims*' practice, including Pseudo-Ptolemy's *Centiloquium*, Tusi's *Si fasl* in multiple copies, Kushyar's *Compendium of Principles in Astrology*, ʿAli-Shah Bukhari's *Fruits and Trees*, Biruni's *Book of Instruction in the Elements of the Art of Astrology*, and at least three copies of Shahmardan-i Razi's *Garden of Astrologers*.[145] His reading tastes also embraced the classics in astronomical theory, such as the Almagest corpus, Tusi's *Memoir*, and Jaghmini's *Epitome*. Last but not least, Müʾeyyedzade's library was well furnished with works on celestial magic and talisman-making, including Ahmad al-Buni's (d. 1225) famous *Shams al-maʿarif* (Sun of knowledge) and Fakhr al-Din al-Razi's *al-Sirr al-maktum* (Hidden secret).[146]

Müʾeyyedzade's intense interest in the astral sciences is all the more striking when we consider that his known extant works do not include any texts devoted to this subject. In the two other fields that stand out in the inventory of books in terms of the number of titles—philosophical theology and poetry—Müʾeyyedzade produced scholarly and literary works as both a respected theologian and an esteemed poet. Upon his return to Ottoman lands, Müʾeyyedzade began teaching in the high-ranking madrasas of Edirne and Istanbul. Throughout the 1480s and 1490s, he was deeply involved in teaching philosophical theology, jurisprudence, and logic, in addition to writing poetry under the pen name "Hatemi" (Signet Maker). Although Müʾeyyedzade's acknowledged erudition and impressive book collection might lead one to expect much more scholarly output, his demanding administrative responsibilities after being appointed as military judge of Anatolia in 1501 and later of Rumelia in 1505 left him little time for academic writing. Among his relatively few extant works, most of which were devoted to jurisprudence, logic, and philosophical theology, is a supercommentary on Jurjani's *Sharh al-mawaqif*, where he displays his expansive knowledge of the astral sciences and refutes Jurjani's attacks on *munajjim*s. Presented to Bayezid II sometime between 1483 and 1502, this text by Müʾeyyedzade takes particular issue with the arguments Jurjani put forth against *munajjim*s and their alleged claims.[147]

Before delving into Müʾeyyedzade's critique of Jurjani, it is essential first to lay out Jurjani's particular stance toward *munajjim*s and their expertise. As a distinguished and widely influential polymath who wrote numerous handbooks and treatises on standard madrasa disciplines as well as lengthy commentaries on Aristotelian logic and *ʿilm al-hayʾa*, Jurjani was undoubtedly one of the pivotal scholars of the postclassical Islamic world, with wide-ranging impact on fifteenth- and sixteenth-century Ottoman scholars.[148] His colossal commentary on al-Iji's *Stations* (on the discipline of Kalam), which he completed in 1404, attained classic status and elicited a wide range of supercommentaries over several centuries across the Ottoman domain and beyond. The text is notably rich in addressing topics related to the astral sciences. Some modern scholars of postclassical Islamic science and learning even argue that Jurjani played a key role in the dissemination of astronomy into madrasas via his *Sharh al-mawaqif*, as well as his theological commentary on Tusi's *Tajrid al-iʿtiqad* (Purification of belief). Jurjani's engagement and familiarity with *ʿilm al-hayʾa* is not in fact surprising, given that he composed three commentaries in the discipline of *hayʾa*, annotating al-Tusi's *Memoir*, Qutb al-Din al-Shirazi's *Gift* (*Tuhfa*), and Jaghmini's *Epitome*.[149]

Jurjani's specific discussion and critique of *munajjim*s are found in the fifth "station," which addresses issues pertaining to God's attributes. For

Jurjani, *munajjims*, whom he associates with the Sabians, are one of the six groups rejecting the belief in an omnipotent God, together with "philosophers" (*falasifa*), "dualists" (*majusis*), and supporters of the Muʿtazilites.[150] Jurjani casts *munajjims* as a group believing in the autonomous power of planets and other heavenly bodies. According to Jurjani, *munajjims* assert that whatever happens in the terrestrial world is dependent on the particular positions of the stars within the zodiac and other stellar configurations. Rather than taking these celestial objects merely as intermediaries, says Jurjani, *munajjims* view them as *mudabbirs* ("that which governs") and believe in their autonomous, creative power. For Jurjani, such claims about *munajjims* are easy to debunk. For instance, he resorts to one of the stock arguments used in medieval and early modern polemics against astrologers and astrology as to the disparate fates of twins.[151] Jurjani questions how it could be possible for twins to have different courses of life while born at the same time and thus under ostensibly identical stellar conditions. Jurjani concludes that, contrary to what *munajjims* strongly hold, a minor change in the configuration of celestial positions does not have any significant impact on terrestrial events.[152]

Müʾeyyedzade engages with Jurjani's treatment of *munajjims* at length, dedicating one-tenth of his entire gloss to this discussion. He first addresses Jurjani's assertion that *munajjims* attribute independent power to celestial entities. Müʾeyyedzade argues that in none of the "books about astrology" (*kutub al-tanjim*) could one ever find a single claim that celestial bodies are the direct, efficient causes of events on Earth. According to Müʾeyyedzade, all authoritative books in the astral sciences consistently state instead that celestial phenomena are merely facilitators for, not creative agents of, manifest effects—similar to the way chilling follows the act of drinking water or eating scammony aids digestion.[153] As Müʾeyyedzade emphasizes, even the most devout individuals hesitate to deny the interdependence of the celestial and terrestrial realms. In fact, this entire cosmic structure and interdependence is a testament to God's unmatched wisdom and a stimulus to faith. To reject it is to verge on impiety and unbelief.[154]

After relaying the core principles of the cosmological underpinnings of *munajjims*' practice, Müʾeyyedzade turns to address specific arguments raised by Jurjani about *munajjims*. On the topic of twins' differing destinies, for instance, Müʾeyyedzade states that what truly matters in astrological computation and interpretation is not the time of birth but rather "the time of conception" (*misqat al-nutfa*).[155] According to Müʾeyyedzade, all the major sources in the astral sciences, which he regrettably leaves unnamed, concur on this key concept. Furthermore, Müʾeyyedzade challenges Jurjani's claim that slight differences in celestial degrees have negligible effects on earthly outcomes. To illustrate his point, Müʾeyyedzade describes a hypothetical situation in which one's ascendant is found in the last degree of one astrological sign, while the ascendant of another person is in the first degree of the following sign. In such an instance, Müʾeyyedzade argues, all associated astrological variables—including the "owner of the triplicity" (*muthallatha*), "terms" (*hadd*), "decans" (*wajh*), prorogators such as the *haylaj* and *kadkhudah*, the "ruling planet of the ascendant" (*al-mustawli ʿala al-taliʿ*), "cardines" (*awtad*), the cusps of the astrological houses, and other pertinent indicators for computation and interpretation—would be entirely different. Müʾeyyedzade even reprimands Jurjani for his lack of expertise in this field, noting that Jurjani frequently misuses and confuses basic astrological terms, such as the exaltation and descension of zodiac signs.[156]

What we have in Müʾeyyedzade's gloss is an invaluable portrait of an Ottoman intellectual from the immediate circle of Sultan Bayezid II sufficiently well-versed in astral lore to venture to criticize a prominent figure of medieval Islamic philosophy and theology for his faulty judgments and disparagement of *munajjims* and their expertise. While we may lack decisive evidence regarding Sultan Bayezid's personal assessment of *munajjims*' expertise, it is plausible to surmise that his perspective aligned closely with that of his lifelong companion, with whom he grew up and rose to prominence, reading the same set of texts and cultivating a similar body of intellectual interests. The influence was probably not merely a one-way street from Müʾeyyedzade to Bayezid II. The sultan, with his remarkable enthusiasm for studying the astral sciences, among other things, must have also had some effect on the formation and sustenance of Müʾeyyedzade's kindred interests.

Messianic Fervor at the Turn of the Tenth Hijri Century

Did the interest in promoting the astral sciences and its practitioners during Bayezid II's time have any connection to the rising tide of messianic discourse in the Ottoman lands on the cusp of the tenth century of the Islamic calendar? Bayezid must have been aware of the messianic currents

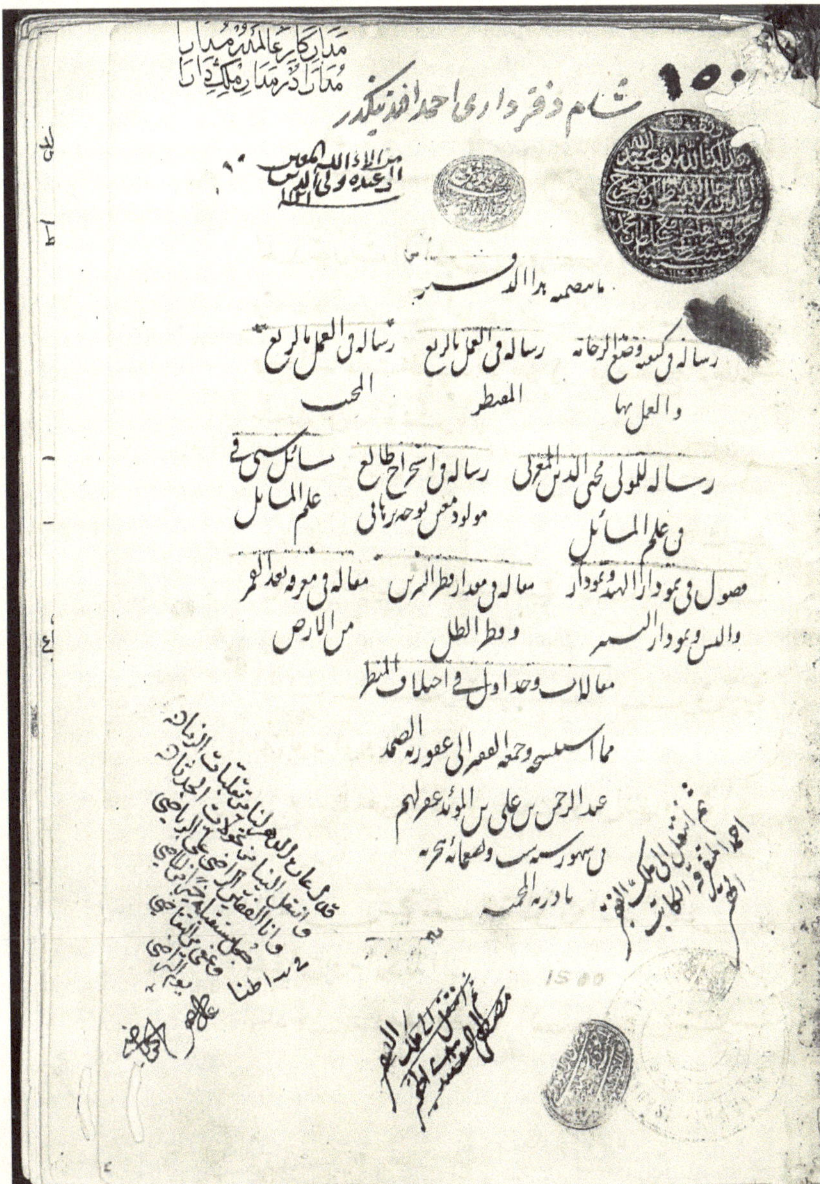

FIGURE 10. Ownership statements by Mü'eyyedzade and Riyazi, among others, on the same miscellaneous compilation of astronomical treatises. Süleymaniye Library, Carullah Ms. 1500, 1a. Reproduced by permission from the Directorate of the Turkish Institution for Manuscripts.

sweeping across the territories he ruled, particularly those triggered by the agents or supporters of the Safavid enterprise that evolved over the fifteenth century from a Sufi order into a messianic political movement, culminating in the formation of the Shiʿi Safavid state under Shah Ismaʿil. In 1492 (898 AH in the Islamic calendar, a mere two years before the turn of the tenth century, marking the closing century of the first millennium in Islamic chronology), Bayezid II narrowly escaped assassination in Albania by a mendicant dervish, likely pro-Safavid, who reportedly shouted before being killed by the sultan's men, "I'm the Mahdi!"[157] Only a year later, in 899 AH/1493 CE, a man arriving in Istanbul from Iranian lands was noted by some contemporary Ottoman chroniclers to have engaged in messianic propaganda, casting himself as the "sergeant of the Mahdi." He was eventually apprehended and executed, according to an eyewitness account, after a fatwa was issued by a group of leading religious scholars of the time.[158]

As both these incidents took place in close proximity to the sultan just before the advent of the tenth Islamic century, Bayezid II must have been aware of the immediate political and ideological resonance of messianic, millenarian currents fueled by Safavid supporters (or others) among or in close proximity to his subjects in Anatolia and Rumelia.[159] Yet it remains difficult to state definitively that Bayezid II espoused these views and utilized them for his own political program. Further, drawing a direct correlation between his extraordinary engagement with the astral sciences and the potentially triggering impact of messianism during his reign demands more stringent evidence than mere conjectures, however suggestive. Modern scholars of late medieval and early modern messianic and apocalyptic currents in Islamic domains sometimes too easily assume a link between the heightened impact of messianic discourses and the simultaneous rise of the occult sciences. While this might be true for specific branches of esoteric knowledge, such as the *ʿilm al-huruf* (science of letters) and *ʿilm al-jafr* (lettrist divination), the textual output by *munajjims*, both in Bayezid II's time and later, rarely engaged with or served as a platform for expressing messianic, millenarian, or apocalyptic claims. Except for a few scattered references, the technical writings of astral experts lack sufficient detail to verify modern scholarly assumptions about a straightforward connection between messianic debates and *munajjims*' activities. Mirim Çelebi's sycophantic remarks about the sultan as the messianic savior of the end times might seem to be a good example, but even in this instance one cannot find any in-depth discussion on the underlying astrological reasoning of such claims.

A more likely source to trace the articulation of such messianic claims is contemporary historical texts written by certain literati and bureaucrats in Bayezid II's close circle. Depending on these authors' literacy in various sciences, these claims could have been couched in astral-scientific discourse. For example, in his *Hasht Bihisht* (The eight paradises), which was commissioned by the sultan himself, the émigré bureaucrat and historian Bidlisi lauds Bayezid II as the messianic renewer of the faith (*mujaddid*), positing his reign as the herald of the tenth Islamic century. As a learned individual well-versed in astrological jargon and techniques, Bidlisi was able to use astrological references extensively to celebrate Bayezid II's rule as ordained by Heaven. Bayezid was exceptional, says Bidlisi, because, in the eyes of the *munajjims*, the sultan's horoscope was superior in fortune to those of all other shahs. Bidlisi goes into further detail, identifying the exact astrological markers of Bayezid's nativity. Accordingly, Venus, the planet symbolizing prophethood and sacred laws, governs the sultan's ascendant. Other important indications drawn from the twelve astrological houses of his horoscope also suggest, so Bidlisi maintains, a steady state of health, a strong natural disposition, and a perfect reign for the sultan. Bidlisi concludes his discussion by asserting that all astrological signs testify to Bayezid II's preeminence over all other rulers in the world.[160]

Aside from Bidlisi, other contemporary history writers commissioned by Bayezid II, including Ibn Kemal (d. 1534), one of the protégés of Müʾeyyedzade and the future Shaykh al-Islam; and Malekzade Ahi (d. after 1486), another Persian-émigré literatus and the writer of an Ottoman dynastic history bearing the title *Shahnama*, also frequently employed the term *sahib-qiran* (master of the auspicious conjunction) to designate Bayezid II in their chronicles.[161] Yet, unlike Bidlisi, they do not offer any astrological justification behind such designation. Perhaps most intriguing in this context is Firdevsi-i Tavil's *Kutbname*, a lengthy history in verse that he composed in 1503 to recount recent Ottoman naval victories against the Venetians. The text's true significance lies in Firdevsi's extensive use of apocalyptic imagery to cast Bayezid II as the prophesied ruler of the age, the *axis mundi* of the time (*qutb al-aqtab*, lit. the "pole of poles"). The notion of the "red apple" (Tur. *kızıl elma*), symbolizing the Ottoman millenarian desire to conquer Rome after having taken Constantinople, also features prominently in the text.[162] The *Kutbname* is certainly not the only contemporary textual evidence for the perpetuation of the millenarian "red apple" discourse around the court of Bayezid II. In an anonymous dream report, likely penned by someone from the frontier ghazi-dervish milieux, the author states that in his dream he saw Sultan Bayezid II sitting next to Seyyid [Battal] Ghazi, the legendary Muslim warrior of the Anatolian frontiers in the 8th century. Seyyid Ghazi then turned to the author and said, "Behold, I have brought Sultan Bayezid, ready for your service. Let him conquer westwards unto the Red Apple and establish the dominion of Islam!"[163]

Bayezid's possible messianic and millenarian preoccupations notwithstanding, these last two examples suggest that the political ambitions, ideological orientations, and cultural aspirations at his court were also shaped within the context of rivalry against contemporary European powers. Contrary to much of the scholarship on Bayezid II's reign, which often portrays his stance vis-à-vis the political and cultural dynamics in Europe as listless and passive, this was far from the case. Particularly during the first two decades of his reign from the early 1480s to the late 1490s, when the Cem Sultan affair escalated into an international crisis, Bayezid II diligently engaged a busy network of spies and informants. These connections acquainted the sultan with the political issues and cultural preferences at major European courts in France, Italy, and Central Europe while simultaneously helping his name circulate in those capitals as both a formidable enemy and a patron of the arts and learning.[164] It is thus little wonder that several artists and humanists in Italy and elsewhere, including Leonardo da Vinci and Michelangelo, considered seeking and even attempted to win Bayezid II's patronage.[165]

Conclusion

Whatever their personal motivations, the exceptional enthusiasm of Bayezid II and his closest companions for the study and promotion of the astral sciences played a pivotal role in the gradual institutionalization of *munajjims*' service within the imperial court through the creation of a permanent position. We should not forget, however, that similar developments were unfolding around the same time for other practitioners of technical knowledge, as manifest in the emergence of the royal corps of architect-engineers and court physicians. It remains a distinct challenge to pinpoint the precise date and dynamics underlying the transition from earlier ad hoc functionaries who received gifts and rewards in return for their service to permanent officeholders obtaining monthly salaries for a routinized set of functions and duties. Envisioning this shift against the backdrop of the Ottoman enterprise's evolution into a bureaucratic world empire with an emerging professional cadre seems to provide a more fitting context. For the creation of a robust Ottoman imperial structure—with its instruments and ethos to administer experts and their domains of expertise—modern Ottoman historiography tends to award credit entirely to the ingenuity of Mehmed the Conqueror and the bureaucratic-mindedness of Süleyman. Yet Bayezid II deserves equal, even more emphatic acknowledgment for his indispensable role in the formation of the early modern Ottoman empire.

The creation of a steady position for *munajjims* at the Ottoman court undoubtedly helped those pursuing the astral sciences in two notable ways. First, it provided relatively stable financial means and granted access to relevant books and specialized instruments kept in the imperial treasury. For example, one of the active or prospective court *munajjims* during Bayezid II's reign submitted a petition to the palace seeking permission to use some items from the treasury, specifically the Ulugh Beg tables, Tusi's commentary on Almagest and an astrolabe.[166] Another set of archival documents from 1504 and 1505 details the items available at the time in the Inner Treasury, including quadrants (*rubʿ daʿire*), celestial globes (*heyʾet topu*), European clocks (*Frengi saʿat*), and several astrolabes (*usturlab*), some of which were preserved in lavish velvet cases.[167] Considering the often prohibitive costs associated with these astronomical instruments, which often features as a theme in the writings of *munajjims*, as we saw in Chapter 1, serving as a court *munajjim* must have alleviated the challenges of material access to tools and books. The other and even more significant benefit of the creation of a permanent unit of court *munajjims* was its implicit role as a school without walls, a milieu where incumbent and aspiring *munajjims* came together and helped with the transmission of their technical expertise across generations. Even though the unit did not have a designated site in the palace, the position itself facilitated the assembly of aspirants in this specific area of expertise.

Despite the permanent presence of the unit of court *munajjims* from the late fifteenth century, royal enthusiasm and favor for the technical expertise of *munajjims* would change for the worse over the next half century, manifest in the reduced number of court *munajjims* staffing the corps during Süleyman's reign, the decline in their remuneration, and a concomitant diminution in the volume and variety of astral knowledge produced. Süleyman's court was certainly attentive to expertise in heavenly matters and foretelling the future, but there was a discernible shift toward preferring the more esoteric and prophetic practice of the lettrists over the more technical and predictive specialization of the *munajjims*. The next chapter will start exploring this change, reconstructing the lives and reflections of *munajjims* from the later decades of the sixteenth century.

FOUR

FORTUNES TURNED

In the second half of the fifteenth century and the first decade of the sixteenth, numerous individuals with diverse bodies of knowledge and expertise were welcomed into the Ottomans' new capital.[1] This was part of the effort by Mehmed the Conqueror and his son, Bayezid II, to transform their polity into a magisterial new empire building upon robust foundations and the know-how of skilled individuals. During these formative, "Ottomanizing" decades of Ottoman imperial formation, groups of émigrés, particularly those from the tumultuous post-Timurid Persianate world, were instrumental in meeting the practical and ideological needs of the burgeoning new empire and filling roles within the imperial administration.[2] They included a wide range of practitioners in various professions, from physicians and scribes to scholars, artists, and *munajjims*. Judging by the names, appellations, extant works, and some biographical and autobiographical accounts of those staffing the emerging unit of court *munajjims*—which, like other similar positions, such as the post of chief physician, chief architect-engineer, or chief artilleryman, becomes archivally visible only during Bayezid II's term—the unit was peopled by experts predominantly of Persianate background.

These émigré *munajjims* contributed to the dissemination of technical astral expertise in the Ottoman realm not only through the books and instruments they brought to the Ottoman capital but also through the indigenous Ottoman *munajjims* they trained, thanks to the opportunities provided to them by virtue of a permanent position at the court. From the second quarter of the sixteenth century, a corps of court *munajjims* who would pen their works predominantly in Ottoman Turkish emerged from the direct disciples or second-generation students of the previous Persianate coterie. As such, the latter played an essential role in passing down the canonical corpus of astral sciences—which is predominantly Persian—to subsequent generations. Yet compared to their precursors at the time of Bayezid II, who received far more recognition and remuneration thanks in particular to the sultan's personal intellectual interest in the astral sciences, *munajjims* of the latter decades of the sixteenth century enjoyed considerably less support and prosperity. Even though their expertise remained in demand and the unit of *munajjims* was preserved in the court bureaucracy, the number of court *munajjims* gradually waned as the century progressed, and their compensation underwent a marked decline, much to the chagrin of certain practitioners.

As in Chapter 2, where the sociocultural and political nuances of *munajjims*' expertise in the Ottoman realm throughout the late fifteenth and early sixteenth centuries were explored via the lives and journeys of two practitioners from the era, this chapter examines the change in identity, compensation, and recognition of court *munajjims* by the mid-sixteenth century through the lens of *munajjims* from Süleyman the Magnificent's time. One might expect a heightened courtly interest in the expertise of *munajjims* during this particular period; after all, modern scholarly literature on the reign of Süleyman, notably the works of Barbara Flemming and Cornell H. Fleischer, have demonstrated the imperial elite's increasing attraction to prophetic, messianic, and apocalyptic discourses in framing and articulating universalist imperial ambitions and a concurrent growth in occult scientific and lettrist expertise.[3] The prevalent impulse among modern scholars and nonspecialists alike to associate *munajjims* and their expertise too readily with other occult practices and practitioners impedes our understanding of why and how *munajjims* were unable to flourish during a period marked by a burgeoning enthusiasm for lettrism and related esoteric knowledge. Evidently, despite Süleyman's and his close confidants' pronounced tendency toward messianic and prophetic fervor—as evidenced by the frequent circulation of key apocalyptic texts and prophecies within the palace, such as ʿAbd al-Rahman al-Bistami's *Miftah al-Jifr al-*

*Jami*ᶜ (Key to the comprehensive prognostics) and the sultan's intimate ties with an expert in the esoteric science of letter divination (namely, Haydar the geomancer)—there was no concomitant increase in the number, prevalence, or activity of court *munajjims*. Counterintuitively, the number of court *munajjims* during Süleyman's reign declined markedly from Bayezid II's era, and their salaries shrank appreciably.

Rather than taking this decline in the number and compensation of court *munajjims* under Süleyman's reign as a token of the Decline in the Ottomans' scientific enthusiasm, a finer-grained perspective would view it in the context of the evolving trends and preferences of those in power regarding their use of experts and fields of expertise. Throughout Süleyman's lengthy reign, there was doubtless a royal interest in heavenly matters, but it was more inclined toward the esoteric and prophetic practice of lettrists than the technical and astronomical knowledge of *munajjims*, a topic treated further in the next and final chapter, where I will contrast *munajjims*' expertise with that of other seers and diviners.

In this chapter, we will examine the stories of *munajjims* known to have served at the court of Süleyman in order to understand what was changing in the extent of the appeal to and recognition of their expertise. Although archival sources note the presence of at least six *munajjims* offering their expertise in different decades of Süleyman's long reign from 1520 to 1566, the uneven nature of extant biographical and autobiographical details regarding the lives of these men leads us to focus on one particular figure: ᶜAli, better known as Riyazi, meaning someone adept in the mathematical sciences (Tur. *ᶜulum-ı riyaziyye*). Riyazi was part of the corps of court *munajjims* from the early 1540s until the early 1550s, by which time the number of court *munajjims* gradually dropped until around 1550, when only two remained. Despite Riyazi's continued employment into the 1550s, he never ascended to the position of chief *munajjim*. Driven by unmet ambitions and unfulfilled desires, he left behind an abundance of autobiographical records that candidly express his disillusionment with the inadequate acknowledgment of his expertise and authority. Furthermore, Riyazi's extant legacy comprises a treasure trove of manuscript records—ownership statements, colophons, and a variety of marginal notes and microscripts—that offer deeper insight into the intellectual interests of a sixteenth-century Ottoman court *munajjim*. To balance the heavy focus on Riyazi in this chapter, we will also take occasional excursions into the lives, works, and experiences of other astral experts from the same period, including Yusuf b. ʿÖmer, Lutfullah (d. late 1540s), and Muvakkit Mustafa. Together, these diverse accounts from the Süleymanic era reveal a less pronounced decline in the stature of *munajjims* and their expertise, even as their counsel remained in demand.

An Ottoman *Mathematicus* and His Sour Times

After Bayezid II was forced by his son, Selim I, to abdicate the throne in 1512, the newly crowned sultan presumably retained the *munajjims* who had served in his father's court. A payment register from 1514, a mere two years after Selim's accession, mentions a distinct body of *munajjims* in the imperial service, but regrettably, the record offers no further clues regarding the identities or compensation of these experts.[4] Selim had one *munajjim* in his princely court in Trabzon, as registered in a separate archival document, who was provided an annual sum of 800 *akçes* for presenting almanacs with prognostications.[5] Lacking an exact reference to his name in the file, it is impossible to trace this *munajjim*'s subsequent whereabouts. Based on the absence of a reference to any *munajjim* in another salary register listing all those people in Selim's retinue while he moved to Istanbul, he likely did not join other court *munajjims* in the capital after Selim's ascent;[6] thus we may surmise that all of the four *munajjims* recorded in the 1514 register had been attached to Bayezid II's court. This supposition is bolstered by two additional archival registers from the early years of Süleyman's reign in the 1520s that record the names of four court *munajjims*—Seyyid İbrahim, İshak, Salman-i ʿAjam, and Yusuf b. ʿÖmer—each of whom had officially begun his service to the Ottomans during Bayezid II's time, as demonstrated in Chapter 2.[7] While Seyyid İbrahim was the son of Khitabi, aka Sayyid Munajjim, Yusuf b. ʿÖmer was a student of Sayyid Munajjim. This generational continuity demonstrates how the institutionalization of *munajjims* at Bayezid II's court started to pay off and shape the composition of relevant cadres in subsequent decades of the sixteenth century. Despite the likely confidential and personalized nature of *munajjims*' business, which might call for dismissing the roster of *munajjims* or at least the chief at each successive accession to the throne, the institutionalization of court *munajjims* ensured their evolution into a cohort providing standardized and impersonal service not drastically affected by a change of ruler.

While there is no known mention of Selim's specific appreciation for the expertise of *munajjims*, several contemporary narrative sources depict him as placing substantial value on astral knowledge. For example, Hakim Shah Muhammad Qazvini (d. after 1523), another émigré natural philosopher from western Iran who served Selim as a court physician upon his arrival in Ottoman lands, bears witness to Selim's interests in the mathematical sciences. In his Persian adaptation of ʿAli Shir Navaʾi's (d. 1501) renowned biographical dictionary of poets, originally written in Chagatay Turkish as *Majalis al-nafaʾis* (Assemblies of distinguished men), Qazvini remarks that Selim inherited from his father a keen interest in and aptitude for the mathematical sciences.[8] Another contemporary work, Keşfi Mehmed Çelebi's (d. 1525) regnal history *Selimname*, vividly illustrates how Selim was swayed by the counsel of *munajjims*. An eyewitness, Keşfi recalls that Selim, intent on launching an attack against the island of Rhodes by 1519, retracted his campaign plans upon hearing—or perhaps reading—foreboding predictions from his court *munajjims* in their almanacs, which indicated that both the start and middle of the year would be highly inauspicious for such a venture.[9] Intriguingly, an extant almanac, presented to Selim in March 1519 by a certain Kemal, contains surprisingly specific prognostications, advocating a delay for any assault on Rhodes for a few months.[10] This overlap between astrological advice in annual almanacs and a contemporary chronicler's testament indicates that Ottoman decision-makers followed and implemented *munajjims*' routine astrological recommendations.

Other chronicles and historical narratives from the period offer further examples of how Selim and his confidants heeded celestial signs and their interpretation by *munajjims*. For instance, Idris Bidlisi, to whom we were introduced in earlier chapters as a bureaucrat and historian of Persianate descent serving in the administration of Bayezid II and Selim I, recounts in his "History of Sultan Selim" that just before the Battle of Chaldiran in August 1514 between Selim's and Shah Ismaʿil's forces, a full solar eclipse occurred in the sign of Virgo. Bidlisi notes that *munajjims*, or as he puts it, "the experts of [the science of] the decrees of stars," reached a consensus by drawing on their intuition (*tahaddus*) and experience (*tajriba*) and declared this to be a clear indicator of "the catastrophe of the fortune of the king of the East of the Earth" and the manifest victory of the western power.[11] Interestingly, in an extant copy of an anonymous Persian almanac from 1514—which must have been presented to the court in early March, months before the Chaldiran campaign—the unidentified *munajjim* notes in the section on the anticipated solar and lunar eclipses of the year that a full solar eclipse was expected to commence 9 hours and 33 minutes into Sunday night, 28 Jumada al-Akhir 920 AH (20 August 1514). The eclipse would occur, the anonymous *munajjim* says, above the horizon in "the Iranian lands" and would be most strongly felt in that area, indicating a calamity for "an individual great of eminence" living in a region encompassing the Euphrates, Mosul, or Persia.[12] In seeming accord with these prognostications, Selim proceeded to inflict a decisive defeat on his eastern neighbor on 23 August.

Around the time Selim ruled the empire and attended to the advice of astral experts, a janissary soldier in the Ottoman military welcomed into the world a son he named ʿAli, who would later rise to prominence as a *munajjim* under his pen name Riyazi. Little is known about the exact date of Riyazi's birth, his childhood, other members of his extended family, or his father Hüseyin's background prior to being recruited as a child levy. Contemporary biographers of poets, such as ʿAşık Çelebi and Mustafa ʿAli, identify Riyazi as an Istanbul-born *kuloğlu*, a term meaning "son of a slave" or "son of a servant," referring to male descendants of Ottoman janissaries and other levied soldiers in the Ottoman army.[13] Biographical dictionaries of poets from the sixteenth century mention numerous *kuloğlus* who, apart from their military roles, frequently participated in the poetic life of the capital and attended various social gatherings held at the mansions of wealthy urbanites. Riyazi indeed developed a passion for poetry, possibly predating his affinity for the mathematical and astral sciences. In addition to his fame as a *munajjim* in the 1540s and 1550s, he was a poet of considerable repute. This aspect of his biography is particularly noteworthy—and useful to the historian—because Riyazi's literary endeavors and skills earned him a spot in contemporary biographical dictionaries of poets. Given the overall dearth of references to *munajjims* in the sixteenth-century biographical dictionaries of scholars such as Taşköprülüzade's *Shaqaʾiq* and its immediate supplements, which exhibit a marked bias against *munajjims* and consistently exclude them from their gallery of learned scholars and wonder-working mystics, we are often left with the scant archival entries that merely list the names and remunerations of court *munajjims*. Riyazi's case thus stands apart in our exploration of *munajjims* from the period. It enables us to combine archival documentation gleaned from payment records with a significant amount of biographical insight provided by his contemporaries, as well as detailed autobiographical remarks scattered through his extant works.

Riyazi's Upbringing and Intellectual Inclinations

Contemporary biographical accounts reveal that Riyazi received the standard madrasa education at the time, likely in his early teens in the mid-1520s. This education was founded on an initial proficiency in Arabic grammar and morphology before gradually progressing into logic, philosophical theology, and other madrasa disciplines, including jurisprudence, hadith, and Qur'anic exegesis.[14] While we lack precise details about which Istanbul madrasas Riyazi attended and the scholars under whom he studied, certain autobiographical notes in his extant writings provide information about the disciplines and texts he explored during his madrasa training. In the lengthy preface of his partial Turkish translation of Ibn Kathir's (d. 1373) comprehensive universal history, which he completed in 1557, Riyazi recounts how he first excelled in the literary sciences and Arabic grammar.[15] He then immersed himself in the study of legal theory, focusing particularly on the book of *Talwih* (The alluding) by the renowned fourteenth-century Persian theologian and jurisprudent al-Taftazani (d. 1390).[16] This phase likely took place in a mid-tier madrasa, as by the first half of the sixteenth century, in the rigidly ranked system of Ottoman madrasas in the core provinces of the empire (i.e., Istanbul, Edirne, and Bursa), a particular middle tier of madrasas were occasionally referred to in Ottoman sources as "*Talwih* schools" on account of their commitment to teaching that specific book.[17] As Riyazi's progress as a student continued within the madrasa system, he branched out into studying substantive law, Qur'anic exegesis, and logic.[18]

Was Riyazi's interest in the mathematical and astral sciences a product of his studies in the madrasas? It seems unlikely. Quite to the contrary, Riyazi's passion for these sciences seems to have flourished beyond and even in direct contrast to the canonical madrasa disciplines. As one of his contemporary biographers noted, Riyazi began to engage with the natural and mathematical sciences, particularly the astral sciences, only after he realized that his "innate disposition" (*tabʿ*) was more attuned to these sciences than to the conventional scholastic path of the madrasas.[19] Curiously, contemporary biographical dictionaries of scholars frequently felt compelled to justify an individual's concentration in the mathematical and astral sciences by attributing it to a peculiar natural temperament. This tendency is perhaps most evident in Taşköprülüzade's widely circulating biographical compendium of scholars, completed in 1561, in which Taşköprülüzade often explained a scholar's dedication to the mathematical sciences with reference to their extraordinary inherent inclination toward these areas of expertise. In his entry on Qadizada al-Rumi, for example, Taşköprülüzade shares an anecdote about a disagreement between Qadizada al-Rumi and Sayyid Sharif al-Jurjani, who briefly maintained a master-student relationship, possibly in Samarkand, during the late fourteenth or early fifteenth century. The discord apparently arose from their contrasting scientific dispositions: Taşköprülüzade notes that Jurjani eventually let go of Qadizada al-Rumi, stating that "mathematics dominated his temperament."[20] Such a portrayal, ascribing an individual's affinity for the mathematical and astral sciences to their personality and character, suggests that by the time Taşköprülüzade composed his compendium, a madrasa scholar's keenness for these sciences was considered unconventional enough to warrant some explanation.

When Riyazi entered the madrasa system as a student, the formal study of the mathematical and astral sciences was not commonplace within Ottoman institutes of higher education. This may stand in contrast with madrasas in certain times and places in the Islamicate world, such as twelfth-century Baghdad, fourteenth-century Cairo, early fifteenth-century Samarkand, or even late fifteenth-century Istanbul, where modern historians of science have uncovered compelling evidence documenting the instruction of texts related to the mathematical and astral sciences, particularly astronomical theory (*ʿilm al-hayʾa*).[21] Ottoman madrasas and their instructors from the 1520s, however, seemed to have a more indifferent, if not overtly adversarial, attitude toward these disciplines. The previous generation of scholars in the later fifteenth century, including figures like ʿAli Qushji, Müʾeyyedzade, and Mirim Çelebi, who had combined their rigor in mainstream madrasa disciplines with a profound interest in natural-philosophical, mathematical, and even astral pursuits, were increasingly being succeeded by scholars who focused exclusively on the teaching of those scholastic disciplines directly linked to judicial training, such as legal theory (*usul al-fiqh*), substantive law (*furuʿ al-fiqh*), hadith, and Qur'anic exegesis.

The dynamics underlying this scholarly sea change are varied and can readily be interpreted within the framework of "Ottoman Sunnitization," a term growingly used by Ottomanists to refer to a range of policies, practices, and processes that paved the way for a more pronounced Sunni orientation among the Ottomans from the sixteenth century.[22] Within the immediate context of changing intellectual trends and epistemological orientations, Sunnitization can be seen not simply as the catalyst but more as the consequence of, or at least a phenomenon concurrent with, the shift in Ottoman scholarly concentrations. The increasing demand for specialization and professionalization in Sunni legal expertise, aiming to cultivate a new generation of judges and jurists who would embody and enforce the Ottoman legal order, evidently held a pivotal role. This need for legal experts was particularly driven by the empire's vast territorial expansion from the 1510s to the 1530s, especially in regions that encompassed the established centers of Islamic learning and holy lands. One must also take into consideration the ongoing Shiʿi Safavid threat, which probably posed the most pressing political and ideological challenge to Ottoman rule in the first half of the sixteenth century.[23] Aside from the growing demand for Sunni legal expertise, one should also note the supply-side factor, that is, the influx of scholars proficient in the traditional madrasa disciplines of hadith and Qur'anic exegesis from greater Syria and Egypt after the Ottoman annexation of the region in 1516–17.[24] Still more research is required to fully understand the exact contours and chronology of the impact that émigré Arabophone ʿulamaʾ might have exerted on the evolution of the Ottoman intellectual landscape.

It is little wonder that a figure such as Katib Çelebi, a bibliophile polymath from the seventeenth century, singled out ʿAli Qushji, Müʾeyyedzade, and Mirim Çelebi, along with Molla Fenari, Qadizada al-Rumi, Hocazade, and İbn Kemal, as constituting a distinctive generation—or two—of scholars in the fifteenth- and early sixteenth-century Ottoman realm who successfully combined *hikma* with Sharia, or in other words, natural philosophy with madrasa science.[25] A similar sentiment was implied by Mustafa ʿAli in his *Künhü'l-Ahbar* (The essence of history), where this displeased madrasa-educated literatus mentions that in the fifteenth century, sultans bestowed much greater favor upon learned individuals with proficiency in various sciences, including *ʿilm al-hayʾa*.[26] In his study on Ottoman scholarly culture, based on three biographical dictionaries from the sixteenth, seventeenth, and eighteenth centuries, Fahri Unan notes that, of the 254 books he identified as composed by Ottoman madrasa scholars in these biographical dictionaries, only 20 are about "rational sciences," with the majority of these works originating from earlier periods.[27] These numbers, as well as the observations of Katib Çelebi on the shift in Ottoman intellectual orientations, reveal that the willingness, capacity, and versatility of fifteenth- and early sixteenth-century Ottoman scholars that enabled them to embrace various types of knowledge, including the philosophical, mathematical, and astral sciences, gradually diminished among later generations.

Katib Çelebi's observations are sometimes downplayed by modern historians of science, who assume that he, as a non-madrasa-educated savant, was an outsider to the entire madrasa system and thus lacking a genuine grasp of its traditions and nuances.[28] However, modern historians of early modern Ottoman scientific and intellectual culture are in no way superior to Katib Çelebi when it comes to accessing and commanding sources or tracing evolving Ottoman scholarly trajectories from the fifteenth to mid-seventeenth century. Katib Çelebi was not only an avid collector and reader of books that amounted to tens of thousands of volumes and spanned various disciplines, both scholastic and otherwise but also had a particular penchant for tracing the biographies and intellectual genealogies of scholars throughout the medieval Islamicate world, including the Ottoman realm of the fifteenth and sixteenth centuries.[29] Furthermore, although his initial training was rooted in the Ottoman chancery rather than the madrasa system, he was by no means a stranger to madrasa circles and their disciplines. After inheriting a substantial sum from his uncle, which allowed him to fully devote himself to intellectual pursuits in his thirties, one of his favorite activities was attending public lectures by prominent madrasa instructors of his day. Given what we know about Katib Çelebi's social and intellectual life, we have no reason to dismiss his critical views regarding the waning—or more accurately, changing—scholarly interest in the natural-philosophical, mathematical, and astral sciences throughout Süleyman's reign.

The diminishing interest in the study of the astral sciences within madrasas from the second quarter of the sixteenth century, which Katib Çelebi diagnosed a century later, is vividly captured in an anonymous archival petition evidently by a high-ranking madrasa instructor dating back to the late 1530s. Written by a professor teaching at one of the *Sahn* schools—

the most prestigious madrasas of the empire by then—this letter in Ottoman Turkish conveys the petitioner's concerns and grievances about his heavy teaching load. At one point in the letter, he mentions that, alongside the standard madrasa disciplines, which encompass both the preparatory sciences—largely focusing on the teaching of the Arabic language—and the legal/religious sciences, he studied astronomical theory, geometry, and the art of using astronomical tools for celestial observation. Curiously, the anonymous petitioner says, these three sciences, which he intriguingly designates as "strange [or uncommon] sciences," failed to garner any scholarly attention among his peers and contemporaries in Ottoman madrasas.[30]

It is important not to interpret such evidence and contemporary observations regarding the gradual diminution of interest in (natural-philosophical, mathematical, and astral knowledge in Ottoman madrasas as confirming Decline-focused modern narratives proclaiming the demise of scientific enthusiasm among the Ottomans from the sixteenth century onward. I suggest steering clear of evaluating the type of scholarly work undertaken in madrasas as a measure and indicator of a "higher" or "lower" scientific standing by modern criteria. Just as the teaching of an astronomy book in a madrasa does not necessarily elevate it to the status of a polytechnic institute of its time, an intensive focus on intricate jurisprudential or exegetical matters in another does not relegate the school to an inferior educational complex. In a similar vein, we should not feel compelled to adhere to anti-Decline narratives that take the sustained and at times amplified scholarly pursuits in certain areas—such as logic or jurisprudence—as conclusive evidence for the enduring vigor of the more technical astral and mathematical sciences in early modern Ottoman madrasas. The main point here is to emphasize the importance of tracking the ever-shifting intellectual trends—inherently influenced by broader political, societal, and institutional dynamics—and to ponder whether these changes might have affected the extent of the appeal of the expertise of any particular group of practitioners, including *munajjim*s.

Riyazi's Early Social Network

Turning our focus back to Riyazi, the main protagonist of this chapter, we find him reaching a decision to embark upon the study of the mathematical and astral sciences, as his natural disposition is reported to have dictated.

This enthusiasm, however, could be nurtured outside the confines of the madrasa. Confronted with the scarcity of educational structures for pursuing these interests, he approached a certain Lutfullah, a court *munajjim* listed on the palace payroll from the late 1520s. The earliest archival mention of Lutfullah dates back to 1527, though the wording of the record—"the customary fee of Lutfullah"—hints that his appointment must pre-date the entry.[31] Lutfullah served as a court *munajjim* from at least the late 1520s until his passing in the late 1540s, ultimately ascending to the position of chief *munajjim*. Only a single work he evidently authored has survived: a detailed birth horoscope in Persian crafted for a certain Mahmud b. Mehmed, who, according to Lutfullah's calculations, was born on 6 December 1489.[32] Lutfullah's name also appears in several manuscripts as a copyist, including a Kandilli Observatory Library copy of the Timurid Sayyid Munajjim's popular astrological treatise *Lata'if al-kalam* (Delicacies of words), which later passed into the hands of Riyazi, as evidenced by some of its marginalia.[33] Lutfullah notes in its colophon that he finished copying the treatise in October 1526.[34] The flyleaf of the same copy also reveals that Lutfullah composed another text to render the Ulugh Beg tables more user-friendly for aspirants to the astral sciences, but the book, *Tashil-i Zij-i Ulugh Beg*, remains undiscovered.[35] Another compilation of texts, now in the Topkapı Palace Museum Library, contains four Persian treatises on various astronomical instruments by Mirim Çelebi, arguably the most distinguished Ottoman astral authority in the first quarter of the sixteenth century. The colophons of these four texts disclose that they were copied in November 1516 by a certain Lutfullah b. Mübarek, who was likely the very same Lutfullah.[36]

Finally, we have Lutfullah's name mentioned in an imperial order dispatched to the judge of Istanbul in 1578, concerning the Istanbul observatory founded by Taqi al-Din. The order stipulated that the judge should retrieve the books once owned by the late Lutfullah on the astral sciences—which were then kept at the chief architect Sinan's *masjid* in Istanbul, as the report says—and send them to the observatory.[37] Since Adıvar's *Osmanlı Türklerinde İlim*, followed by İzgi's *Osmanlı Medreselerinde İlim*—and many others relying on these two works—modern scholars have circulated the mistaken supposition that the Lutfullah in question is Molla Lutfi, who was executed in 1495 on charges of heresy.[38] This assumption has been uncritically accepted owing to the convenience of linking Molla Lutfi with Taqi al-Din and thus perpetuating the narrative of "visionary" Ottoman scientists/scholars executed by traditional ʿulamaʾ. It appears to be the case, however, that the reference to the *munajjim* named Lutfullah concerns the courtly practitioner of the astral sciences with the name Lutfullah, from the first two decades of Süleyman's reign. Viewed from a different angle, this episode of modern scholarly disregard for Lutfullah the *munajjim*, conflating him with Molla Lutfi, is also a powerful testament to the oblivion into which Ottoman *munajjims* fell.

The difficulty inherent in reconstructing the life stories of *munajjims* because of the limited availability of biographical and autobiographical sources is also evident in the case of Lutfullah. We do not know where he was born and raised except that he was one of the court *munajjims* in Istanbul from the 1520s through the 1540s and that he tutored Riyazi, and possibly others, in the astral sciences. That his extant writings—the horoscope he composed and the texts he copied—are all in Persian suggests a possible Persianate background for him, or at the very least, close interactions with one of the Persianate-émigré *munajjims* at Bayezid II's court, perhaps Mirim Çelebi. The specifics of Riyazi's initial encounter with Lutfullah and the beginning of their master-disciple relationship remain obscure. Contemporary biographical sources that provide snippets of Riyazi's life indicate that following his father's passing, Riyazi became the stepson and protégé of a wealthy financial bureaucrat, Kara Balizade. Renowned for the social gatherings he organized and the support he provided to skilled individuals, especially those "men of skills coming from Arab and Persian lands," Kara Balizade likely facilitated Riyazi's introduction to the elite circles of the capital by the early 1530s.[39] Given his position in the financial branch of the bureaucracy, Kara Balizade would also have been instrumental in presenting Riyazi to İskender Çelebi (d. 1535), the empire's chief financial administrator in the late 1520s and early 1530s. With his ostentatious wealth and power, İskender Çelebi was known for not only financially supporting but also helping promising young men, whose ranks included Riyazi himself, in other ways. İskender Çelebi's court—a meeting place for poets and other learned and qualified individuals—may have provided the conducive environment Riyazi needed to cultivate his new intellectual interests and build networks, potentially including his connection with Lutfullah.[40]

Interestingly, in one of his extant writings, Riyazi reveals that he participated in the Ottomans' eastern campaigns of 1533–35 against the Safavids. In his later translation of Ibn Kathir's history, written nearly three decades after this campaign, Riyazi makes a fleeting reference to his presence in the advancing army as a janissary.[41] Given his heritage as the son of a janissary soldier, it may be the case that Riyazi had the prerogative to join military campaigns within the janissary units and to claim war spoils. For instance, an archival register from 1526 listing the salaries within janissary units cites a certain ʿAli b. Hüseyin among the *nan-horegan* (lit. "bread-eaters"), a term denoting the sons of current or former janissaries who received modest material and financial support.[42] While it remains a matter of speculation to equate our Riyazi with the ʿAli b. Hüseyin noted in the 1526 janissary register, Riyazi's sparse autobiographical reflections, penned thirty years later, suggest that as of the early 1530s he had not yet embraced the role of court *munajjim*. In these reminiscences, he recalls the time he spent in Baghdad during the campaigns once Sultan Süleyman decided to stay there for the winter. Riyazi recounts that he visited the shrines of the seventh Shiʿi imam, Musa ibn Jaʿfar al-Qazim (d. 799), and his grandson, the ninth imam, Muhammad ibn ʿAli al-Jawwad (d. 835).[43] He also notes a special visit to the tomb of Nasir al-Din al-Tusi, whom he considered a paramount inspiration after he began his career as a practicing *munajjim*.[44]

Books Riyazi Read

Tusi's *Zij-i Ilkhani* was one of the first books Riyazi acquired after he decided to delve into the astral sciences. Approximately thirty extant manuscripts, dispersed across various manuscript libraries worldwide, from the University of Pennsylvania's Kislak Center for Special Collections and the Bibliothèque nationale de France in Paris to Leiden University's Oriental manuscript collection and several libraries in Turkey, evince traces of Riyazi, sometimes as owner, sometimes as copyist, and sometimes as annotator of the book in question. These manuscripts comprise a diverse yet specialized array of works, mostly covering mathematics, history, literature, and notably, the astral sciences. Not every manuscript offers clues to the specific dates and locations where Riyazi put his hands on the book, but the years from 1540 to 1560 stand out as the time when he was most avid. Among the books owned by Riyazi that pertain specifically to the domains of *munajjims*' expertise—which constitute over half of the entire collection of the manuscripts bearing his marks—a copy of the Ilkhanid tables, currently located in Diyarbakır in southeastern Turkey, contains his earliest dated statement of ownership. A brief note on the last folio states that Riyazi possessed the book in 1541–42.[45] At that time, Riyazi was probably still not a court *munajjim*, as he used no such title in the ownership statement. The earliest instance where he refers to himself as a court *munajjim* is a compilation of four short texts on astrolabes and other astronomical subjects that he notes he copied in May 1543.[46] The same self-designation of Riyazi as a court *munajjim* appears frequently in texts he possessed between 1544 and 1553.[47]

The *Zij-i Ilkhani* was not the only text from the Tusian corpus Riyazi studied. Tusi's *Tahrir* (Recensions of the Almagest) and Nisaburi's commentary on the Ilkhanid tables also feature in the collection of extant manuscripts once held by Riyazi.[48] This treasure trove of textual fragments, unearthed by means of thorough-going detective work, enables our partial reconstruction of the intellectual journey and career of a sixteenth-century *munajjim* who not only mastered virtually all branches of the astral sciences, from *hay'a* and *zij*es to different genres of astrology and celestial magic, but also demonstrated a keen interest in other realms of knowledge, most notably history, genealogy, medicine, and philosophy. Thanks to these fragments, we can establish Riyazi's possessing copies of the pseudo-Ptolemaic *Centiloquium* (*Kitab al-Thamara*) and Ghaznawi's *Sufficiency of Learning in the Art of Astrology*;[49] reading Ibn al-Shatir's (d. circa 1375) astronomical treatise, *Nihayat al-su'l fi tashih al-usul* (An end to questioning in the rectification of first principles);[50] correcting information in a copy of Biruni's compilation of historical chronologies, *Al-Athar al-Baqiyya ʿan al-Qurun al-Khaliyya* (The remaining signs of centuries past);[51] leaving his expository remarks on a copy of Ulugh Beg's *Zij*, Abu'l-Fida's (d. 1331) *Mukhtasar Ta'rikh al-bashar* (Concise history of humanity), and the *Kitab al-bariʿ fi ahkam al-nujum* (Outstanding book on the judgment of stars) of ʿAli b. Abi al-Rijal (d. after 1037), commonly known in medieval European circles as Albohazen Haly.[52] In addition, in various notes, Riyazi mentions owning other, yet-to-be-discovered works, among them Fakhr al-din al-Razi's popular treatise on celestial magic, *Al-Sirr al-maktum fi mukhatabat al-nujum* (The hidden secret in the proclamation of the stars); Riyazi writes that he began a translation of the *The Hidden Secret* but could not complete it.[53]

Riyazi's comprehensive literacy in the mathematical and astral sciences as well as various other disciplines was duly acknowledged by his contemporaries. In the eyes of his contemporary biographers, he was a true "polymath" (*camiᶜ-i ᶜulum*), proficient in a dazzling array of sciences, including the arts of casting talismans and crafting magic squares, and known to be well read in the Ptolemaic corpus and Euclidian geometry, spherical astronomy, natural philosophy, celestial magic, logic, theology, genealogy, and chronology.[54] His contemporaries regarded his expertise in the astral sciences so highly that one of his biographers, ᶜAşık Çelebi, asserted in punning rhymes that even Ibn al-Shatir, the acclaimed fourteenth-century Arab astronomer, did not possess half the proficiency of Riyazi, while Abu Maᶜshar, the most renowned astrologer in the entirety of Islamic history, was not even "a tenth part of a tenth" his equal.[55] The same witness also claims that, so proficient was Riyazi in the astral sciences, he could execute intricate calculations on the movements of celestial objects and decipher their corresponding influences on the terrestrial realm with extraordinary accuracy and speed. Another late sixteenth-century biographer, Kınalızade Hasan (d. 1607), whose father was likely to have been personally acquainted with Riyazi, hailed him as the "Jamshid al-Kashi of the time" and "Aristotle of the age."[56]

Books Riyazi Wrote and Patrons He Approached

As Riyazi's self-chosen epithet implies, he shared his contemporaries' convictions about his unmatched erudition in this domain. This self-confidence surfaces, to varying degrees, in all five of Riyazi's extant works. The first two, written in the first half of the 1550s, reflect a period when Riyazi was still hoping to gain the recognition that he felt was his due. These earlier works include a tract he wrote no later than June 1550 to report an auspicious time for beginning construction of the Süleymaniye Mosque complex and another short treatise on talismans—specifically, astral magic—crafted at the request of Sultan Süleyman to ward off the bubonic plague. During this period, Riyazi seems to have had closer ties to the imperial court in Istanbul, even appearing in payment registers of the time as one of two court *munajjims* along with Yusuf b. ᶜÖmer.[57] Nevertheless, his earlier compositions failed to bring him the fame, fortune, and comfort he thought he deserved. His subsequent extant works, far more extensive than the former two, include an Ottoman Turkish translation of Ibn Kathir's popular Arabic chronicle, dedicated to one of Sultan Süleyman's sons, Prince Bayezid, in 1557, and a philosophical treatise presented to Grand Vizier Semiz ᶜAli

Pasha (d. 1565) in 1562. In these later texts—and especially their sweeping prefaces—Riyazi reflects upon his waning career, reasserts his erudition and expertise, and voices his deepest resentments about the times in which he has had to live. As such, these texts offer valuable insights into the early modern Ottoman lexicon of expertise and the fiercely competitive terrain of Ottoman men of knowledge as perceived by those actors.

As he highlights in the introduction to his last known treatise, *Kuvvet-i Kudsiyye* (Sacred powers), a philosophical compendium in Turkish on the faculties and strength of the human psyche, there was scarcely any field of expertise that escaped Riyazi's attention. He spent—or perhaps wasted, as he might have said—his entire life mastering diverse branches of knowledge and engaging with the intellectual elite of his time.[58] He never seems to have tired of speaking about his intellectual journey, which began with a foundation in Arabic grammar and the study of madrasa disciplines, studying several important ontological and etymological conundrums in philosophical theology and natural philosophy pertaining to God's nature, the limits of knowledge, and the eternality or createdness of the universe. He then progressed to the mathematical and natural sciences and studied arithmetic, mathematical equations, land measuring and surveying, geometry, astronomical theory, celestial observation, and astral sciences, including the arts of creating almanacs, using *zij*es and other astronomical instruments, and making astrological judgments.[59] As was common to many other *munajjim*s, Riyazi's scientific pursuits also extended to medicine, driven by his desire not to be dependent on the remedies of what he considered the "clueless physicians" of his day.[60]

By the second half of the 1550s, Riyazi seems to have had more time to contemplate the course of his life. He was beset by a profound sense of frustration as the result of unfulfilled aspirations and a once-promising career left in tatters. He perceived himself as the victim of a series of afflictions and injustices inflicted upon him by some unnamed, villainous individuals. In a somber retrospective, Riyazi conceded that his extraordinary erudition and intellectual prowess had ultimately come to naught. This sorrowful realization was poignantly captured in a verse of his own composition, one of numerous such examples that he artfully interwove throughout his translation of Ibn Kathir's chronicle, as well as in his other extant works. The lines eloquently encapsulate Riyazi's sentiments regarding his unparalleled skills and knowledge and his deep-seated sense of disheartenment stemming from the lack of recognition and recompense he believed he rightfully merited:

> Having acquired myriad virtues, most refined
> My mastery and skillfulness, perfected
> Yet banished to a corner, desolate
> From recognition and favor, separate
> *Though in my superiority, solace found*
> *Yet, where does reverence for the experts abound?*[61]

Despite Riyazi's prevailing sense of defeat and persistent bitterness from the mid-1550s, he was still in dire need of patrons—people who might reward him out of an affinity for books and learning at large, and especially a penchant for the philosophical and mathematical sciences. He chose to seek the favor of Prince Bayezid, over his brother Selim, the two remaining sons of the aging Sultan Süleyman, who were on the cusp of open conflict.[62] In the spring and fall of 1557, Riyazi presented to Prince Bayezid his two-volume partial translations of Ibn al-Kathir's fourteen-volume universal history. In the preface of the first volume, completed in the spring, Riyazi justified his preference for Prince Bayezid as a patron by alluding to the prince's interest in reading history and literature, subtly contrasting it with Selim's inclination toward what Riyazi regarded as the more frivolous pursuits of music and dance.[63] Riyazi was accurate in his assessment: Bayezid was known for his support of learned people in his princely court in Kütahya, commissioning translations of Persian and Arabic literary classics into Turkish.[64] Conversely, his younger brother, Selim, the future sultan, gained a reputation for hosting drinking parties and indulging in pleasure.[65] Riyazi's translations of Ibn Kathir were, indeed, commissioned by Prince Bayezid. However, only two years after Riyazi endeared himself to the prince, his bet proved ill fated. In 1559, Bayezid suffered defeat at the hands of Selim's forces, backed by the still-reigning Sultan Süleyman, and the prince sought refuge in Safavid Iran under the protection of Shah Tahmasb (r. 1524–76). Bayezid was garroted in the Safavid capital of Qazvin in 1561 by agents of his father with the consent of the Safavid shah.

Having jeopardized his readmittance to the imperial court in the capital by becoming the client of a contesting prince who had fallen out of favor with Sultan Süleyman, Riyazi targeted the grand vizier Semiz ʿAli Pasha as his next prospective patron. This choice was not random. Semiz ʿAli Pasha was known for his keen interest in books, especially those on the astral sciences.[66] He also had a unique collection of mechanical clocks, a relatively new technology imported from Europe that had started to attract the attention of ruling elites. Just a few years before Riyazi attempted to approach him, Semiz ʿAli Pasha had welcomed another expert in astral sciences, Taqi al-Din, who eventually completed a treatise on mechanical clocks by relying on Pasha's rich collection.[67] Riyazi must have held similar aspirations when, in October 1562, he presented his concise philosophical treatise to the grand vizier. The questions as to whether Riyazi succeeded in reaching out to Semiz ʿAli Pasha and receiving his benevolence and whether he had the opportunity to acquaint himself personally with Taqi al-Din, the future founder of the Ottomans' first observatory, remain shrouded in uncertainty owing to the scarcity of clues in the available sources.

Tracing Riyazi's life after he made his bid at the grand vizier's court presents an even more formidable challenge. The latest date encountered in the extant manuscripts bearing evidence of Riyazi is 1563–64.[68] ʿAşık Çelebi, who finished his biographical dictionary in 1568, records that at the time he wrote his relevant entry, Riyazi was residing in Ottoman Hungary as the beneficiary of a generous land grant (*zeʿamet*), which would have offered an annual income of 20,000 to 100,000 *akçes*.[69] This snippet of information becomes particularly intriguing in light of Tibor Halasi-Kun's study on a certain *zeʿamet* recipient named ʿAli Çelebi from Buda, who passed away in 1587.[70] While definitive evidence identifying ʿAli Çelebi and Riyazi as one and the same remains lacking, the 1588 inventory of Ali Çelebi's surviving possessions, which include an astrolabe and around 120 books—among them a treatise on astronomical observation listed as *Rasadname*, at least four copies of perpetual calendars (*ruznames*), a scroll addressing some form of celestial magic (*niranjat*), a copy of a treatise on dream interpretation, another copy of a book on fortune-telling (*falname*), a copy of Zakariya b. Muhammad al-Qazwini's (d. 1283) *Ajaʾib al-makhluqat* (Wonders of creation), several medical treatises, and various historical texts and compendia of poetry—lends considerable weight to this tantalizing hypothesis.[71] If the ʿAli Çelebi of the 1588 inventory was the same person as Riyazi ʿAli, it could then be inferred that Riyazi likely benefited from the benevolence of Grand Vizier Semiz ʿAli Pasha, who bestowed upon him a handsome land grant on the Central European frontier.

Riyazi's Fall from Royal Grace: His Treatise on Talismans

That Riyazi was at Prince Bayezid's provincial court after having served as a *munajjim* to Sultan Süleyman in Istanbul raises an interesting question: What prompted him to ultimately withdraw from the imperial court by the mid-1550s and set out in search of alternative patrons? In the latter half of the 1540s, Riyazi was on the Topkapı Palace payroll as one of two court *munajjims*, together with Yusuf b. ʿÖmer. Extant archival registers from the period document that he received a meager daily stipend of 6 *akçes*, while Yusuf was paid 12 *akçes*. When measured against the remuneration afforded to court *munajjims* half a century ago during the reign of Bayezid II, who were, on average, earning more than 30 *akçes* a day, the sums disbursed to the top two *munajjims* of the empire during Süleyman's so-called Magnificent Age were nothing short of an affront to the significance of the service they provided. Further comparisons with the earnings of contemporaneous experts within the court lay bare the fact that *munajjims* were clearly not among the priorities of Süleyman's administration. For example, the chief physician of the time, Seydi ʿAli b. Kaysuni (d. after 1548), was the beneficiary of eighty *akçes* a day, nearly seven times more than the then-chief *munajjim* Yusuf b. ʿÖmer's allowance, and the chief architect-engineer, Sinan (d. 1588), was receiving fifty-five *akçes* per diem, more than four times the chief *munajjim*'s stipend.[72] Riyazi's predicament, however, was decidedly more acute, for not only did he receive merely half the wage of his sole colleague in the corps of court *munajjims*, but his income also paled in comparison to those in professions requiring far less technical and textual knowledge, such as scribal apprentices or marble cutters, who were making between 6 and 10 *akçes* a day.

ʿAşık Çelebi's detailed portrayal of Riyazi in his biographical account suggests that Riyazi's treatise on talismans, purportedly written at Sultan Süleyman's behest to propose a method for repelling the bubonic plague, marked a turning point for him, as it failed to garner the royal attention Riyazi so ardently desired.[73] This display of royal disregard, combined with the imperial court's already tepid appreciation and compensation for his expertise and services, seemed to be the last straw breaking Riyazi's back, pushing him to relinquish his role as a court *munajjim* and even to cease submitting annual almanacs to the palace. Indeed, none of the almanacs Riyazi crafted have survived to this day. This absence extends to an additional twenty books he mentions having composed. The longer list of Riyazi's irretrievably lost or yet-to-be-discovered compositions includes works on genealogy and history, such as *Ümmü'd-dünya* (The mother of the world), *İskendername* (The Alexander legend), and *Tabakatü'l-ʿalem ve derecatü'l-ümem* (The classes of the world and the ranks of people); biography: *Tabakatü'l-şuʿara* (The classes of poets) and *Terceme-i Vefeyatü'l-ʿAyan* (Translation of the obituaries of eminent men); talismans and celestial magic: *Tılsımat-ı nuraniyye* (Luminous talismans), *Kenz-i mutalsam* (A talismanic treasury), and *Kunuzu'l-metalib* (Treasures of desires); and even physiognomy: *Mirʾatü'l-insan* (The mirror of man)—all together demonstrating the breadth of his erudition.[74]

Not all five of his extant works have a colophon recording a definitive completion date, but examining the contents of these works and arranging them chronologically indicates that Riyazi must have composed his treatise on talismans sometime between 1550 and 1556. We know that no later than June 1550, Riyazi, in his ongoing capacity as a court *munajjim*, presented the sultan with a surprisingly lengthy report containing a horoscope that delineated the most auspicious moment to commence construction of the Süleymaniye Mosque complex. This intriguing report, structured as a booklet, will undergo a thorough examination in the next chapter, where it will be contrasted with other contemporary texts of prophetic and prognostic significance authored by a diverse array of future-knowers beyond the realm of *munajjims*. The remainder of this chapter will be dedicated to a closer examination of Riyazi's treatise on talismans. Vividly illustrating *munajjims*' need for precise astronomical data and serving as an open call to Süleyman to support new research in the astral sciences, this short treatise on talismans illuminates the complex relationships between scientific experts and empires in the sixteenth century. Additionally, the text's peculiar afterlife renders it an enthralling instance of pseudepigraphy, as the treatise's authorship eventually became disassociated from Riyazi and circulated under the name of Ibn Kemal, a celebrated madrasa scholar from the early sixteenth century.[75]

The text is recognized and cataloged in modern bio-bibliographical sources simply as *Risale-i tılsım* (Treatise on talismans), with at least twelve extant manuscripts dispersed among various libraries.[76] This profusion is intriguing, as all his other surviving works have survived in only one copy. The relative popularity of the text stems from its circulation under the name of Ibn Kemal, one of the most prestigious and prolific religious scholars in the early modern Ottoman intellectual sphere, whose full name was Şemseddin Ahmed b. Süleyman (aka Kemalpaşazade, 1469–1534).[77] The earliest dated copy of the treatise dates back to December 1596 and bears a title written in a bold red hue: *This is the treatise of talismans by Ibn Kemal.*[78] The text's epilogue further records the name of its author as Ahmed b. Süleyman. These details suggest that within a mere generation or two after Riyazi, the text began to circulate under Ibn Kemal's name. This inadvertent or perhaps deliberate misattribution can be understood more easily given the fact that Ibn Kemal is known to have authored an extensive Arabic treatise on the plague, titled *Rahat al-arwah* (The comfort of souls), which prescribes, among other remedies, spiritual and esoteric techniques, including the frequent use of amulets and magic squares inscribed with prayers and formulae.[79]

In the absence of any reference to Riyazi within the text, how can we determine that he was indeed the treatise's original author? Two distinct passages in the treatise refer to two separate, more comprehensive works that the author claims to have compiled, addressing various applications of talismans in greater detail: the *Kenz-i mutalsam* (Talismanic treasury) and the *Kunuzu'l-metalib ve rumuzu'l-gara'ib* (Treasures of desires and signs of curiosities).[80] Neither of these titles, however, can be found in modern catalogs, bio-bibliographic works, or Ottoman-era encyclopedic collections, with the exception of a brief mention of the former in Katib Çelebi's *Kashf al-zunun*. Even here, Katib Çelebi did not attribute it to Riyazi.[81] The only text attributing the authorship of these two works to Riyazi is his own Turkish translation of Ibn Kathir's chronicle, wherein he cites both *Kenz-i mutalsam* and *Kunuzu'l-metalib* as his original compositions.[82] Moreover, the specific wording and phrases Riyazi uses in his Ibn Kathir translation, whenever the discussion touches upon the uses of talismans, match almost verbatim the language of the talismanic treatise.[83] Riyazi seems so proud of his more comprehensive works on the topic that he suggests his readers consult these works if they wish to explore more examples of the uses of talismans.

Risale-i tılsım commences with a passage proclaiming that Riyazi composed it in response to the sultan's request to learn more about the principles of the science of talismans, their diverse types and benefits, and their specific application in combating the bubonic plague—a persistent and vexing issue in the sixteenth-century Ottoman landscape.[84] Unless this was a mere authorial trick devised to justify the work's composition, Süleyman might in fact have made such a request, considering the profound impact of the plague's recurrent outbreaks. Although the bubonic plague had been an issue during earlier phases of Ottoman history, the empire's rapid territorial expansion after 1517 on all fronts, along with the concomitant population growth and swift urban transformations in various cities, especially the capital, introduced new and diversified vectors of infection. The territories governed by the Ottomans in the sixteenth century experienced numerous waves of plague epidemics, with one of the most severe outbreaks occurring in the 1550s and lingering for the subsequent two decades.[85]

Unlike other plague treatises that offer detailed prescriptions, Riyazi's account lacks diagrams, symbols, letters, prayers, recipes for suffumigation, and other such elements one usually encounters in treatises on talismans.[86] Instead, it bears a resemblance to a modern researcher's grant proposal, conceptualized and submitted to a funding institution (in this case, the sultan), succinctly outlining the proposed utility of the project and promising to develop it further upon the reception of support. The treatise is structured into a general introduction, three short chapters, and a brief epilogue, with each chapter further divided into three subsections. Throughout the text, Riyazi often discusses only the overarching theoretical underpinnings of talisman-making without proposing specific treatments for particular cases. On the rare occasions when he does offer anecdotal experience, he often concludes abruptly, directing the reader to his two other works mentioned above.

In the general introduction to the treatise, Riyazi briefly discusses the nomenclature and conceptual nuances of various magical practices that elicit certain effects in the terrestrial realm, always in alignment with the eternal will of God. The first chapter is devoted to explicating, through concrete historical examples, the diverse phenomena that affect the sublunary world. These include terrestrial forces such as magnets and the Yada stone, a substance believed to bring rain and control nature in traditional Turkic cultures, as well as celestial forces encompassing specific stars and planetary configurations.[87] Consistent with many other *munajjims* who never questioned the concept of Divine Providence, Riyazi asserts the pious premise that God alone endows each celestial object with distinct qualities and human beings' limited comprehension is incapable of fully grasping their underlying mechanisms. He openly contests the arguments of those who deny celestial influences over the terrestrial realm, since he believes, based on "external proofs" and "experiential knowledge," it is unreasonable to reject the heavens' influence on the sublunary world.[88] In concluding the first chapter, Riyazi transitions into discussing talismans and illustrates their successful use in both the pre-Islamic and Islamic eras through brief anecdotes gleaned from various historical accounts, including the chronicle of Ibn Kathir that he would later translate into Turkish.[89]

In the second chapter of the treatise, Riyazi provides a more intricate astrological discussion to elucidate the knowledge and expertise required for crafting talismans. The astral sciences, Riyazi writes, provides the theoretical foundation for talisman-making, for the efficacy of talismanic images and motifs is dependent upon the extent to which stellar powers are transferred to sublunary elements. It is therefore of paramount importance to select the appropriate planets for the operation at hand, calculate the astrologically auspicious moment (i.e., the ascendant) when the relevant celestial forces are dominant, and deploy natural substances that intrinsically align with the qualities of the chosen planets. Riyazi draws an analogy between the science of talismans and the science of medicine, asserting that just as medicine heals the human body, talismans serve to cure the "body of the earth."[90] He then proceeds to enumerate some of the prerequisites a proficient practitioner of talismans must possess, the foremost of which is excelling in the astral sciences. He reiterates that the theoretical framework of the art of talismans is predicated solely upon the astral sciences. For the first time in the treatise, he also briefly mentions the importance of using astronomical instruments to achieve astrological precision, which, in his words, is feasible only with the financial backing of sultans and rulers.[91]

The discussion on the importance of precise astronomical instruments and data for more effective use of talismans pervades the remainder of Riyazi's treatise. Toward the end of the second chapter, for instance, he contends that imprecise calculations of celestial degrees impair the efficacy of the manufactured talisman.[92] He recommends the use of large astrolabes and quadrants bearing detailed markings in degrees and minutes to facilitate the computation of exact celestial positions. The same recommendation is also featured in the third chapter, where Riyazi probes deeper into the process of producing talismans. He emphasizes that tin and copper—metals corresponding respectively to Jupiter and Venus, the planets believed to be linked to bubonic plague—are essential in crafting efficacious talismans against the disease.[93] The talismans to be made from these metals should be as tall as human beings, says Riyazi, and the manufacturers must know not only what to inscribe on them, including the names of heavenly spirits and other formulae, but also when to write them, that is, astrologically favorable moments. In order to comply with all the underlying astrological principles related to the timing of sculpting talismans and completing inscriptions, the mathematical degree of the ascendant needs to be calculated precisely with the help of astronomical instruments. Here Riyazi reemphasizes how imprecise calculations of degrees diminish the talismans' effectiveness and thus advocates for deploying sizable astrolabes and quadrants. Such instruments are not always within easy financial reach of practitioners, as he stresses; only through the patronage of the ruler of the age can practitioners acquire them and properly pursue their celestial endeavors.[94]

Riyazi revisits the necessity of royal support for the acquisition of precise astronomical instruments in the epilogue of the text as well. He even adds a brief autobiographical passage recounting, per usual, how he has been interested since his early youth in obtaining knowledge across various disciplines. More recently, he writes, he has delved into the intricacies of natural philosophy, specifically the astral sciences.[95] From this foundation, he then moved on to the art of talismans, studying its theoretical principles along with its practical applications. However, as he admits, he has been unable to execute the necessary operations, owing to his lack of adequate practical experience as well as technical and financial support. If the sultan were to assist him in taking the essential practical measures, Riyazi believes he could promptly produce an efficacious talisman for repelling the bubonic plague.

In this unsurprising appeal for royal support, Riyazi addresses another critical issue related to cutting-edge research and technology in the astral sciences. He asserts an urgent need for a new observational program and a revised handbook of astronomical tables, as all available *zijes*, including even the most up-to-date Ulugh Beg tables, fail to satisfy practitioners' requirements for precision. As Riyazi emphasizes, temporal calculations of celestial phenomena such as eclipses or planetary conjunctions, which were usually completed by drawing on these tables, often contradicted personally observed data, leading those unaware of the tables' defects to blame *munajjims* for miscalculation and ultimately the inaccuracy of predictions. Therefore, a new systematic observational program needs to be launched to rectify the existing data and parameters found in the circulating *zijes*. Undertaking such a costly endeavor would be impossible, however, without the financial backing of the sovereign. If the sultan were to demonstrate his benevolence in supporting such a venture, even the most inconsequential and ineffectual person, in Riyazi's words, may soon emerge as a "new Aristotle or Plato."[96]

But Riyazi's plea for royal support and insistence on the pressing need to compile new astronomical tables seems to have fallen on deaf ears. Subsequently, we do not find Riyazi serving at Süleyman's court nor have any efforts to initiate systematic celestial observations on the part of later Ottoman learned practitioners been found until the mid-1570s, when another esteemed astral expert by the name of Taqi al-Din, hailing from Egypt and Syria, managed to convince Süleyman's grandson, Sultan Murad III, and several influential statesmen in his entourage of the necessity of constructing an observatory in Istanbul to revise the data of existing *zijes*. One of Riyazi's contemporaneous biographers, Kınalızade Hasan, declares that if Riyazi had received royal support, he would have been immortalized as one of the most eminent astral experts of his age.[97] Kınalızade Hasan offers no further clues as to why Riyazi failed to garner such support even when he was among the court *munajjims*, but a revealing detail is provided later by Mustafa ʿAli, who in the entry on Riyazi in the biographical section of his chronicle attributes the nonfulfillment of Riyazi's proposal to the religious sensitivities and apprehensions of certain statesmen. The ruling elites, according to ʿAli, did not green-light Riyazi's proposal to erect human-sized talismans for fear that such objects might "infringe upon the beliefs of the populace."[98]

Here we see another instance of ambivalence about the practice of *munajjims*: Riyazi's expertise was initially sought for its potential utility, but he was eventually given the cold shoulder for offering a solution that was potentially in conflict with religious sensibilities. Mustafa ʿAli describes the controversy as being about the creation of human-size talismanic objects. While more research needs to be undertaken to trace the spectrum of attitudes toward talismanic objects of different types and sizes in early modern Ottoman society, a similar ambivalence toward their reception and use seems to have been common. Prevailing sentiments do not seem to have always been hostile toward surviving talismanic statues from Greek and Roman antiquity, such as the famous Serpent Column located at the Hippodrome of Constantinople; their mythical histories and putative protective properties were often integrated into Ottoman historical narratives about the foundation of the city of Istanbul.[99] Moreover, talismanic shirts and amulets with undeniable talismanic qualities were frequently used by various segments of society for their presumed protective power.[100] Yet, these talisman-friendly examples notwithstanding, there seems to have been little to no desire for making talismanic statues of the sort Riyazi proposed, for reasons that appear to be related more to the objects being statues rather than to their being *talismans*. For instance, when Grand Vizier İbrahim Pasha (d. 1536) brought several statues to Istanbul after the conquest of Buda in 1526 and displayed them in the garden of his newly established palace overlooking the Hippodrome, accusations of impiety against him from some Ottoman contemporaries quickly ensued.[101]

We lack any direct accusation of impiety toward Riyazi, other than Mustafa ʿAli's remarks mentioned above. However, it seems plausible to argue that his proposal was dismissed by the ruling authorities on the grounds of religious sensitivity, a rejection that surely offended and further alienated Riyazi from practicing his profession at the imperial court. Even prior to this ultimate dismissal, he likely had concerns about whether or not to continue his service, given his lowly status in the court hierarchy, the diminishing salaries allocated to court *munajjims* during Süleyman's reign, and the already meagre amount of scholarly esteem ascribed to his expertise. While it is difficult to conjecture how his career might have unfolded had circumstances been otherwise, the early modern Ottoman history of science—particularly the history of the astral sciences—might well look very different had Riyazi found the support and recognition he felt was his due.

Riyazi's Peers under Süleyman's Rule

Riyazi's evident frustration may distort our view in this chapter focusing on the status of *munajjms* and the demand for their expertise during Süleyman's time; after all, this perspective is derived from the subjective experiences of just one practitioner, who was characterized by some of his biographers as "naturally irascible."[102] In addition, it is clear that, even if *munajjims* did not enjoy extraordinary favor at the time, their standard services were still actively sought after. Nonetheless, Riyazi can also be viewed as the most—or potentially the only—outspoken member of the community of *munajjims* of his era, who likely faced comparable challenges related to the diminishing appreciation and esteem for their expertise and the evolving preferences of patrons. Just because similar autobiographical narratives with emotional outbursts are not available in the extant writings of Riyazi's peers does not mean that other practitioners of the astral sciences experienced favorable circumstances. In fact, a close reading of their writings reveals instances where certain considerations and challenges regarding the recognition of their expertise are implied, however subtly.

Consider the case of Yusuf b. ᶜÖmer es-Saᶜati, with whom Riyazi worked, perhaps not always on collegial terms, for nearly a decade as the two functionaries of the corps of court *munajjims*. Among all the court *munajjims* in the sixteenth century, Yusuf boasted the lengthiest tenure. The earliest records about him date from spring 1504, when he was a junior member of the astral contingent at the court of Bayezid II, while the latest trace of him comes from 1560, when he presented Sultan Süleyman with an annual almanac replete with unusually rich references to a rather mystical and esoteric interpretation of astrological practice.[103] Over his service, spanning more than half a century, Yusuf must have presented more than fifty *taqwims* and other short reports, but unfortunately only nine of these almanacs have survived. Beyond his surviving *taqwims*, modern bio-bibliographical catalogs attribute to him a short treatise in Arabic on timekeeping, *Jamiᶜ al-miqat* (The compendium of timekeeping).[104] A copy of this treatise resides in a *majmuᶜa* (multiple-text manuscript) owned by Riyazi, suggesting that the members of the body of court *munajjims* were actively engaged in reading one another's works.[105]

Of Yusuf's extant almanacs, seven date between 1511 and 1531 and the remaining two are from 1554 and 1560. In the extensive register of payments and gifts distributed during the last decade of Bayezid II's rule, Yusuf appears numerous times, sometimes under the name Sinan: in Ottoman textual culture, the names Sinan and Yusuf were often used interchangeably for individuals named Sinan al-Din Yusuf. He is documented on more than fifteen occasions as the recipient of cash and robes in recognition of his regular service as a monthly-salaried court official submitting annual almanacs. He was even granted, on 18 May 1511, an additional 1,000 *akçes* to cover the costs of his impending visit to his hometown to reconnect with his loved ones, though the location is not specified.[106] This special royal treatment at the end of Bayezid II's reign, when Yusuf was merely one among several court *munajjims*, may be related to Yusuf's having been a student of Khitabi, the chief court *munajjim*, presumably making him one of the most promising young *munajjims* of the day. Yet despite his prestigious scholarly lineage and long tenure at the Ottoman court, Yusuf had no known student.

While Yusuf served as a court *munajjim* from the onset of the sixteenth century, he also held the role of timekeeper at the Fatih Mosque complex in Istanbul, at least for a certain period, which explains why he was known rather as *saʿati,* which literally means someone competent in measuring the time or hour. Yusuf's earliest surviving almanac, dated March 1511, features a signature in Arabic identifying himself as "the timekeeper in the new mosque complex of Sultan Mehmed."[107] His later almanacs were signed exclusively as "*munajjim* at the imperial court."[108] Without more archival records and (auto)biographical details, we can only speculate whether he kept his job as a timekeeper at the Fatih Mosque complex while serving as a *munajjim* at court in the 1520s and 1530s. Taking his signatures as the sole evidence, one might infer that being a court *munajjim* began to hold more personal significance, leading him to prefer that designation when signing all his later *taqwims*. It should be added, however, that prestige concerns operated differently for different practitioners of the astral sciences. For example, a contemporary indigenous Ottoman expert in the astral sciences originating from Bursa, Necmeddin b. Seyyid Muhammed, consistently preferred to identify himself as a timekeeper in his autographs, even though extant archival records disclose that he too was a court *munajjim.* Necmeddin served as a timekeeper, first in the Old Mosque in Edirne during the 1520s and later in the Fatih Mosque complex in the 1540s and 1550s. During those years, he was also listed among the court *munajjims*, yet all the signatures in his five extant almanacs, spanning from 1526 to 1553, reveal that Necmeddin preferred calling himself a *muwaqqit* rather than a court *munajjim.*[109]

Like all his professional colleagues, Yusuf went unmentioned in sixteenth-century biographical dictionaries of scholars. Since Yusuf left no trace of literary or poetic endeavors, he was never included in the contemporary biographies of poets, a privilege Riyazi enjoyed. Therefore, we are without insights regarding his experiences and possible reception among

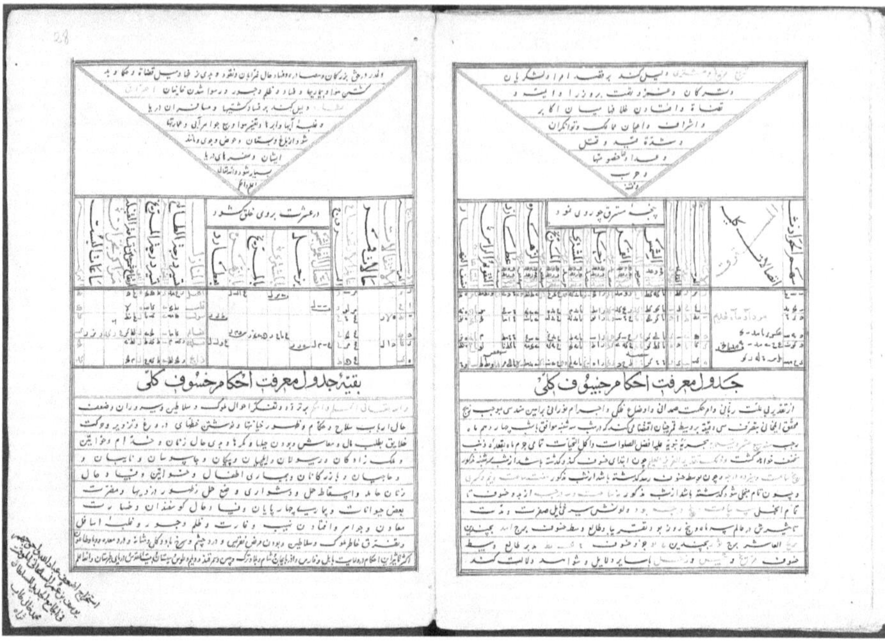

FIGURE 11. Yusuf b. ʿÖmer's almanac for 1511, bearing his signature as the timekeeper of the Fatih Mosque. Topkapı Palace Museum Library, Ms. A 1960, 27b–28a. Reproduced by permission from the Directorate of National Palaces.

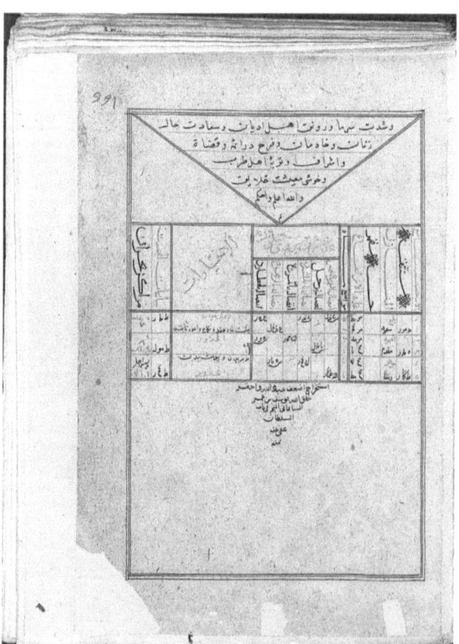

FIGURE 12. Yusuf b. ʿÖmer's almanac for 1517, bearing his signature as a court astrologer. Topkapı Palace Museum Library, Ms. Revan 1711, 291a. Reproduced by permission from the Directorate of National Palaces.

his contemporaries. Nine of his *taqwims* have been found to date, but his extant almanacs, adhering to the conventions of the genre, are devoid of any autobiographical details. Still, these almanacs, which cover a remarkable chronological span from the early 1510s through 1560, provide some clues to how Yusuf may have had to slightly tailor the contents of his almanacs, likely in response to the trending topics and popular references of the day. While his gradual transition from writing in Persian to Ottoman Turkish and his shift from using the Ilkhanid to the Ulugh Beg tables echoes the overall pattern in Ottoman literary and scientific production from the late fifteenth through the mid-sixteenth century, his latest surviving almanac, produced in March 1560, contains certain aberrant features that merit closer examination.[110]

Unlike his eight other surviving almanacs, in which Yusuf adopted the standard sequence of *taqwim* contents—beginning with the calculation of the ascendant at the moment of year-transfer and proceeding with detailed astrological prognostications and calendrical information—the 1560 almanac intriguingly allocates two pages to a detailed exploration of the theory of "hidden spirits" (*rijal al-ghayb*). Originating in early Sufi circles to describe the influence of the invisible saints believed to offer aid to the inhabitants of the world and later refined by Ibn al-ʿArabi (d. 1240) in his conceptualization of sainthood, the *rijal al-ghayb* theory delineates a hierarchy of spiritual beings, presided over by the *axis mundi,* or pole (*qutb*), and subdivided into different ranks.[111] While the names, numbers, and functions of these ranks vary across different sources, the hidden saints were generally considered to appear on specific days of each month from certain directions, aiming to assist travelers, soldiers, sailors, or others in need. As such, the doctrine gained significant traction and found practical application across various social milieux in early modern Ottoman society. Hence, it would not be surprising for a *munajjim* to add such universally appealing and useful content to his almanac; indeed, references to hidden spirits are frequently found in extant Ottoman almanacs from the fifteenth century, when the *taqwims* had yet to be standardized and still more closely resembled anthologies of different sorts of useful information. However, by the sixteenth century, these elements, including historical chronologies and charts as well as descriptions of various divinatory tools, were largely purged from almanacs.[112] Yusuf himself seems to have duly followed these conventions in all eight of his extant almanacs from the first half of the sixteenth century, yet in the 1560 *taqwim* he provides an extensive discourse on the *rijal al-ghayb* theory, explicitly referring to Ibn al-ʿArabi—whom he proclaims the master of the verifiers and the pole of the gnostics—and illustrating the congruence of the 366 hidden saints with the total number of days in a (solar) year.[113] Yusuf's decision to dedicate space to this theory, entirely absent from his earlier surviving almanacs, raises the question of why he did so. Perhaps he was introduced to Ibn al-ʿArabi and his *rijal al-ghayb* theory only later in his career. Or possibly, observing that Ibn al-ʿArabi and his prophetic authority had currency among the sultan and his inner circle—which is evident in the circulation of apocalyptic texts within the palace produced by staunch adherents of the Ibn ʿArabian tradition, such as ʿAbd al-Rahman al-Bistami and Remmal Haydar—Yusuf might have sought to capitalize on it to enhance the prestige of his almanac and himself.[114]

Other astral scientists at the time also faced the necessity of calibrating the scope of their services and expertise to align with the expectations, preferences, and level of knowledge of their intended audience. This sometimes induced frustration, as they felt constrained by the mismatch between what their knowledge entailed and what they thought their audiences were prepared to comprehend. An illustrative example is found in the writings of an Istanbulite timekeeper, Muvakkit Mustafa, arguably the most prolific astral expert during Süleyman's reign. Modern catalogs attribute to him around twenty-five works on timekeeping and the use of various astronomical instruments, the vast majority of which were composed in Turkish in the late 1520s and 1530s.[115] Given the substantial number of surviving copies of his accounts across different centuries—his manual on the astrolabic quadrant alone appears to exist in more than one hundred copies—Muvakkit Mustafa was pivotal in the Turkish vernacularization of technical astral knowledge in the early modern Ottoman world.[116] Of his numerous treatises, his *Tuhfetu'z-zaman ve Haridetü'l-avan* (The gift of time and the pearl of the moment), completed in 1547, epitomizes his entire scholarly output, encapsulating virtually everything he had previously written on the general principles of the science of *hay'a*, geography (then deemed a component of *ʿilm al-hay'a*), timekeeping, various astronomical instruments, astrology, and even talismans.[117] While contemporary scholarship on Muvakkit Mustafa, driven by a brief reference in an early seventeenth-century biographical account, posits that he held the position of chief court *munajjim* prior to Taqi al-Din's appointment in 1571, I have not been able to locate a corroborating archival record. The same scholarship also proposes that Muvakkit Mustafa was a student of Mirim Çelebi, but this also needs conclusive evidence.[118]

In one specific section of his *Gift*, Muvakkit Mustafa voices serious concerns and complaints about the public's lack of interest in the technical expertise of timekeepers. This particular discussion revolves around the question of how sunset time is calculated and what the appropriate moment would be to start performing evening prayers and/or breaking the fast during Ramadan. Referring to the prophetic tradition that suggests expediting the call to prayer and breaking the fast as soon as the sun sets, Muvakkit Mustafa emphasizes the importance of not delaying until after the setting of the sun.[119] However, timekeepers' proper and scientifically calculated sunset time was evidently met with skepticism by the public. For *muwaqqits*, the setting of the sun was a mathematical and astronomical problem for computation, and the moment of sunset did not require the full visible disappearance of the sun. Nevertheless, "the majority of the public," he says, had a different understanding, preferring to wait until the onset of complete darkness.[120] In an attempt to find a middle ground, many timekeepers would postpone the announcement of the time of sunset by a few minutes. However, even at this moment—improperly and impiously late in the timekeepers' opinion—the sky would not seem dark enough to the masses, who were quick to label the timekeepers as ignorant, oblivious of their own lack of knowledge about the "intricacies of timekeeping."[121] Muvakkit Mustafa found it particularly disconcerting to observe some "scholarly types" (*ehl-i ʿilm*) of his era aligning with the general public instead of adhering to the expert guidelines of the timekeepers. Thanks to such individuals, says Muvakkit Mustafa, timekeepers were left frustrated and heartbroken.[122]

Along with the perceived necessity of modifying the content of their regular submissions in response to the changing tastes of decision-makers and frustration with the inadequate acknowledgment and utilization of their technical expertise by both patrons and the public, certain court *munajjims* besides Riyazi also voiced explicit grievances and appeals pertaining to the monetary aspects of their professions. For example, a petition from April 1575, penned by the chief court physician, conveys a request made by a certain Bali, who was one of the court *munajjims* at the time.[123]

Bali was apparently receiving 10 *akçes* per diem, a sum insufficient, as the chief physician underlines, for his living expenses and incommensurate with the extent of his services. The chief physician attests in his petition to Bali's competence in his science and prompts the members of the Imperial Council to remember that previous court *munajjims* had received 40 or 50 *akçes* a day. He forcefully argues that Bali is in dire need of an adjustment in his salary. Subsequently, a decision was made to enhance Bali's daily compensation by five *akçes*. We cannot definitively ascertain, in the absence of (auto)biographical details, whether this 50 percent raise in salary satisfied Bali. Nevertheless, this petition clearly shows that in the second half of the sixteenth century, Riyazi was not the only *munajjim* discontented with his compensation. Additionally, and perhaps more significantly, the petition indicates that by the 1570s, the corps of court *munajjims* had lost its administrative independence and the task of overseeing its internal modus operandi, particularly the appointments and promotions of court *munajjims*, had been delegated to the chief court physician.

Conclusion

By juxtaposing Riyazi and other courtly practitioners of the astral sciences during Sultan Süleyman's lengthy reign with those treated in the previous two chapters who worked during the reign of Bayezid II, we can trace and reconstruct the evolving dynamics of astral-scientific patronage within the Ottoman court from the late fifteenth to the mid-sixteenth century. This change encompassed a gradual transition from the recruitment of Persian-émigré astral scientists to the cultivation of homegrown Ottoman experts and from a reliance on Persian in the composition of relevant texts to the growing use and prestige of Ottoman Turkish as a scientific language. This juxtaposition also illustrates the shift from a more generous disbursement of funds for *munajjims* to a tightening of available resources, as well as the drift from a closer intellectual engagement with the expertise of *munajjims* to relative indifference to their knowledge.

Süleyman's so-called magnificent and millenarian reign, especially the first part, from the 1520s to the 1550s, which was spent undertaking an ambitious plan of territorial expansion and ideological propagation, is often celebrated in contemporary literature as the zenith of the empire and the pinnacle of court-commissioned endeavors in the occult sciences in a spirit of lettrist-prophetic zeal. However, the imperial appetite for supporting the experts of the astral and mathematical sciences and the genuine intellectual interest of the imperial elites in their knowledge were on a decidedly downward trajectory that needs to be spelled out and explained. This is encapsulated with remarkable lucidity in Riyazi's story, which serves as an important reminder to adopt a more nuanced perspective and a more finely tuned periodization when inquiring into the place of the occult sciences in sixteenth-century Ottoman political and intellectual culture. The next chapter will probe more deeply into how the prognosticative power and technical expertise of *munajjims* in producing predictions became overshadowed by the prophetic authority of other types of experts involved in the culture and business of forecasting during the Süleymanic and post-Süleymanic eras.

FIVE

OCCULT RIVALRIES

Who had the authority and expertise to predict the future in the early modern Ottoman world? To address this question, it is necessary first to clarify what we mean by "predict" and "future," as these concepts can swiftly slip into ambiguity, particularly in their late medieval and early modern contexts, where the modern, post-eighteenth-century understanding of time that conceives of the future as an anticipated, malleable, and "open" entity, as Reinhard Koselleck puts it, had yet to prevail.[1] In the medieval and early modern past, predicting encompassed a broad spectrum of phenomena and expertise, ranging from formulating well-thought-out plans for implementing certain measures in regard to forthcoming events to employing scientific and arcane methods to forecast the unknown.[2] In the interest of analytical clarity, we may categorize the former as "projection" and the latter as "prognostication," "divination," or even "prophecy." We should always bear in mind, however, the subtleties and contingencies that sometimes clearly demarcated these notions and at other times made their boundaries fuzzier. Moreover, terms in the original sources rarely reveal a semantic distinction between these analytical categories.[3]

To take an example of projection, we may consider the practice of tax farming in early modern Ottoman finance, which required "projecting" the estimated income of a revenue-generating asset—whether agricultural land, mines, or customs revenue from a commercial product—and subsequently auctioning it off to prospective buyers as a method of internal borrowing.[4] Likewise, in the realm of architectural construction, projection involved selecting a suitable space for a planned building and designing a model prior to initiating the actual construction.[5] In contrast, prognostication was a distinct form of prediction involving the interpretation of visible or hidden signs and portents—celestial, terrestrial, bodily, or otherwise—in order to set expectations and gain insights into the future or the unknown (Ar. *al-ghayb*).[6] The time frame of such prognostications could range from the immediate future to distant horizons, depending on the purpose of the act, the actors, and the scope of the prognosticative tools employed.

The early modern Ottoman world has bequeathed to us an impressive gallery of sources that reveal the widespread use of prognosticative tools by educated elites and laypeople alike. From astrology, dream interpretation, and physiognomy to geomancy, bibliomancy, and other divinatory arts, each practice drew upon various bodies of knowledge and generated its own unique corpus of textual artifacts.[7] Consider, for instance, the practice of physiognomy, which sought to predict someone's character and destiny by taking as portents their bodily and facial features. This practice was of particular use in different manifestations of Ottoman slavery, such as the regular collection of *devshirme* boys from among the Christian Orthodox subjects of the empire and the selection of domestic servants in slave markets.[8] Another divinatory practice, bibliomancy, was a standard feature of courtly parties and scholarly gatherings, as well as other occasions, where revered books—typically the Qur᾿an but also esteemed poetry collections such as Rumi's (d. 1273) *Masnavi* or the *Divan* of Hafez (d. 1390)—were opened at random and the first encountered passage or verse interpreted as an omen for the future.[9] Each of these practices depended on a distinct type of expertise, accompanied by varying claims to authority on the part of their practitioners.

Modern scholarship on the history of these practices in the Islamicate world tends to lump them together, studying them *en masse* under the catch-all term "occult sciences."[10] While the convenience and utility of this conceptual approach for accessibility and comprehension are understandable, it also risks obscuring or flattening important differences among these areas of expertise and their practitioners and thus impeding our ability to trace moments and actors in competition and conflict across differing claims of authority. After all, the politically contested terrain of the future was a busy playground where numerous individuals with varying claims of expertise operated—sometimes side by side, sometimes in keen rivalry with one another—each asserting their peculiar forms and foundations of authority and expertise.

Among the various actors with diverse epistemic grounds of expertise, *munajjim*s emerged almost as a household name, coming to be regarded as the supreme, if not the sole, authority in the realm of foretelling the future. This reputation evolved to such an extent that in modern Turkish, Persian, or Arabic parlance, the connotation of the term has been expanded—or maybe distorted—to encompassing any and all types of fortune-tellers, seers, magicians, and soothsayers, regardless of the scope, methods, and objectives of their business. As this book occasionally reiterates, however, these deeply ingrained modern conceptions in regard to *munajjim*s stand in the way of a more accurate understanding of the scope of their expertise and experiences within the particular social and intellectual context of the medieval and early modern eras.

Just as *munajjim*s were not the sole authority in forecasting the future, the scope of their engagement in this sphere was also markedly different from that of other prognosticators, including geomancers, dream interpreters, and mystics claiming to possess some arcane form of prophetic power. Contrary to modern scholarly assumptions that often ascribe to *munajjim*s all-encompassing expertise in predictions of the future and take for granted that they were preoccupied with prophetic, apocalyptic, and millenarian considerations, as well as other long-haul forecasts, extant textual materials from Ottoman *munajjim*s present a different picture. If *munajjim*s held any authority over contemplating and computing the future, it was, in fact, more narrowly centered on determining and recommending auspicious moments, that is, assessing whether undertaking a venture at a designated moment in the near future would accord well with the temporal order that was considered divinely ordained.

This chapter examines various textual examples of forecasting, some penned by *munajjim*s and others by competing actors in the domain of predicting the future in order to elucidate the peculiar characteristics and boundaries of their respective domains of expertise. While a majority of these examples are derived from the time of Süleyman, reference to materials from earlier and later periods will also be made, demonstrating the enduring and ubiquitous nature of certain practices, as well as the subtle distinctions between them. Süleyman's reign from 1520 to 1566 is of particular significance, given the heightened currency of prophetic lore and the growing expression of millenarian and apocalyptic concerns at the time, as lavishly documented by Barbara Flemming and Cornell H. Fleischer in their pioneering studies.[11] The rising tides of messianic and apocalyptic currents are often thought to have promoted and enhanced the discourse produced by future-knowers, but little attention has been paid to unearthing the possible nuances and competitive interactions between those experts. Contrary to common belief, *munajjims* were only infrequently involved in the production and circulation of apocalyptic discourses. Nevertheless, with their ability and authority to interpret both imminent and remote futures through heavenly portents, they were subject to denunciation and demonization, particularly on the part of certain geomancers and prominent Sufi shaykhs who claimed exclusive authority over predicting the future based on their privileged knowledge of letter divination (*ʿilm al-jafr*).[12] Employing expertise as a framework thus enables us to trace and reconstruct the ways in which ostensibly analogous types of experts and branches of knowledge, often categorized under the umbrella terms "occult sciences" or "esoteric knowledge," were in fact engaged in vigorous competition for social capital and epistemic recognition.

The Hours Holding Promise

An essential facet of *munajjims*' expertise in measuring and interpreting future time involved determining astrologically auspicious moments for their patrons to embark on an activity. In this domain, *munajjims* enjoyed all but unrivaled authority. Early modern Ottoman chronicles and historical narratives, much like contemporary or earlier historical sources from the Islamic world, abound with references to *munajjims*' designation of astrologically favorable moments. These references in narrative sources are often expressed in a rather careless manner, not specifying the identity of the particular *munajjims* or the exact time computed for the occasion in question. The periodic insertion of standard remarks in historical narratives, stating that an imperial endeavor—be it setting out on a military campaign, launching a newly built ship at sea, or constructing an important edifice, including sacred spaces—commenced at an "[astrologically] auspicious hour" (Tur. *bir vakt-i saʿd* or *saʿat-i mesʿud*) raises the question whether these statements reflect historical fact or are simply rhetorical conventions. Fortunately, the Ottoman archives provide us, once again, with an impressive set of records documenting how this practice was routinely exercised by court *munajjims*. At least a few hundred surviving archival documents, predominantly from the seventeenth to the nineteenth century, make it abundantly clear that determining auspicious hours was *munajjims*' primary task and exclusive expertise, to the extent that it, along with the routinized submission of annual almanacs, justified their enduring presence in the administrative hierarchy for many centuries. Although certain Ottoman rulers and other decision-makers occasionally expressed concerns about heeding *munajjims*' advice on auspicious moments, in general the ruling authorities actively sought and dutifully complied with the court *munajjims*' recommendations.[13]

A typical *munajjim*'s report preserved in the Ottoman state archives would take the form of an easily discardable, page-long memorandum, presumably crafted in response to an earlier order dispatched to the court *munajjims* by the ruling authorities—usually the sultan himself but at other times the grand vizier or another relevant individual or administrative body with decision-making capacity. Oftentimes, a horoscope chart showing twelve astrological houses in the projected auspicious hour appears at either the top or bottom of the document. As the *munajjims*' priority was to communicate the most favorable time for the matter in question, they dispensed with lengthy explanations, scientific justifications, and daunting references to authoritative sources. Owing to the ephemeral nature of their material existence as single-sheet documents, as well as their occasionally confidential content possibly necessitating their swift disposal, the number of reports that have vanished over time is likely far greater than the ones that have survived. Nevertheless, the multitude of extant reports, especially from the later periods of Ottoman rule, enable us to document and trace the practice in the Ottoman court on the basis of substantial archival evidence. One such treasure trove resides in the Topkapı Palace archives, cataloged under register number 1298. Housing over fifty distinct reports spanning the fifteenth to the nineteenth century, this register—likely bundled much later by modern archivists, in view of the thematic coherence of the documents—serves as a testament to the enduring allure of the *munajjims*' expertise in providing future-oriented counsel.[14]

The earliest report in the register originates from the late fifteenth century, pertaining to a request made by one of the "viziers" (*Başa* [*sic*] *hazretleri*) or perhaps even the grand vizier, in March 1486, on the eve of an imminent military campaign, probably against the Mamluks.[15] A year before, Bayezid II had begun to take a more aggressive stance against the Mamluks, setting out on a campaign that resulted in the Ottoman invasion of Cilicia in the summer of 1485.[16] According to this report on the most auspicious hour, the unnamed *munajjim* recommended "sunrise" (*gün toğa tururken*) on Tuesday, 20 Rabi al-Awwal, in the year 891 AH (26 March 1486) for the army to begin its march. He explained—and also illustrated with a floral horoscope—that at that precise time, the Sun would occupy the first house, exalting in Aries, with Jupiter residing in the tenth, Venus in the second, Mars in the eleventh, and the Moon in the ninth, among other celestial positions. These celestial alignments, said the *munajjim*, would foreshadow a decisive Ottoman victory, forcing their adversaries to sue for peace. Several of the governors in the unnamed rival polity would meet their demise, and others, including a figure of prominence, would be taken captive. The *munajjim* went on to predict that these portents would start manifesting from the 8th of Jumada al-Awwal (12 May 1486) and extend through the 10th of Jumada al-Akhir (13 June 1486) or perhaps as late as the 25th of Rajab (27 July 1486). Concluding his Turkish report in Arabic, the *munajjim* posited that the conquest of certain territories in Arabia within the ensuing four years would be highly probable should the campaign commence at the time of his recommendation.[17]

Some undated reports in the same register likely date back to the late fifteenth and early sixteenth centuries, given their composition in Persian and their notable similarity in idiom and penmanship to the works of previously mentioned *munajjims* at Bayezid II's court. However, the majority of the reports in this folder, as well as in other collections in the Ottoman state archives, date from the eighteenth and nineteenth centuries. This predominance is not due to the practice being more popular in later centuries but likely because the majority of relevant documents from earlier centuries simply could not survive, written as single-page documents without copies being recorded to a register. That the preponderance of these reports date from later centuries underscores how the practice of *munajjims* as "ritual specialists" retained its role in the social and political life of the Ottoman court, even at a time when concerns surrounding *munajjims* and their expertise gained additional layers amid shifting definitions and implications of what (modern) science entails and who qualifies as a scientist.[18]

A noteworthy distinction between earlier and later reports of auspicious hours can be found in the heightened precision of the latter's designated times, which is likely a natural consequence of the growing use of mechanical timepieces that allowed for minute marking.[19] This is not to suggest, however, that *munajjims* of earlier centuries were not familiar with using hours and minutes in their temporal markings—quite the contrary, as experts in the precise measurement of celestial degrees and hence markers of time, *munajjims* in the fifteenth and early sixteenth centuries were also capable of designating auspicious moments in terms of hours and minutes. Yet in the comparative absence of mechanical timepieces, their target audience mostly lacked the tools to adjust their temporal rhythm based on precise temporal advice. That most people, even the empire's elite, still lived their lives without precise timekeeping technology is surely the reason for the anonymous *munajjim*'s rather imprecise recommendation of "sunrise" in the above example from the late fifteenth century. Later cases, however, consistently pinpointed the exact hour and minute, and occasionally even the second, of the time proposed.[20]

The time window within which *munajjims* were expected to designate an auspicious hour and report it to the court was far from unbounded. Indeed, their task was to identify the nearest favorable date so as not to incur any delays in the ongoing preparations for the event. In other words, *munajjims* were typically not in a position to decide whether a particular event would transpire or not. Their responsibility instead lay in determining an appropriate starting time for the pre-planned occasion, one that would synchronize and harmonize with the heavens and God's temporal order. In some instances, when the initial recommendation could not be observed because of unforeseen setbacks in the actual preparations, the *munajjims* were asked to designate an alternative auspicious hour. For instance, in a memo from 1788, the chief court *munajjim* and the deputy *munajjim* were asked to propose a different hour for the planned military campaign toward Bulgaria, as the initially proposed time had proven too early for the completion of necessary preparations.[21] In other cases, when an operation had to commence without a *munajjim*'s report, the *munajjims* were consulted later to verify if the date on which the operation had begun would prove favorable.

For example, a report from February 1765, written by the third *munajjim* in the office, provided an interpretation of whether a military campaign toward Georgia's having been launched on 19 February 1765 would bode well or ill for the ongoing campaign.[22]

Aside from the memos of *munajjims* that have serendipitously survived in the Ottoman archives, the court *munajjims*' routine practice of submitting reports to convey their computations and recommendations for auspicious moments can be traced through other forms of archived correspondence. Consider, for instance, the Hatt-ı Hümayun collection, which consists primarily of documents and reports dispatched to the sultan by the grand vizier's office for sultanic deliberation in the eighteenth and nineteenth centuries.[23] This collection houses numerous examples of *munajjims*' *responsa* on auspicious hours, summarized in reports presented for the sultan's approval. What these documents reveal, in fact, is an impressive bureaucratic stream and network of correspondence between different offices in the central administration, simultaneously involving a multitude of actors. In a telling example from the time of Sultan ʿAbdülhamid I (r. 1774–89), one document informs the sultan of two separate computations made by the chief court *munajjim* and the second *munajjim*, respectively, for the occasion of the symbolic ceremony bestowing the royal seal upon the new grand vizier.[24] Apparently, the grand vizier's office had requested reports from the office of court *munajjims*, to which the two highest-ranking court *munajjims* responded separately, though regrettably these actual reports have not survived. A discrepancy arose between the chief *munajjim*'s computation of Thursday, the 24th of Rabi al-Akhir, 7 hours and 4 minutes after sunrise, and the second *munajjim*'s designation of 5 hours and 51 minutes after sunrise on the same day. Confronted with this discrepancy between two expert opinions, the grand vizier's office turned to the sultan for the ultimate decision, who appended the following note to the upper portion of the document, remanding the decision back to the grand vizier's office: "In truth, the domain of the unknown (*ghayb*) can be known only by God alone, but whoever among those two *munajjims* shows a higher level of expertise in their science, then act according to his computation."[25] Alas, we lack further evidence to trace whose expertise and computation ultimately held sway. Still, the document is a strong indication that even holding the title of chief did not make one the supreme expert of that practice, as expertise was constantly negotiated and contested.

Aligning with the Cosmos: The Foundation Horoscope for the Süleymaniye Mosque

The last example above, concerning the divergent calculations and suggestions of two court *munajjims* with regard to the same question, reveals the fertile ground these sources present for tracing rivalries within the same field of expertise. In examining the discrepant computations of *munajjims* and their implications for articulating competing claims of professional expertise, one of the surviving accounts of Riyazi, whose life and tribulations were explored in the previous chapter, provides further insight. Composed prior to the construction ceremony for the Süleymaniye Mosque complex in June 1550, Riyazi's report, which recommends an auspicious time at the behest of the aging Sultan Süleyman, is a priceless artifact.[26] Apart from the composition date that establishes the text as one of the oldest surviving examples of Ottoman electional astrology, Riyazi's account—crafted as a separate book rather than a single-page memo and divided into nine chapters, each presenting comprehensive scientific explanations to justify the author's recommendations and predictions—illustrates how a self-assured scientific expert asserted his erudition and superiority vis-à-vis his competitors within the same profession. Moreover, given that Riyazi's proposed date for commencing construction matches the actual date of the foundation ceremony of the Süleymaniye Mosque complex, as recorded in contemporary historical narratives and inscriptions found on the portals of the mosque, the text bears witness to the political and practical recognition accorded to *munajjims*' recommendations concerning astrologically auspicious hours, even when such recommendations pertained to religious purposes and sacred spaces.

In stark contrast to the typical, single-page *munajjim* memos of auspicious hours preserved in the Ottoman state archives, Riyazi's report—penned in response to Sultan Süleyman's open "call for proposals" from court *munajjims* before the construction ceremony for his planned mosque complex—arrived in the form of a twenty-five-page booklet now housed in the Süleymaniye Library in Istanbul.[27] The sole surviving copy is inscribed in a clear, legible *naskh* script, complete with painstakingly added diacritical marks to aid in reading it correctly. Verse markers and section headings are rendered in red ink, while the titulature used for Süleyman shines in gold. The manuscript is bound in brown leather, and at the center of the binding lies an embossed circular medallion. Both the medallion and decorative lines adorning the corners of the binding are stamped in gold. Taking into account the physical qualities of its cover, folios, and script, the surviving manuscript appears to be the presentation copy delivered to the palace.

A particularly striking aspect of the report's composition as a booklet is the presentation of its contents in nine separate chapters, each delving into Riyazi's intricate scientific reasoning and predictions concerning the fortunes of the edifice. Unlike the *munajjim* memos preserved in state archives, which informed their target audience merely of the suggested auspicious time without elaborating on its scientific basis, Riyazi's booklet serves as a bravura showcase for his exhaustive knowledge of the astral sciences. In addition to his painstaking references to the teachings of several authorities in the Islamicate practice of electional astrology, from Kushyar to Ptolemy—which probably would be of little interest to the sultan himself—the text occasionally features samples of Riyazi's original poetry. These verses served to limn his persona as a self-confident and bumptious yet disenchanted expert in the astral sciences. For instance, even at the beginning of his account, where he explains how his calculation is superior to that of his peers, Riyazi incorporates one of his verses expressing unease with his own era, a time when other learned individuals and self-proclaimed astral experts—whom he considers utterly incompetent—enjoy greater rewards:

> Riyazi's heart, weighed down by fortune's ruthless hand,
> Ignorant minds misconstrue his plight, misunderstand.
> His fame, as radiant as the sun at its peak,
> Yet those blind to stars' secrets, more favor they seek.[28]

There is little doubt that the entire account was composed as a reminder and an invitation for the sultan to reevaluate and appreciate Riyazi's erudition and expertise. Before concluding his treatise, for instance, Riyazi revisits his favorite theme, reiterating how numerous calamities have recently befallen him and a multitude of excruciating obstacles and adversities have impeded him from living a blessed life.[29] In a peculiarly self-aggrandizing manner, Riyazi emphasizes his intellectual merits and expresses his discomfort—as he does in many of his other extant works—in the face of the erosion in his time of what he portrays as the meritocratic ethos and proper order of previous decades:

How strange! Riyazi in twenty-four sciences is profound,
Yet in this realm, no wealth or property he has found.
Curiously, his own superiority is to blame,
For the humiliation cast upon his name.[30]

In light of the particular historical and personal circumstances surrounding Riyazi's account, his pronounced resentment was not entirely unfounded. Although the text lacks a colophon to document its exact composition date, it was likely penned in the first half of 1550, before the construction ceremony for the Süleymaniye Mosque complex in June. By that time, as we saw in the previous chapter, Riyazi was one of the two court *munajjims* listed in the palace payment registers.[31] Riyazi was receiving a mere six *akçes* a day, whereas his colleague Yusuf b. ʿÖmer was given twelve *akçes*, according to the extant archival register of payments that dates back to just a year before the likely composition date of the account. A contemporary witness, ʿAşık Çelebi, shares in his biographical dictionary of poets a slightly mocking poem by Riyazi that alludes to the dynamics of his recruitment as one of the two court *munajjims*, along with Yusuf b. ʿÖmer, after the death of the chief court *munajjim* Lutfullah:

Today I, a humble speck, was sought
By the sun of knowledge, radiantly fraught.

Lutfullah, the chief *munajjim*, now gone
His salary remains, no claim upon.

Who in the imperial court now stays
Like a [Jamshid al-]Kashi, to stargaze?

I replied, ʿÖmer's son and I alone
In this world, as *arse* and *head*, we're known.[32]

In the absence of definitive evidence shedding light on the precise nature of the relationship between Riyazi and the incumbent chief court *munajjim*, Yusuf b. ʿÖmer, any attempt to interpret this poem as Riyazi's insinuation—casting himself as the head and Yusuf b. ʿÖmer as the arse—would only be speculative. However, the unusual length of his report communicating the auspicious hour for construction of the Süleymaniye Mosque complex provides additional clues that become coherent if the fiercely competitive nature of relations between experts within the same profession is taken into consideration. In fact, his possible resentment might explain why Riyazi decided to write such an unusually long report in the first place, filled with occasional personal comments and samples of poetry, rather than simply presenting a concise memo outlining his recommended date for commencing construction. Riyazi probably took Sultan Süleyman's initial call for proposals as an opportunity to educate and persuade the court of his unrivaled expertise in the astral and mathematical sciences, thereby elevating himself above all other contemporary *munajjims*. The vigorous efforts he put forth to distinguish himself from his peers are particularly evident in the first two chapters of the report, to which we now turn.

Riyazi opens his treatise with a concise explanation of the reasons for its composition. Accordingly, Süleyman, whom he extols with specific epithets that echo the ensuing contents of this tract—such as "the sultan of the Two Holy Cities," "the caliph of the two mosques," "the destroyer of idols and the houses of unbelief"—had requested an auspicious hour to lay the foundation of a colossal new Friday mosque in the capital.[33] Upon hearing this royal request, Riyazi sat down and made the necessary calculations to determine the most auspicious hour to begin the mosque's construction. His wording here does not specify the type of instruments he used—whether a *zij*, an astrolabe, or another astronomical tool.[34] He simply states that after diligently analyzing planetary aspects and detecting the relative positions of the celestial objects according to Istanbul's latitude, he designated a propitious hour, which was 4 hours and 12 minutes after sunrise on Thursday, the 26th of Jumada al-Awwal 957 AH, when the ascendant would be in the twentieth degree of the sign Leo. Riyazi concludes his opening remarks by emphasizing that the auspiciousness of his recommended time was immediately acknowledged by "all other *munajjims* and almanac-makers."[35]

Upon determining the exact date and time for his choice, Riyazi proceeded to compose the horoscope chart for his proposed moment. Intriguingly, in contrast to the prose section in the introduction, where he wrote down that the ascendant should be in the twentieth degree of Leo, Riyazi made a subtle change when denoting the degree of the ascendant on the chart here. In their notations, *munajjims* and mathematicians in the Ottoman and Islamicate world predominantly used the *abjad* system—an alphabetic numerical system that assigned different numerical values to letters in the Arabic alphabet and their combinations. For instance, the number 20 was denoted by the letter ک, its assigned numerical value, and 22 was written as کب, combining 20 (ک) and 2 (ب).[36] Where the first astrological house is marked in his horoscope chart, Riyazi quite unusually wrote the degree of Leo as کاا, signifying twenty-two, as if he had initially written ک and later added the two *elifs* to arrive at 22. It remains unclear why he did not simply add a ب to make it 22 (کب). The exact reasons and dynamics of his textual adjustment notwithstanding, Riyazi subsequently explained in his second chapter why the ascendant should be at the 22nd degree rather than the 20th degree of Leo as recommended by other *munajjims*.

In the first chapter, Riyazi explains the rationale behind selecting Leo as the basis for his calculation. Drawing extensively from Abu'l-Mahamid al-Ghaznawi's relatively popular treatise, *Sufficiency of Learning in the Art of Astrology*—a copy of which was in his possession, as mentioned in the previous chapter—Riyazi asserts that when a royal enterprise is at stake, *munajjims* should consider a zodiac sign associated with royal people, the two luminaries (the Sun and the Moon), and the two remotest planets (Jupiter and Saturn).[37] The signs ruled by Saturn—Capricorn and Aquarius—would be inappropriate for this specific venture, as the former is a tropical sign, inauspicious for matters related to construction, while the latter is overall an unfavorable sign. Similarly, the signs ruled by Jupiter would be irrelevant from the standpoint of a construction project. Moreover, Riyazi explains, during that particular season, the ascendants of these two signs would appear only at nighttime, thus rendering them improper as options for commencing construction. Cancer, another tropical sign ruled by the Moon, is also inappropriate for this type of election. Consequently, the only viable choice is Leo, a fixed and royal sign befitting imperial construction. Riyazi maintains that all other authorities in the astral sciences, including Kushyar, are in complete agreement regarding the validity of Leo in designating an auspicious moment for royal construction activity.[38] In another accurate reference to Ghaznawi, Riyazi contends that the nativity of the querent—in this case, the sultan himself—is equally crucial, as it should harmonize with the horoscope of the designated hour. As Riyazi and many other practicing *munajjims* of the time likely concurred, with each person's being born under distinct astrological conditions, the most auspicious time to hold an event would vary for different individuals. The selection of an auspicious moment would not prove successful unless it was compatible with the natal chart of the querent. Riyazi does not provide a comprehensive account of Süleyman's nativity here, except for saying that the align-

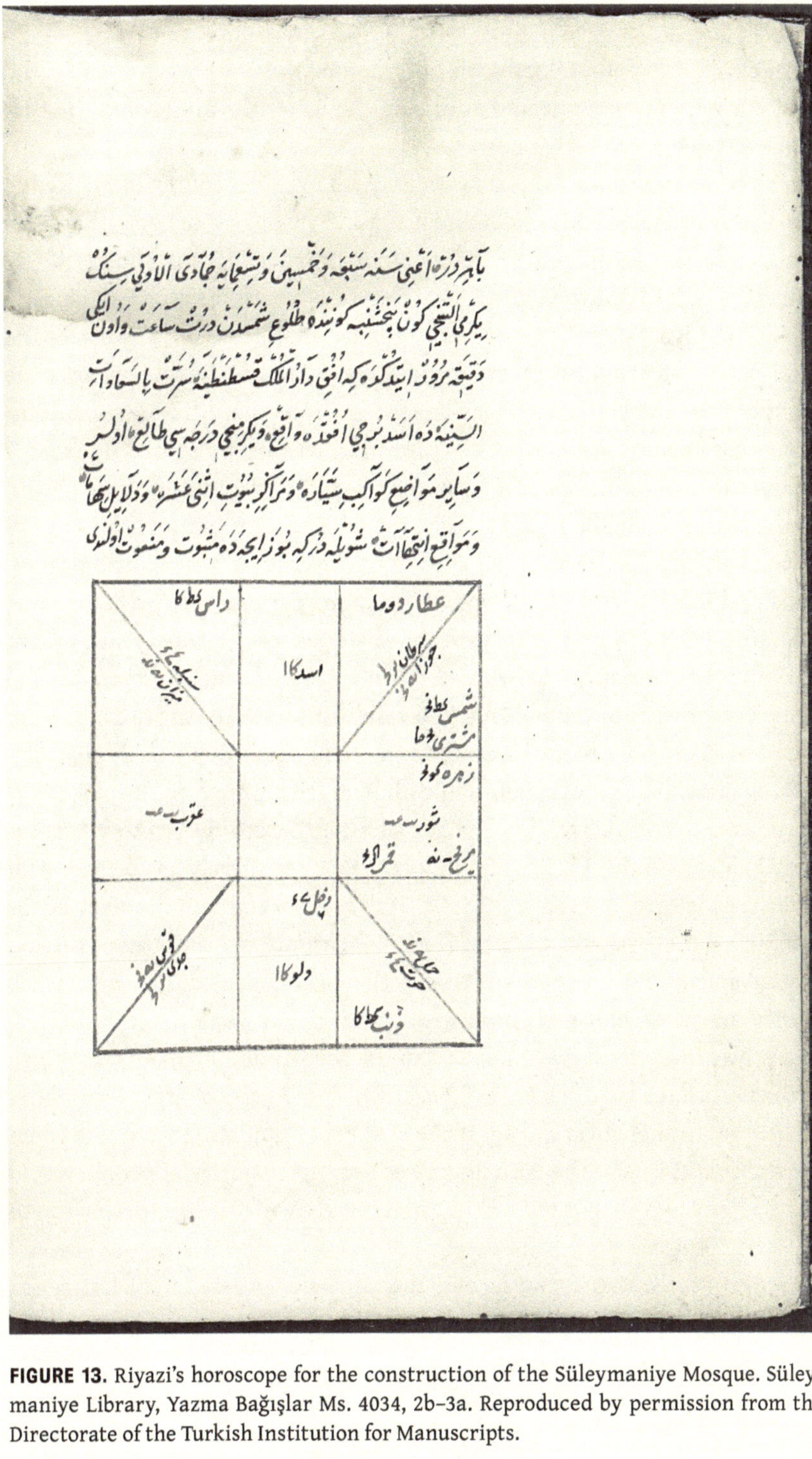

FIGURE 13. Riyazi's horoscope for the construction of the Süleymaniye Mosque. Süleymaniye Library, Yazma Bağışlar Ms. 4034, 2b–3a. Reproduced by permission from the Directorate of the Turkish Institution for Manuscripts.

ment of celestial bodies during his birth conforms favorably to the celestial configuration at the projected auspicious hour for beginning construction of the mosque.[39]

In the second chapter, Riyazi expounds upon his reasoning for adjusting the degree of the ascendant, setting it as the 22nd degree of Leo rather than the alleged consensus reached by contemporary *munajjims* and almanac-makers reportedly favoring the 20th degree of the same sign.[40] Other *munajjims*, in Riyazi's opinion, tended to believe that different degrees of the same astrological sign generate more or less the same effects, whereas he highlights a number of nuances that stem from the differences in individual degrees of the sign. As he states, just as various zodiac signs have distinct terrestrial influences, different degrees within a particular sign yield divergent outcomes and should, thus, be interpreted accordingly. For Riyazi, this rule had been established in treatises imparting Babylonian secrets and in some works written by "Muslim philosophers."[41] He explains that after consulting several books of earlier authorities, particularly the teachings of Teucros of Babylon—known in the Islamic world under the name Tankalusha or Tangalusha—he resolved to rectify the ascendant to the 22nd degree of Leo, as he came to the realization that this degree, signifying piety and sound belief, would align much better with the objectives of the construction.[42] No further explanation is provided as to when exactly he came to this realization or whether it was merely a typographical error that led him to suggest the 20th degree of Leo at the outset of his tract. By directly quoting a passage in verbatim from the Arabic version of Tankalusha, Riyazi contends that if the 20th degree of Leo were to be chosen, as other *munajjims* proposed, the portents would be unfavorable. Not only is the 20th degree malefic, but the corresponding fixed star is also ominous—a fact that other *munajjims*, whom he refrains from explicitly identifying, simply overlooked. If the mosque's foundation were laid when the ascendant was at the 20th degree of Leo, the majority of people would be reluctant to frequent the mosque and might even develop an aversion to the space.[43]

Riyazi's discourse in the second chapter holds greater significance for the insights it presents into the existence of professional competition and the clash of expertise among practicing *munajjims*. Regrettably, due to a dearth of documentary evidence concerning the calculations of other contemporary *munajjims* for the event in question, we are unable to verify

Riyazi's claims about the existence of recommendations for the projected time differing by a mere two degrees in the sign of Leo. In clock time, this difference in the degrees of zodiac signs amounts to slightly less than a ten-minute discrepancy.[44] In other words, accepting Riyazi's statement—which is at this time the sole evidence at hand—would mean that the other *munajjims* must also have proposed the exact same day (i.e., Thursday, 26 Jumada al-Awwal 957 AH/12 June 1550 CE) but a different hour, one marginally earlier than Riyazi's suggested four hours and twelve minutes after sunrise. The question then remains: Which projected hour was ultimately taken into consideration for the commencement of construction—Riyazi's or that of the other *munajjims*?

Throughout the latter portion of the second chapter, Riyazi, with a distinctly self-promoting tone, asserts that his explanations regarding the need to rectify the ascendant's degree by two degrees eventually convinced his peers to revise their initial positions and acknowledge Riyazi's "mastery" (*fazilet*) to an even greater extent than before.[45] His remarks here evoke an image of a milieu in which all astral experts convened and openly discussed the ideal auspicious hour for the commencement of construction. In fact, in his Turkish translation of Ibn Kathir's chronicle, completed seven years after his horoscope account, Riyazi recounts that following Süleyman's order to the "experts of the [science of the] stars" to designate an auspicious hour for the construction ceremony, they convened in the presence of the statesmen and computed a moment.[46] Riyazi was, apparently, not present during this initial meeting. Upon his arrival, Riyazi says, he examined the designated horoscope and discovered that even though the auspicious moment selected was apt for the business of construction, it would only lead to misfortune for Sultan Süleyman, as owner of the planned mosque. Riyazi then adjusted the horoscope accordingly, and his modifications were met with unanimous approval from the other astral experts.

Accurate as Riyazi's later recollections may be, there is little proof for the presence of such a collegial and cooperative scientific environment in archival documents, manuscript sources, or narrative sources. On the contrary, as the aforementioned archival document from the late eighteenth century also suggests, even the chief *munajjim* and the second *munajjim* did not always operate as a collaborative team, at least occasionally recommending different auspicious hours for the same event. Thus, Riyazi's remarks appear to serve more to distinguish his own prestige and scientific superiority than to provide an accurate representation of his specific professional milieu; otherwise, he would not have inserted the previously quoted verse of his own—the one that speaks about his sun-like fame in contrast to the ignorance of others—immediately following this section.

Fortunately, we have a number of contemporary sources that document the date of the construction ceremony, albeit less precisely than the time designated by Riyazi in the only extant foundation horoscope for the mosque complex. Foremost among these sources is the inscription that still stands over the mosque's portal. This inscription, composed by the Shaykh al-Islam Ebu's-suʿud Efendi and inscribed by the famed calligrapher Hasan Çelebi (d. 1594) records that the construction began in the last ten days of Jumada al-Awwal 957 (circa June 1550).[47] Moreover, in the semi-autobiographical memoirs of the Ottoman chief architect Sinan, who was in charge of the construction of the mosque complex, the starting date is also given as Jumada al-Awwal 957, though he does not provide a specific day, let alone an hour, in the accounts he dictated almost thirty years after the ceremony.[48] The information from mosque inscriptions and the chief architect is further corroborated by the remarks of the empire's chief bureaucrat at the time, Celalzade Mustafa (d. 1567), who not only witnessed the construction of the complex in the capacity of chief chancellor but also ascribed to it a lofty imperial and ideological significance.

In his monumental chronicle recounting the pivotal events of Süleyman's reign, Celalzade devotes a separate, extensive chapter to the Süleymaniye Mosque complex, framing it, as Kaya Şahin says in his monograph on Celalzade Mustafa, as the "most articulate cultural statement of Ottoman imperial Sunnism."[49] According to Celalzade, after the construction site was determined by architect-engineers and the foundations excavated, "the experts among learned people of almanacs and stars," whom he did not explicitly identify, designated an auspicious hour.[50] The foundation was laid accordingly on Thursday, 27 Jumada al-Awwal 957. There appears to be a one-day discrepancy between the narrative of Celalzade and the computation of Riyazi, but in reality both refer to the same day: Thursday. In the perennial challenge of converting dates from the Islamic calendar to the modern Gregorian calendar, modern experts in calendar conversion recommend using the weekday as the principal reference point rather than the number of days in the given Hijri month. Each month in the Islamic lunar calendar has either twenty-nine or thirty days. While *munajjims* had their own astronomical conventions and computations, laid out in annual *taqwims*, for determining the first day of each lunar month and the exact number of days—twenty-nine or thirty—within it, the public's prevailing method used firsthand observations of the lunar crescent attested by individuals deemed trustworthy in the society, including religious scholars and local judges.[51] If the new crescent was sighted on the twenty-ninth day, the new month would commence on the following day; if not, the new month would begin after the thirtieth day. In that sense, the same weekday might be denoted by a *munajjim* as the Xth day of the month and by others as the (X + 1)th or (X – 1)th day. As no extant copy of a *taqwim* has survived from the year in which construction of the Süleymaniye Mosque commenced, we cannot verify the *munajjims*' designation of lunar months for that year. The crucial point here is that regardless of the one-day difference in the dates Celalzade and Riyazi provide, it must be the same Thursday at the end of Jumada al-Awwal of 957 AH (likely 12 June 1550).

By examining the information gleaned from these three key contemporary testimonies and comparing it with the date suggested in Riyazi's horoscope and the one presumably advised by other *munajjims*, it becomes clear that the imperial authorities did indeed heed *munajjims*' temporal guidance for the construction ceremony of the Süleymaniye Mosque complex. As for the question of whether the imperial authorities adopted Riyazi's adjusted recommendation or that of other *munajjims*, who allegedly suggested the moment when the ascendant would be in the 20th degree of Leo, a definitive answer remains unreachable, given that the difference in the suggested times was only a few minutes. Still, there is no question that despite the waning intellectual interest in the expertise of *munajjims* during Süleyman's time, their counsel as ritual specialists was routinely taken into serious consideration—even for significant religious events such as the construction of an imperial mosque.

The conjunction of astrology and religion or *munajjims* and mosques may strike some modern readers as wrong-headed, given that the prevailing narrative about *munajjims* often portrays them as impious astral determinists and ungodly sorcerers from a strictly religious standpoint. How, then, could their astrological advice be taken as guidance for a religious undertaking? Although Riyazi's report may stand alone in the entirety of Islamicate history as the sole surviving horoscope of an auspicious hour for the construction of a sacred space, it was not an isolated incident. Chronicles and historical narratives from both earlier and later periods, originating from the Ottoman and other Islamicate realms, reveal additional instances where *munajjims*' expertise was sought to pinpoint astrologically favorable moments for the construction of mosques. For instance, the late sixteenth-century historian and bureaucrat Mustafa ʿAli cites the Mamluk-era polymath Suyuti's (d. 1505) statement that the building of the al-Azhar Mosque in Cairo began at an auspicious time recommended by *munajjims*.[52] Likewise, the early seventeenth-century Ottoman chronicler Mustafa Safi's eyewitness account of Sultan Ahmed I's reign (r. 1603–17) notes that the foundation of the Sultan Ahmed Mosque in the capital was laid on 8 October 1609, at an auspicious moment computed by "the masters in astrology and the learned men of *zij*es and almanacs."[53] Only when we recognize *munajjims* as experts in the computation and interpretation of time, tasked with finding the precise moments that would be in perfect harmony with the heavens and God's temporal order, does it become clear how their expertise was appealing, even for sanctified endeavors such as mosque construction. The remainder of Riyazi's text offers additional clues as to how a strictly astrological text could serve as a forum for expressing lofty Sunni Islamic references and conveying predictions about the future of the planned imperial mosque. In the third chapter, for instance, he draws on Kushyar's teachings to expound upon eight specific planetary positions that he anticipates will align at his suggested moment, all of which, he contends, foretell a glowing destiny for the mosque. The first position concerns the Moon, which Riyazi calculates to be at its exaltation in the tenth house at his chosen moment, in conjunction with Venus. According to Riyazi, this means that the lofty ceiling of this noble edifice will resemble the tenth celestial sphere and its courtyard will look like the ninth degree of paradise. The other seven specific positions Riyazi explicates in the chapter all portend that, once constructed, the mosque will become a distinguished place of worship with the finest architectural qualities. From its elegant façade and sturdy columns to its luminous windows, the mosque will attain the status of the al-Aqsa Mosque and, more important, *al-Bayt al-Maʿmur* (The flourishing house), which, according to most Qurʾanic commentators, is the heavenly counterpart of the Kaʿba.[54] The same portents also imply, in Riyazi's view, that the planned Süleymaniye Mosque will overflow with devout people, much like the Plain of Arafat on the Hajj, and that the saintly men of the unseen realm will perpetually grace its presence, like the Kaʿba. Those who will perform their prayers in the mosque—and in fact, regarding the mosque's anticipated popularity among worshippers, the astrological signs are all positive according to Riyazi—will find themselves prosperous and thriving.[55] Likewise, the fifth and sixth chapters of the tract are devoted to a discussion on the importance of harmony among four separate horoscopes: the horoscopes of the construction, of the nativity of Sultan Süleyman, of his accession to the throne in 1520, and of the year-transfer of the most recent year, which should be 1550. For Riyazi, all of these horoscopes align harmoniously, although he does not tabulate and verbally detail the latter three. As he emphasizes, such harmony is a rare occurrence, further signifying that once completed, the Süleymaniye Mosque—again likened to the Ka'ba and the legendary Iram of the lofty pillars—will be robust, enduring, and immortal.[56]

In addition to the fortunes of the mosque and its divinely ordained significance, Riyazi's tract also offers predictions, or perhaps reflections, concerning the prowess of its architect and craftsmen, as well as the potential completion date of the construction. For example, in the eighth chapter, dedicated to elaborating on the mastery and expertise of the architect and the artistry and skill of the craftsmen, Riyazi quotes Aphorism 59 of the pseudo-Ptolemaic *Centiloquium* verbatim to claim—without explicitly naming Sinan—that the celestial configuration of the designated moment clearly signifies the architect's unique acumen and originality in crafting his architectural plan, which Riyazi thinks will ultimately earn Sinan the sultan's favor.[57]

In the ninth and final chapter, Riyazi attempts to determine the possible completion date of the construction. Through his calculations he arrives at two conclusions: according to the first, the construction would take five years, three months, and twenty-three days, while according to the second, it would take four years and nine months. The discrepancy between the two, he explains, is due to the fact that the first calculation takes as its starting point the moment the foundations are to be dug, whereas the second begins with the actual commencement of construction activity. In any case, as he explicitly states, all of these (astrological) indications and calculations are mere "approximations" (Tur. *takribi*).[58] He concludes by sharing his hope that the construction of the complex will be completed within five years. Given that construction actually lasted more than seven lunar years—much to the chagrin of the sultan—Riyazi's final chapter provides conclusive evidence that the tract was composed well before the mosque's completion and likely prior to the beginning of construction.

Riyazi's mention of approximate estimation in the final chapter of his horoscope tract reflects the prudent and probabilistic approach typically employed by *munajjims* in limiting their visions of the future to the confines of their discipline. As discussed in more detail in Chapter 1, concerning *munajjims*' self-reflections on the layers and limitations of their expertise, they were acutely aware of what their predictions could and could not entail Yet there were other types of experts whose claims about and authority in foretelling the future were based on different epistemic foundations and techniques. Among such figures were geomancers, dream interpreters, practitioners or enthusiasts of letter divination (Ar. *ilm al-jafr* and *ʿilm al-harf/huruf*), and prominent mystics or holy men who asserted prophetic authority, either through one or several of these techniques or through more enigmatic, intuitive clairvoyance. In stark contrast to the rather moderate, generic, and technically grounded forewarnings and predictions of the *munajjims*, the pronouncements about the future made by these latter groups were much bolder, more provocative, and hence more palpable and intelligible to their audience. Indeed, it was in their esoteric and prophetic discourse, rather than the generic and technical predictions of the *munajjims*, that the apocalyptic, millenarian fervor of the sixteenth century found its most elaborate expression. Their textual output was also notable for its copious remarks deriding *munajjims*, which leads us to employ expertise as a framework in which to examine conflicts between different seers of the future.

Seers beyond the Stars: Whose Expertise Reigns Supreme?

The history of Ottoman seers of the future is as old as the history of the Ottomans themselves. Dreams—and those who interpret them—play a pivotal role in the foundational narratives that recount the prophesied ascension and expected universal dominion of ʿOsman's tribe. In the most famous rendition of the well-known dream narrative, ʿOsman, the namesake founder of the Ottoman enterprise, experienced a prophetic dream while he stayed one night in the abode of a holy man named Ede Bali, his future father-in-law:

> Then sleep won him over, so he lay down and rested. He saw that among (his acquaintances) was a highly esteemed shaykh, whose many miracles were well-known . . . His guest house was never vacant and ʿOsman Gazi himself used to come from time to time and be a guest of this holy man. As ʿOsman Gazi slept he saw that a moon arose from the holy man's breast and came to sink in ʿOsman Gazi's breast. A tree then sprouted from his navel, and its shade compassed the world. Beneath this shade there were mountains, and streams flowed forth from the foot of each mountain . . . [When ʿOsman awoke, he went and] told the story to the holy shaykh, who said, "ʿOsman, my son, congratulations, for God has given the imperial office to you and your descendants, and my daughter Malkhun shall be your wife."[59]

There is little doubt that this foundational dream narrative, different variations of which may readily be found in the chronicles of ancient and medieval Near Eastern civilizations, circulated among Ottoman ruling elites and literati from the fourteenth century.[60] Recorded in writing for the first time in the late fifteenth century, it served specific political agendas about the sovereignty and legitimacy of the Ottoman dynasty.[61] Yet in the centuries following this story's alleged events, prophetic dreams and their interpreters remained a constant component of the Ottoman sociopolitical and cultural fabric. Not only do the chronicles and other narrative sources, particularly those written or circulated within Sufi circles, abound with dream narratives, but the Ottoman state archives, too, house myriad individual documents from the fifteenth through the early twentieth century submitted by individuals from all walks of life who wished to share their prophetic dreams regarding the fortunes of the incumbent or impending sultan.[62] An implicit—or sometimes pronounced—expectation of these petitioners was that these narratives would yield social capital and possibly immediate monetary rewards for the dreamer.[63]

One recurrent theme of these narrated dreams is the imminent triumph of the reigning sultan over his domestic or foreign adversaries. The reports frequently communicate their authors' unwavering confidence in the sultan's and empire's bright future, enhanced by all the divine signs in the dream. Yet there were also certain conditions related to the actions the sultan must undertake, as often outlined in the dream. In this regard, many of these dream narratives embed a veiled warning about the course of future actions, subtly woven into a prophetic discourse. For instance, a dream report from the late fifteenth or early sixteenth century, presented to Sultan Bayezid II by a man identifying himself as the deputy of a certain İskender, recounts a visionary experience where the Prophet Muhammad told him that Bayezid was under the Prophet's protection and Bayezid's current state of affairs tended toward mental and physical strength.[64] As the author of the dream narrative emphasizes, the implications are clear: Bayezid should remain steadfast in his adherence to the Sharia and uphold justice. In doing so, his reign would be blessed with prolonged prosperity, and his soldiers would achieve unending victory.

These dream reports sometimes came along with their own interpretation, as exemplified by the dream communicated by İskender's deputy. Other times, they were so overt that they required no interpretation, such as one dream account penned by a woman named Şerife Hatun. In this brief archival document, she candidly articulates her desire for a house as recompense for sharing her dream, wherein she envisioned the sultan immersed in a state of abundant joy.[65] Yet there existed an array of other dreams that were significantly more enigmatic, including those experienced by the sultans themselves. These visions, cloaked in mystery, demanded the expertise of a skilled interpreter. In such instances, to whom did they turn for counsel?

While there was no discrete profession devoted exclusively to the interpretation of dreams, historical records do frequently refer to a distinct group of individuals known as *muʿabbirs* (from Arabic "interpreter"; in Ottoman Turkish, "dream interpreter"). *Muʿabbirs* should not be viewed, however, as holding a professional role akin to that of *munajjims*, who were reading heavenly portents, or physicians, who were interpreting bodily portents, both of which required the acquisition of an extensive amount of textual knowledge, as Riyazi never tired of pointing out. Rather, these oneiromancers were thought to possess innate, intuitive skills for unlocking the enigmatic riddles presented in dreams. This fairly instinctual, rather than technical, authority is key in this context, given that this sort of expertise could not be obtained merely through book learning: there was no established training, school, or delineated method for acquiring and honing this craft. Yet, similar to other divinatory practices, the art of dream interpretation generated its own textual corpus, most notably in the form of dream-interpretation manuals. Evidence is scant, however, as to whether dreamers or interpreters regularly consulted these written manuals at the actual moment of interpretation. Instead, we find dreamers more frequently seeking the guidance of Sufi shaykhs or other revered mystics in a nonbookish fashion.[66] Perhaps the most renowned and well-documented example of such a relationship is that of Şüca^c Efendi (d. 1588) from the Halveti Sufi order, who established and maintained an intimate association with Sultan Murad III.[67] As the preferred interpreter of Murad III's dreams, visions, and vocal descriptions of other mystical and aural experiences, Şüca^c Efendi held a privileged place close to the sultan for over fifteen years from the time of the latter's princedom in Manisa. This relationship between the two has been documented, at least from one perspective, thanks to a unique compilation of Murad III's dream logs and brief entries of visionary and mystical experiences, totaling nearly two thousand separate records.[68] Despite this wealth of information and the privilege of having access to an Ottoman sultan's inner experiences—or to his utterances, to be more precise—we lack documents that shed light on the shaykh's interpretive process.

Sultan Murad III's interest in communicating his dreams and other mystical experiences to a Sufi shaykh was an inextricable part of his fascination with the esoteric domains of knowledge at a time when the first Islamic millennium's end (October 1592) was drawing close.[69] In light of some of his dream records, wherein he fashioned himself as an almost divine figure, befriending the "hidden spirits" (Tur. *ricalü'l-gayb*) and "saints of God" (Tur. *evliya'ullah*), as well as occasionally hearing God directly speaking to him, Murad III harbored grand political and ideological aspirations of casting himself as the expected messianic ruler before the arrival of the apocalypse.[70] Such lofty aspirations did not go unnoticed by contemporary literati, who were no strangers to the prevailing apocalyptic fervor. Mustafa ʿAli, for instance, composed at least three texts that capitalized on and further stimulated Murad's preoccupation with his cosmic significance before the impending end. In one of these short tracts written in 1584, Mustafa ʿAli toyed with the numerological significance of the number twelve and drew parallels between the twelve months of the year, twelve signs of the zodiac, twelve hours in a day, twelve imams—including the awaited twelfth, the Mahdi—and twelve good characteristics of Murad himself, who was fortuitously the twelfth Ottoman sultan. Thus, Mustafa ʿAli surmised that Murad was predestined to live for ten cycles of twelve, reaching the ripe age of 120.[71]

Sultan Murad III's generous patronage of the observatory, built in Istanbul in the mid-1570s with the initiative of the incumbent chief *munajjim* Taqi al-Din, has been linked to the burgeoning messianic and apocalyptic sentiments during his reign.[72] While the dynamics and actors underpinning this ambitious project are only partially understood, it is almost certain that the venture was not birthed unprompted by Murad but midwifed through the initiative of Taqi al-Din, who eventually sold the sultan on his idea of carrying out new, systematic research on celestial spheres.[73] As has already been discussed in previous chapters, the post-thirteenth-century Islamicate East teemed with incessant complaints and demands from astral experts about the need to refine the astronomical data contained in the *zij*es. Some of these experts were fortunate enough to harness royal patronage and conduct systematic celestial observations in well-equipped observatories. As an aspiring expert in the astral sciences facing the perennial problem of discrepancies and inconsistencies in the computation of celestial movements and events, Taqi al-Din no doubt had the same desire. Unlike some of his professional predecessors in the Ottoman realm, such as Riyazi, who had formally made the same request to Sultan Süleyman two decades earlier, Taqi al-Din was lucky, at least at the beginning, to have found favor with Murad III. Whether or not the sultan's preoccupation with the imminent apocalypse factored into his support remains a question unanswerable by the current state of the evidence, yet this supposition seems highly plausible, given similar instances and manifestations of Murad's genuine concern with his cosmic role in the End Times.[74] The ultimate decision to fund the construction of an observatory may also have been affected by Murad III's desire to prove himself a generous benefactor of learning and learned people, as well as to promote his rule as an ambitious, absolutist monarchy capable of launching high-flown projects.[75] Moreover, intellectual endorsements of Taqi al-Din and his project on the part of Grand Vizier Sokollu Mehmed Pasha (d. 1579) and the royal tutor Hoca Saᶜdeddin (d. 1599) likely further convinced the sultan to allocate generous funds.[76] It should also be added that Hoca Saᶜdeddin even took a personal scientific interest in the project. As Taqi al-Din notes in one of his unfinished *zij*es, certain lunar observations he completed in October 1576 were undertaken at Hoca Saᶜdeddin's residence, conveniently located near the observatory in the Galata district of Istanbul.[77]

The compilation of Murad III's dreams, visions, and mystical experiences reveals a sovereign profoundly committed to a messianic-apocalyptic persona, with a keen interest in the science of letter divination (*ʿilm al-jafr*), particularly the works of a certain shaykh from the Bayrami branch of the Halveti Sufi order, İlyas b. ʿİsa Saruhani (d. 1559/60), whose profile will be introduced below.[78] A similar millenarian fascination is not readily discernible in the extant oeuvre of Taqi al-Din, however. Avner Ben-Zaken, in a somewhat contrived fashion, has attempted to make a case for Taqi al-Din's apocalypticism in view of a few references to the talismanic qualities of mechanical clocks in the preface of his treatise on these objects.[79] However, not only do these passing references bear little relation to apocalyptic considerations, but Taqi al-Din's other writings also reveal little about his engagement with this discourse.[80] This lack is hardly surprising, for in the extensive survey of the textual corpus composed, copied, and consumed by fifteenth- and sixteenth-century *munajjims*, not a single *munajjim* has been identified as having been deeply engaged with messianic and prophetic considerations.

In any case, Taqi al-Din's proximity to the sultan and his inner circle did not last long; even if he had played a role in stimulating the sultan's already rooted esoteric interests, he would scarcely have had time to cultivate them. Only a few years after he initiated his systematic celestial observation program, alongside a group of colleagues and assistants—a scene most eloquently captured in the famous illustration of the Istanbul observatory—Taqi al-Din's observatory was demolished, sanctioned by a fatwa from the then Shaykh al-Islam, Kadızade Şemseddin.[81] High-ranking members of the ʿulamaʾ within or close to Kadızade's circle were so angry with Taqi al-Din that they wanted him to be executed on a charge of heresy and unbelief.[82] The grand vizier, Koca Sinan Pasha (d. 1596), who seems to have been aware of the plans of these high-ranking ʿulamaʾ, recommended to the sultan in a private report that Taqi al-Din should vacate the capital and be sent to Egypt.[83]

The exact reasons for the hostility to Taqi al-Din among the leading ʿulamaʾ are not known with certainty; however, it is improbable that astrological pursuits per se were the principal or sole reason for their grievances. Before and after the establishment and subsequent dismantlement of the observatory, court *munajjims* regularly rendered their astrological service to both patrons and the public without facing significant opposition. Moreover, only two years after the observatory was demolished,

Murad III commissioned a madrasa scholar, Mehmed Suʿudi Efendi (d. 1591), to translate an astrological and divinatory manual, which was published as *Metaliʿu's-saʿade* (Book of felicity) and dedicated in two copies to the sultan's two daughters.[84] Had the ʿulamaʾ's aversion to Taqi al-Din been grounded in his practice of astrology, we would expect to find similar instances of aggressive hostility both earlier and later. This, of course, is not to suggest that the ʿulamaʾ openly embraced astrological pursuits. In fact, instances where traditional ʿulamaʾ extended sympathy and endorsement to astrological endeavors were more anomalous than routine. Still, Taqi al-Din's case is distinctive in that it concerned the construction of a novel edifice that housed a group of at least fourteen practitioners equipped with skills and colossal tools that were, at best, regarded with suspicion and, at worst, perceived as detrimental to public piety in the same way that Riyazi's proposal to erect human-size talismans was perceived twenty years earlier. There may also have been something specifically irritating in Taqi al-Din's character and demeanor. Koca Sinan Pasha's same report, for instance, hints at Taqi al-Din's arrogant and obstinate nature, which may have further incited the ire of the ʿulamaʾ.[85]

This most emblematic episode in the history of Ottoman science is often narrated with reference only to the zealotry and narrow-mindedness of the Ottoman ʿulamaʾ. Yet it appears that Taqi al-Din's contemporary adversaries extended well beyond the high-ranking ʿulamaʾ and included some preeminent Sufi mystics endowed with esoteric and spiritual power. For example, in a letter dispatched to the sultan after Taqi al-Din left the capital, İbrahim-i Kırımi (d. 1593), a Halveti shaykh of a branch different from that of Şücaʿ Efendi, expressed his gratitude to the sultan for expelling Taqi al-Din, whom he decried as a "devil among men and an abominable monk."[86] Before the final decision was made to level the observatory, Sultan Murad had also sought the counsel of his own shaykh and dream interpreter, Şücaʿ Efendi, inquiring in a brief memo, "Oh, my dear soul! We want to discontinue and dismantle the observation. What would be your order, my soul?"[87] Regrettably, Şücaʿ Efendi's exact response remains undisclosed. Considering the ultimate fate of the observatory, however, together with the remarks by his fellow Halveti shaykh, we can conjecture that his sentiments about Taqi al-Din were far from cordial.

The outright hostility to Taqi al-Din demonstrated by those Halveti shaykhs in close proximity to Sultan Murad III, for reasons that remain somewhat nebulous in the relevant sources, is just one example among several instances where mystics and Sufi shaykhs are documented as berating *munajjims*. In fact, within the Ottoman cultural milieu, and to a certain extent in the broader Islamicate world, some of the most vehement criticisms directed at *munajjims* originated from the shaykhs of various Sufi orders and other individuals deeply steeped in mysticism. This may sound somewhat startling, given certain established convictions in modern scholarship about Sufi thought and practice as the quintessence of Islamic esotericism. How could those mystics—who are typically thought of as the adepts par excellence of occult or esoteric knowledge—target *munajjims*, who are often counted as fellow practitioners of the occult sciences? Was it not the case that these two groups pursued analogous interests guided by shared principles?

These questions are actually rhetorical, as their answers have already been woven into the fabric of this book's central themes: expertise and authority. That is to say, there is little surprise in the marked Sufi disdain for *munajjims*, as the expertise and authority *munajjims* purported to hold in regard to interpreting and predicting the future appeared as a direct challenge to many mystics vis-à-vis their own claims to authority and expertise. Consider the following example drawn from a hagiography written by an Ottoman dervish in the late sixteenth century about the founder of his Gülşeni order, İbrahim Gülşeni (d. 1534). The work recounts a time when İbrahim Gülşeni was still serving the Aqquyunlu court of Sultan Yaʿqub (r. 1478–90), prior to his eventual relocation to Ottoman domains.[88] As the story goes, Sultan Yaʿqub's cousin, Bayandur ibn Rustam, had decided to attack Yaʿqub's forces. Those *munajjims* in the service of the sultan, who were receiving hundreds of thousands of *akçes*, as the account specifically highlights, interpreted the astrological signs of the moment as indicating Yaʿqub's impending defeat. Upon seeing Gülşeni still accompanying the sultan, these *munajjims* could not help but ask: "Why did you not leave, given the evident forecast of his defeat based on our calculations? As his *munajjims*, we are his servants; hence, we had no other option but to stay!"[89] In response, Gülşeni informed them that what he discerned in the "divine astrolabe" (*usturlab-ı ilahi*) was not the sultan's downfall but his triumph and the death of Bayandur. According to Gülşeni, this would also corroborate the saying of the Prophet: "All *munajjims* are liars."[90] Upon seeing the *munajjims* steadfastly backing their predictions, Gülşeni posed a challenge to them: "If your predictions prove false, would you agree to a cut in your salaries?" The *munajjims*, taken aback, implored him to refrain from such a proposition. Witnessing the *munajjims*' recoiling, Gülşeni remarked, "Were your practice grounded on certainty, you would not have harbored any doubts about your predictions and interpretations."[91] As events unfolded, İbrahim Gülşeni learned that Sultan Yaʿqub defeated Bayandur and that his prediction—rooted in the "divine astrolabe" of his spiritual intuitions rather than the technical expertise of *munajjims*—was indeed the accurate one.

We cannot take at face value such remarks and anecdotes embedded in hagiographies and narrative sources—indeed, in any textual artifact of the past. Needless to say, this seemingly insignificant anecdote serves the purpose of illustrating Gülşeni's spiritual prowess as a Sufi master, yet it also provides colorful insights into how competing claims of epistemic authority were laid out by men of competing fields of expertise in the early modern courtly context. The story of Gülşeni and the *munajjims* is not the sole instance in which the authority of Sufi shaykhs and dervishes to speak about the future through more esoteric, intuitive means—"the divine astrolabe," so to speak—came into conflict with the technical expertise of *munajjims* who made use of manufactured, external means and instruments (e.g., the standard astrolabe) to interpret the future course of events. For example, İlyas b. ʿİsa Saruhani, the Sufi shaykh of the Bayrami branch of the Halveti order, whose extensive writings on prophetic letter divination enjoyed widespread readership among Ottoman subjects, including Sultan Murad III, derided *munajjims* in his popular *jafr* treatise *Rumuzu'l-künuz fi'l-cifr* (The secret signs of the treasury in letter divination).[92] This enigmatic work, copied and widely circulated in notably divergent renditions, features Saruhani's prophecies concerning the destiny of the Ottoman Empire and its ruling elites, including sultans, viziers, royal tutors, judges, jurists, provincial governors, Sufi shaykhs, and others through the year 2035 AH—or until the year 3000 in certain copies. These prophecies were based on the lettrist principles of *ʿilm al-jafr*, which assume a direct link and intricate system of correspondences between letters, numbers, the four elements, and a variety of respective phenomena. Saruhani, however, often leaves these connections unexplained and reports his prophecies in a straightforward manner, without necessarily providing their underlying lettrist explications. For instance, when he speaks in the tenth chapter about the possible celestial and terrestrial oddities that will take place in the year 3000 AH, he enumerates a number of prophecies, from massive earthquakes in Istanbul and the Black Sea region to famine and drought in Egypt, but he presents no justification from a lettrist perspective. Still, in Saruhani's view, the science of *jafr* holds supreme authority and thus is more reliable than the arts of the *munajjims* and geomancers. What is incumbent upon the sultans and princes is to devote their attention solely to Sufi shaykhs and "holy men" and completely disregard the utterances of *munajjims* and geomancers.[93] As he puts it, these individuals, in contrast to true men of wisdom, lack the divine qualities of *karama* (charisma, or miraculous power) and *walaya* (spiritual authority): all the things they proclaim are nothing more than insubstantial illusion and nonsense.[94]

Although his *Rumuzu'l-künuz* contains only a few references to celestial phenomena, the overall lettrist outlook of Saruhani, evident in some other works authored by or attributed to him, such as the *Kava'id-i Teshirat* (Rules of enchantments) and *Mücerrebat* (Things repeatedly experienced), postulates an inextricable connection between the celestial and terrestrial spheres.[95] Everything from planets, days, the four elements, names of individuals, and various other phenomena are interconnected in a complex weave, enabling those adept in the lettrist principles to discern aspects of nature, predict and manipulate the future, and comprehend the supernatural. This lettrist cosmology can be traced back to the teachings of Ibn al-ʿArabi and ʿAbd al-Rahman al-Bistami, which laid out an interpretive system of correspondences among the seven planets, twelve zodiac signs, twenty-eight lunar mansions, and numerous heavenly bodies, all believed to be designed by God and administered through the mediation of spirits, angels, and hidden saints. Each zodiac sign is thought to be connected to a particular angel, and each of the seven planets is associated with a prophet, exerting an influence over seven millennia and seven weekdays. Within this elaborate system of numerical and lettrist correspondences—which some scholars have defined as "mystical astrology" or "esotericist astrology"[96]—there is no real need for, nor any deep familiarity with, the technical astronomical and mathematical knowledge possessed by *munajjims*.

While it is hardly surprising to find a shaykh of a Sufi order well-versed in lettrist cosmology attacking *munajjims*, the expansion of Saruhani's critique to encompass geomancers calls for further examination. Unlike the expertise of *munajjims*, which relied on elaborate astronomical and mathematical computations of celestial positions as a prerequisite to their astrological interpretations, geomancy shared a much closer relationship with—indeed, was profoundly influenced by—lettrist cosmology. The divinatory art of geomancy (*ʿilm al-raml*; lit. "the science of sand" or "divination by earth") was and remains a divinatory practice conducted by constructing and subsequently interpreting a design known as a geomantic tableau.[97] Comprising sixteen positions, each occupied by four lines of a geomantic figure, this tableau is the central component of the geomancers' art, analogous to an astrologer's horoscope. In its original form, these figures were created by sketching lines of dots in the sand. However, by the medieval era, this practice evolved to include drawing the dots on a piece of paper. Each geomantic figure has a distinct meaning for the fortune seeker, applicable to every imaginable facet of human experience, from love and death to health and professional careers.[98] Drawing upon associations with the four elements, seven planets, twelve zodiac signs, and their corresponding characteristics, the art of geomancy shares striking similarities with the "mystical astrology" of the lettrists. Unlike *munajjims*' astrology, however, geomancy requires no complex astronomical calculations or the utilization of astronomical tools, including *zij*es. The contrast between these two methods of divination underscores the diversity of the avenues of expertise and authority at the time, lending further complexity to our understanding of these historical subjects and dynamics.

Geomancers vs. *Munajjims*

Geomancy and geomancers are indeed crucial for understanding competing claims of authority and expertise during the reign of Süleyman. Around the time when Saruhani started to reach a popular audience through his tracts on lettrist divination and Riyazi was striving to secure royal favor by showcasing his unmatched proficiency in the celestial sciences, an adept of the art of geomancy and lettrism named Haydar emerged as a close confidante and advisor to Sultan Süleyman.[99] Having migrated from Safavid Iran, likely in the early 1530s, Haydar swiftly secured a position within the inner sanctum of the palace owing to his peculiar expertise. Details of his biography remain unknown, including his background prior to his arrival in the Ottoman domains as well as the exact processes and intermediaries that helped him find a seat at the sultan's table. From the late 1530s through the 1560s, however, Haydar unquestionably operated as a clandestine advisor to the sultan, orchestrating grandiose ideological assertions of Süleyman's messianic rule while also providing concrete policy recommendations, all founded upon his expertise in the art of geomancy and letter divination. Haydar's extant writings, including a fifty-five-folio treatise and a handful of scattered archival documents in the form of confidential reports, bear witness to Süleyman's firm belief in his role, at least during the first three decades of his rule, as the "Last World Emperor," the true universal conqueror and "Master of the Conjunction" (*sahib-qiran*), destined to reign supreme at the prophesied end of history.[100] These textual artifacts also provide insight into how an expert in a specific divinatory art participated in imperial policy-making processes and how he distinguished his area of competence from others.

Haydar's comprehensive report, likely submitted to the palace in the early 1540s, is a prime example of how expertise in geomancy and lettrism differed from that of the *munajjims*—in terms of both their epistemic pillars and the scope of policy recommendations. At first glance, the report appears similar to the annual prognostications in *munajjims*' almanacs, where the fortunes of the sultan and various societal groups were predicted for the new (solar) year, traditionally commencing with the Sun's entry into the sign of Aries in the spring equinox. In fact, even on the report's opening page, Haydar explicitly mentions the arrival of the blessed moment when the Sun transitioned into Aries, a time he has seized upon as an opportunity for presenting his in-depth predictions about Sultan Süleyman, all rooted in geomancy and lettrism.[101] As he constantly reiterates in the text, the beginning of the report highlights Süleyman's unique status as "the cosmos's Master of the Conjunction" [of Jupiter and Saturn] and "master of the age" and "the pole of poles" of the present apocalyptic era, which, he declares, is the end of the Age of Saturn and the beginning of that of the Moon.[102] Süleyman's divinely ordained stature in this doomsday is ensured, says Haydar, by the signs found in his horoscope that no other world conqueror, including Faridun and Jamshid, the legendary kings of ancient Iran, has possessed.[103] Such references to astrologically relevant notions and phenomena notwithstanding, the text does not qualify as a piece of astrological work, nor does the expertise Haydar asserts show any serious technical familiarity with *munajjims*' domain of knowledge.

A salient example demonstrating Haydar's unfamiliarity with the astrological expertise typically seen in the works of *munajjims* is his treatment of Süleyman's horoscope. Throughout the book, Haydar recurrently highlights Süleyman's unique ascendant, which he thinks set the sultan apart from all preceding and contemporary rulers. However, these references are noticeably bereft of horoscope charts or detailed technical explications concerning the relevant astrological houses and celestial degrees. In fact, the text wholly lacks any elaborate or even superficial engagement with the jargon and techniques intrinsic to the astral sciences. As Haydar constantly reminds his primary audience—namely, Sultan Süleyman, whom he personally addresses in a notably direct and intimate manner throughout the text—all his judgments and predictions are based on "the proofs of the geomancy of the Prophet Daniel" and "the science of lettrism," a discipline that he says draws its authority from the esoteric knowledge of ʿAli b. Abi Talib.[104] Haydar briefly notes that Süleyman provided him with several books on the science of lettrism for his consultation, including one by Ibn al-ʿArabi.[105] In no part of this lengthy report does Haydar cite any work or authoritative figure specifically associated with the domain of the *munajjims*.

Haydar's extensive prognostications in the report, anchored in his expertise in geomancy and the science of lettrism, also diverge notably from the astrological predictions of *munajjims*. Unlike *munajjims*' written forecasts, which typically refrained from directly mentioning a particular individual or event and employed a discourse of predictive uncertainty rather than prophetic certainty, Haydar's assertions are often both forthright and specific. For instance, he delves deeply into military and political affairs on the eastern frontiers, specifically within the Safavid territories, where he expects—and names explicitly—the impending death of Shah Tahmasb no later than 1557–58 and the defeat of the Safavids at the hands of the Shaybanid Uzbeks. In Haydar's view, the Uzbeks, despite their Sunni orientation, would pose an even graver threat to the Ottomans than the Shiʿi Safavids, owing to their relentless Tatar character and aggressive nature.[106] But his strikingly specific predictions were not just about the Ottomans' Muslim adversaries on their eastern frontiers: he also touches upon the Ottomans' relationships with Christian powers, including the French, Habsburgs, and Portuguese. He even makes specific policy recommendations, one of which concerns capturing Bahrain: in his view, taking Bahrain would enable the Ottomans to control Hormuz and take more aggressive action against the Portuguese in the Persian Gulf, as well as delivering to them Bahrain's abundant mines and wealth.[107] Aside from these predictions and specific policy recommendations on international affairs, Haydar offers extensive commentary on Süleyman's health and general well-being: for instance, the sultan need not be overly concerned about his foot pain, which is simply related to his known problem of gouty arthritis (*nikris*).[108] He also foresees that Süleyman will soon finish two poetry collections before starting a third. All the hidden signs Haydar uncovered, thanks to his expertise in geomancy and the science of letters, indicate that Süleyman was destined to live to the age of 90—though he hopes it might extend to 100—and that all his sons, eagerly waiting to ascend to the throne, would be left disappointed.[109]

The influence of Haydar and his particular expertise on Süleyman likely persisted until the sultan's final days. Evidence of this enduring influence is not limited to the handsome land grants Haydar received until the late 1560s, as recorded in palace payment registers.[110] Several archival documents in the form of geomantic reports, preserved in the palace, also attest to this lasting relationship. One such document, probably drafted in 1560 amid the ongoing contest between Prince Bayezid and Prince Selim—Süleyman's two remaining sons vying with each other to assume the throne when Süleyman (then in his sixties) died—shows Haydar attempting to soothe the sultan by means of geomantic prophecy. As Haydar mentions in the report, Prince Bayezid had taken refuge at the court of Shah Tahmasb, which was evidently a source of great distress for Sultan Süleyman. It seems clear, based on Haydar's comments in this report, that he was busy at the time submitting such reports to the sultan on a regular basis; his most recent report, he notes, was delivered "last Wednesday" to the *agha* of the Gate of Felicity.[111] Since Haydar placed great emphasis on secrecy, it is no surprise that most of his reports have not survived. At the beginning of his lengthy report, for instance, he explicitly requests, *à la Mission Impossible*, that once the sultan is informed of the report's predictions, the document should be "put out of sight and destroyed, lest anyone become aware of its contents."[112]

Aside from ensuring the confidentiality of his recommendations and predictions, Haydar also seems concerned in his reports to stress the utmost epistemic authority of his particular expertise in comparison to others with concomitant authority over foretelling the future, including *munajjims*. The aforementioned report he submitted on the case of Prince Bayezid, for example, is replete with remarks emphasizing the supremacy of geomancy and the science of letters, a fact that had already been affirmed, he argues, by the Prophet Muhammad. To bolster his stance, Haydar cites several hadiths ascribed to the Prophet. One such hadith, according to Haydar, claims that the science of geomancy is a miracle bestowed upon prophets, specifically seven of them, starting with Idris and culminating with Daniel.[113] In another instance, Haydar cites a purported dialogue between the Prophet Muhammad and Ibn ʿAbbas (d. 687/8), one of the cousins and early companions of the Prophet. According to the tale, Ibn ʿAbbas one day came to Prophet Muhammad and said, "O messenger of God! Today, I saw a man drawing lines and [fore]telling the unseen (*al-mughayyabat*). People gathered around him; some of them called him as a liar, and others believed him. What would you say about them?" In his response, the Prophet said: "The science of lines [i.e., geomancy] is prophetic knowledge granted to prophets . . . and the learned concur that the science of *raml* is a miraculous gift bestowed upon seven prophets."[114] Not revealing the sources of these hadiths, Haydar proceeds to quote a Qur'anic verse from the Surah al-Ahqaf (Sand dunes; 46:4): "Bring me a scripture [revealed] before this or a [remaining] trace of knowledge, if you are truthful."[115] He does not explain, however, aside from the title of the chapter, how the verse relates to his discussion on the prophetic authority of geomancy and the science of lettrism.

The document that perhaps most clearly manifests Haydar's assertion of the superiority of his skills and expertise over those of *munajjims* is a letter now preserved in the Topkapı Palace archives. Although the letter is undated and unsigned, it was one of the latest such documents Haydar presented to Sultan Süleyman. Striking similarities between the wording in this letter and phrases from Haydar's other known writings—such as his self-reference as being a "loyal devotee" (*ʿaşık-ı sadık*) and his portrayal of Süleyman as the ultimate embodiment (lit. "seal" [*hatem*]) of the "sahib-qiran" and the chosen one endorsed by saints and invisible intermediaries—strongly attest to Haydar's authorship.[116] The letter seems to have been intended not to offer predictions and judgments on a specific matter but rather to remind the sultan of Haydar's superior skills in the sacred art of geomancy and lettrism and how dedicated he remains as he continues to put this expertise in Süleyman's service. Even though in the letter he seems disconcerted by his adversaries' animosities toward him, he displays confidence in the sultan's sustained support, a sentiment echoed in his other works, where he frequently conveys his gratitude to Süleyman.[117] Still, Haydar cannot help but accentuate the unparalleled nature of his expertise in his craft—and, in fact, in the entire gamut of what he calls "uncommon science[s]" (*ʿilm-i garaʾib*). He highlights how his proficiency and work had garnered wide acclaim, spanning the globe from India and China to the countries of the north and west, making him a uniquely proficient practitioner. For our purposes, the most intriguing part of the letter is where he underscores the prophetic essence of the art of geomancy, which, he says, was acknowledged by the Prophet Muhammad and reiterated in severable reliable books on Qur'anic exegesis, such as the *Kashshaf* (The revealer) by al-Zamakhshari (d. 1144). In order to denigrate *munajjims*' expertise as opposed to his own craft, Haydar even notes that the Prophet Muhammad did not approve of the astral sciences but held "[their] science" in higher esteem.[118]

Conclusion

Predicting the future was not the monopoly of any specific profession or group in the early modern Ottoman realm. A remarkable array of individuals, all claiming their own unique epistemic authority and expertise, participated in the culture and business of prediction using distinct sets of tools and discourses. Lumping them together under the catch-all terms "occult scientists" or "practitioners of esoteric knowledge," no matter how convenient and useful these may be, risks obscuring some essential differences between various experts, pertaining to the foundations of their epistemic authority, the scope of their expertise, and the precise purposes of their practice. By looking beyond these umbrella terms and focusing on separate bodies of experts, we can perceive the actual nodes of competition between those individuals with diverse claims to expertise. With expertise as a lens, we can also track the ups and downs in the appeal of a specific body of practitioners in a particular form of expertise.

Munajjims were only one of several types of experts whose insights into the future and unknown were relied upon by the patrons and public in the Ottoman world. Unlike comparable actors who lacked regular institutional support in the form of a courtly bureaucratic unit, *munajjims* remained steadfast fixtures in the social, intellectual, and political fabric of the early modern Ottoman realm. Owing largely to the limitations inherent in the epistemological foundations of their discipline—which were acknowledged by many *munajjims*, as detailed previously—their practice was focused mainly on the presentation of brief reports on auspicious hours and the submission of annual almanacs with imprecise and general astrological prognostications. They likely offered further counsel and expressed more specific predictions by naming names verbally on the spot, which has left us, as historians, with the basic conundrum of how to apprehend what is undocumented. Moreover, information concerning their engagement in more prophetic methods of foretelling the future is difficult to find in any of the extant textual and anecdotal evidence of court *munajjims*' practice. That type of expertise was predominantly the domain of groups other than *munajjims*, whose textual output sheds light on the competitive sociopolitical landscape of predicting the future and offers further insight into the contemporary negative perceptions of *munajjims*.

Incorporating such alternative forecasters of the future and their materials into our portrayal of *munajjims* in the long sixteenth century thus enables us to track more accurately the changing tides of attraction toward different bodies of expertise and currents of discourse. Apocalyptic concerns and millenarian considerations undoubtedly appealed to certain segments of the Ottoman ruling elites and literati in the Islamic calendar's tenth—and expected final—century, most notably during the reign of Sultan Süleyman. However, *munajjims* were not the principal promoters of these beliefs; in fact, the escalating preoccupation in Süleyman's court with crafting and propagating the image of the Ottoman sovereign as the divinely chosen world conqueror and the last universal ruler, who would usher in eternal justice before the impending end of the universe, prompted various individuals to put themselves in the forefront, proclaiming that they knew and could decipher the hidden signs and meanings through their mastery of lettrism. Even though the references in their writings occasionally contain celestial symbols and astrologically relevant materials, these specialists and others who based their authority on the rather esoteric and intuitive science of letters could not, and indeed should not, be categorized as *munajjims*.

CONCLUSION

ISTANBUL OBSERVATORY *and the* AFTERMATH

The famous image of the Istanbul observatory depicting Taqi al-Din and his associates at work with various astronomical instruments graces the covers of numerous books in multiple languages on the intellectual history and history of science in the Ottoman and Islamicate worlds.[1] The overuse of this depiction tells a tangled tale. On the one hand, it shows the efforts to counter the still-widespread narrative of the Golden Age of Islam followed by an irrevocable decline. The portrayal of sixteenth-century Ottoman astral experts hard at work suggests that scientific enthusiasm persisted and even grew in the postclassical era and that the Ottoman and Islamicate world continued to produce scientists of a high caliber, comparable to Copernicus, Tycho Brahe, or Galileo. While the importance of studying intellectual and scientific pursuits in postclassical Islam cannot be overemphasized, the assertion that the sciences continued to flourish in the Ottoman world in the sixteenth century and beyond has occasionally led to unfounded claims. These claims are particularly pronounced among a new generation of Western narrators of the global history of science, who show excessive eagerness to trace and underscore non-Western contributions to the so-called Scientific Revolution—perhaps driven by a motivation to redeem many decades of a pronounced Eurocentric bias in the historiography of science. For example, a recent and otherwise useful narrative of the early modern global history of science makes grandiose claims for sixteenth-century Ottoman scientific production—"At the same time that Nicolaus Copernicus was causing a stir in Europe, astronomers and mathematicians in the Ottoman Empire were entering their own Renaissance"—without providing any supporting evidence.[2]

On the other hand, ironically, the frequent use of the image of the Istanbul observatory works against its triumphalist intentions and evokes instead a sense of decline, a perception that is equally widespread among contemporary historians. If the sciences and scientists continued to flourish in the early modern Ottoman and Islamicate world from the sixteenth century, one might then ask why alternative images depicting scenes of scientific practice are not more common. Is this due to a dearth of sources and images, or does it reflect a reluctance on the part of modern historians to seek out alternatives—or perhaps laziness? If the reason is the former, what does this scarcity of visual evidence tell us about the standing of the sciences and scientists in early modern Ottoman culture and society? Besides these hypothetical questions, the famous image of the Istanbul observatory, along with the surrounding story that contains the interference of religious scholars leading to its destruction, has been deployed by several historians antithetically, using it to mark a decisive moment of downturn. In these accounts, the image epitomizes the final, albeit short-lived, triumphal period of scientific production in Ottoman and Islamicate history before its ultimate decline in the face of the emerging new sciences and technologies in western Europe. Had the observatory not been destroyed, according to this view, the early modern Ottoman Empire might have kept pace with its western European counterparts.[3]

Let us consider for a moment a counterfactual early modern Ottoman world in which Taqi al-Din encounters no opposition from the members of the ʿulamaʾ and his observatory continues to operate under sustained royal patronage. Could we then confidently assert that this hypothetical early modern Ottoman social order would have exhibited as much scientific and technological vigor as contemporary western Europe? How realistic would this be, considering the notable dearth of certain means and media—unprecedented wealth accumulation from colonial exploitation, the growing institutionalization of scientific vocations through learned societies and academies, and movable-type printing and its transformative effects—all of which contributed in different measures to the changing definitions

FIGURE 14. Taqi al-Din's observatory in Istanbul. From Seyyid Lokman, *Shahinshahnama*. Istanbul University Rare Books Library F. 1404, 57a. Reproduced by permission from the Istanbul University Library and Documentation Department.

of science and its practice in western Europe? The point here is not to perpetuate the narrative of decline or to justify characterizations of the early modern Ottomans (read: Muslims) as culturally backward and inferior; we need to move beyond such categorizations. Nonetheless, this caveat should not deter us from frankly acknowledging that from the sixteenth century, if not earlier, differing material and capital conditions, along with evolving intellectual priorities and developments, led to divergent, nonlinear paths of progress across Eurasia.

Addressing what was lacking and why in a specific historical context, rather than focusing on what was present and how, poses a risk for historians. After all, it is challenging to ascertain whether those phenomena or entities that are presumed absent were actually nonexistent or simply have not been preserved or discovered. Consider, for instance, Harun Küçük's claim in his *Science without Leisure* that between 1660 and 1732, Ottoman Muslims in Istanbul "produced virtually no theoretical astronomy, no natural philosophy, and no anatomy."[4] Küçük bases this assertion almost solely on catalog information in modern inventories of Ottoman scientific production, not by browsing manuscripts themselves, which thus makes it an open target for critique.[5] Nevertheless, historians should—and do—have the liberty and responsibility to question why the elements they perceive as lacking in a particular historical context are absent. When such assertions are taken more as thought-provoking intellectual exercises than definitive conclusions drawn from exhaustive archival research, new avenues of research can then open up that encourage paying closer attention to factors that may have previously been overlooked by scholars, including material, financial, societal, or institutional conditions, rather than vague and essentialist cultural explanations.

Is it possible to affirm, in light of material, financial, and societal circumstances, that the early modern Ottoman world could have matched the scientific and technological advances of contemporary western Europe if Taqi al-Din had not been ousted and the Istanbul observatory had remained operational? Instead of speculating on how hypothetical scenarios might have unfolded, let us revisit Taqi al-Din's story and explore the specific social, cultural, and political context within which his aspiration to establish an observatory and conduct new research on celestial motions was articulated. Doing so will also enable a more accurate comparison of the case of Taqi al-Din (and the Ottomans) with other medieval and early modern observatories and scientific endeavors.

Born in Damascus in 1526 into an established scholarly family, Taqi al-Din began his studies in the madrasas of greater Syria and Egypt, mastering Arabic grammar, jurisprudence, hadith, and Qur'anic exegesis.[6] His curiosity about the mathematical sciences and mechanical arts seems to have been piqued early on, as attested by autobiographical vignettes found in some of his later treatises, but the full extent of this early interest remains obscure.[7] After completing his education in the 1540s, he began teaching in the madrasas of Egypt, which were gradually coming under tighter control by the centralized Ottoman scholarly establishment. Likely driven by a desire to establish closer ties with the high-ranking scholar-bureaucrats in the capital, who had executive authority over teaching and judiciary appointments, he traveled to Istanbul in the early 1550s with his father. Taqi al-Din attended lectures and social gatherings (*majlis*) of several key scholar-bureaucrats, including Shaykh al-Islam Ebu's-suʿud Efendi. As Taqi al-Din notes in an extant archival petition, he was serving, during the Ottoman campaign against the Safavids between 1552 and 1554, as a private tutor to the high-ranking administrator ʿAli Pasha, expecting to secure an appropriate madrasa appointment. Initially, he received no offers, but he was subsequently granted the Shayhuniyya madrasa in Cairo with a daily salary of 25 *akçes*. After spending some time in Egypt, first teaching at the Shayhuniyya and then at the Surgatmishiyya (which offered 30 *akçes* a day), Taqi al-Din returned to Istanbul and reconnected with ʿAli Pasha, who by then had become the grand vizier (he served in this capacity until 1561). With ʿAli Pasha's support, Taqi al-Din received a letter of appointment for a higher-ranking madrasa in Edirne with a salary of 40 *akçes* per day, but he declined it owing to financial and emotional concerns for his family back in Egypt. Instead, he accepted a position in the judiciary in Cairo, receiving 150 *akçes* a day as the deputy to the chief judge of Egypt.[8]

Taqi al-Din's stay in Egypt in the early 1560s facilitated his transformation into an expert "stargazer" (Tur. *rasıd*—the scholarly sobriquet he preferred), with the ambitious goal of conducting celestial observations. Although he already had an interest and some background in the study of the mathematical sciences, it was likely in the 1560s that Taqi al-Din fully immersed himself in the Persianate and Timurid traditions of the astral and mathematical sciences, thanks to his serendipitous encounter with ʿAbdülkerim b. Kutbuddin, the newly appointed judge of Egypt and a direct heir to Mirim Çelebi and ʿAli Qushji. ʿAbdülkerim was a learned man who had been encouraged by his father to study the mathematical and astral sciences and to prepare instruments for celestial observations. He had also inherited from his father a large corpus of treatises on these disciplines by ʿAli Qushji, Jamshid al-Kashi, and Qadizada al-Rumi. While serving as deputy to ʿAbdülkerim, Taqi al-Din seized the opportunity to further cultivate his enthusiasm for the astral sciences.[9] We know that he began conducting observations in Egypt, likely in the 1560s, which he believed could be perfected in a full-fledged observatory with larger instruments providing greater precision.[10] His next trip to Istanbul in the late 1560s or early 1570s was intended to convince the ultimate decision-makers to build such an observatory. According to the eyewitness account of Seyyid Lokman's *Shahinshahnama* (The book of the King of Kings), completed in 1581, Taqi al-Din first paid a visit to Grand Vizier Sokollu Mehmed Pasha[11] and then had a meeting with Hoca Saʿdeddin, an eminent scholar and Murad III's private tutor. At first, Saʿdeddin was skeptical toward Taqi al-Din's request, but after two or three additional interviews and careful consideration, Hoca Saʿdeddin was persuaded to present the proposal to the newly enthroned sultan—not merely by submitting a written petition but "by obtaining an audience for his own gracious self."[12]

With his background, training, and long-term professional service as a madrasa instructor and member of the judiciary, Taqi al-Din does not seem to fit our earlier definition of a *munajjim*, but there is no conclusive evidence of his practicing astrology before his eventual move to Istanbul to establish an observatory. Although reports from near-contemporary sources state that he was appointed as the chief court *munajjim* in 1571, three years before the enthronement of Sultan Murad III, there is no archival confirmation of this appointment, nor have any almanacs or astrological texts from the period been discovered that decisively verify Taqi al-Din's service in this capacity.[13] Therefore, the claim by biographer ʿAtaʾi (d. 1635) that Taqi al-Din assumed the position of chief court *munajjim* in the early summer of 1571, following Muvakkit Mustafa's death, is subject to doubt; it seems more plausible that Taqi al-Din was granted the title after receiving Murad III's authorization to establish an observatory.

In any case, regardless of the precise date of his appointment, the designation of Taqi al-Din as the chief court *munajjim* indicates that the pool of reliable indigenous Ottoman astral experts in the capital at the time was quite limited; otherwise, a learned figure from Egypt serving in the judiciary would probably not have been recruited for the position. Taqi al-Din's dazzling skills in a wide array of mathematical and natural sciences must have played a role in his integration into the imperial center's administrative ranks. Still, his selection for the position over any incumbent practitioner within the corps of court *munajjims* is yet another clue to, and perhaps a consequence of, a diminishing interest in the astral sciences in the capital throughout the long reign of Süleyman.

The case of Taqi al-Din, a proficient practitioner of the astral sciences who required royal patronage to realize his costly scientific dreams, exemplifies the experiences of astral experts across the broader Eurasian context down to the seventeenth century. As we have already seen, Taqi al-Din was not the first individual within the Ottoman—or the wider Islamicate—domain to propose establishing a new observatory and conducting fresh observations to revise existing tables. The history of observatories and observational enterprises in the medieval Islamicate world may be viewed through the same patronage dynamics: between research-minded astral experts and sponsoring sovereigns who supported these endeavors because of their personal intellectual interests, ideological considerations, and the pragmatic benefits. Not all *munajjims*, of course, nurtured aspirations for research or were capable of undertaking such projects. Some medieval astral experts, such as al-Qabisi (d. 967), attempted to establish typologies for *munajjims*, stating that only a few could qualify as "complete astrologers," capable of preparing tables through their own direct observations.[14] It is hard not to find assertions of authority and expertise behind such remarks and efforts to create typologies. Even were such typologies accurate, it is impossible to determine what percentage of *munajjims* in the Ottoman and broader Islamicate realms actively sought to conduct new research. Nevertheless, we know that several *munajjims* formally asked to undertake new astronomical observations and revise existing tables. Taqi al-Din was one of the luckiest ones in the wider Islamicate context and the only fortunate one in the Ottoman realm, at least in the beginning. His request was given the royal green light, and he received generous land revenues from various Ottoman territories to run his project.[15]

Taqi al-Din's case is also typical when compared with some of his better-known contemporaries in Europe. Around the same time Taqi al-Din established an observatory in Istanbul with royal support and conducted new research to track celestial motions, Tycho Brahe (d. 1601) was afforded similar opportunities by the Danish king Frederick II (r. 1559–88) to run astronomical observations at Uraniborg, the observatory Brahe established on the island of Hven (now Ven in Sweden) in the 1570s. Some scholarly ink has been spilled on the curious similarities between these two men of science, including the instruments they used and invented, their astrological interpretations of the same comet in 1577, and the new planetary data and models they aspired to produce.[16] What is often overlooked in these useful comparisons, however, is the fact that Tycho Brahe's activities on Hven also ceased following the death of his patron, Frederick II, in 1588. The new young king, Christian IV (r. 1588–1648), who was far less passionate about the astral sciences than his father, gradually reduced Tycho Brahe's funding. Consequently, Brahe left the island in 1597 because of the financial and political challenges of maintaining his operations under Christian IV's rule. He eventually moved to Prague to operate under the patronage of the Holy Roman emperor Rudolph II (r. 1576–1612), who was famous for his penchant for astrology and alchemy. After Brahe's departure, Uraniborg fell into physical disrepair and had been largely dismantled by the mid-seventeenth century.[17]

While the demolition of the Istanbul observatory is often cited as a prime example of the Ottoman decline in science and technology from the sixteenth century, the dismantling of Uraniborg has never elicited a similar declinist view. This discrepancy is hardly surprising, though, given the fact that systematic celestial observations in western Europe continued unceasingly after the sixteenth century, whereas the Ottoman world would not house another observatory until the second half of the nineteenth century. Even Tycho Brahe himself continued his scientific endeavors after leaving the island of Hven and settling in Prague. Indeed, the seventeenth and eighteenth centuries saw the proliferation of astronomical observatories in various European cities, including Leiden (1633), Paris (1667), London (1675), and Berlin (circa 1700).[18] Enhancing the precision of available planetary data was still the primary motivation underlying these observations, yet the introduction of telescopes into astronomical observation from the early seventeenth century led to a gradual epistemological change in the conduct of research and assessment of precision. Another noteworthy difference between observatories established before and after the seventeenth century is their mode of funding and operation. Although royal patronage remained an indispensable source of support for many observatories, the increasing involvement of universities, along with learned societies and academies, helped these enterprises gain a more institutional, impersonal, and collective character, thereby ensuring their longevity.[19]

Aside from the lack of continuity for observational enterprises in the early modern Ottoman world, another significant reason the story of Taqi al-Din and the destruction of his observatory is often associated with decline is that it fits conveniently into a science-versus-religion polemic that has been widely espoused in the vast historiography of Ottoman-Turkish modernization/westernization. According to the standard narrative, Taqi al-Din's "visionary" "scientific" endeavors were brought to a halt by restrictions imposed by "fanatical" religious scholars. The narrative is sometimes even embellished with mythical elements, such as the claim that the ʿulamaʾ objected to the observatory because they thought this "violated the privacy of angels in the sky."[20] Yet for the proponents of this narrative, this incident was neither the first nor the last in Ottoman-Turkish history where the reactionary ʿulamaʾ vehemently opposed "innovative," "unorthodox" intellectuals. The aforementioned episode about the fifteenth-century scholar Molla Lutfi executed on ʿulamaʾ's charge of heresy; the apocryphal account regarding the alleged seventeenth-century polymath Hezarfen Ahmed, who is naively believed in Turkey to be the first aviator in world history to have made a successful flight with artificial wings; and the general portrayal of the nineteenth- and twentieth-century ʿulamaʾ as "guardians of tradition," obstructing progress and full-fledged modernization, are all framed in similar terms—analogous to the Catholic Church's portrayal in comparably polemic accounts of Galileo's career.[21] Taqi al-Din's case is particularly emblematic, as it involves a building thought to have operated like a modern research center.

For those who tend to view Taqi al-Din's observatory as having been like a modern scientific research institute, the astrological considerations behind Taqi al-Din's—and indeed, any late medieval *munajjim*'s—observational enterprise are too insignificant to highlight. However, as previously explored in Chapter 1, systematic celestial observations at medieval observatories were closely linked to revising planetary data that *munajjims* needed in order to make more accurate temporal and spatial calculations. These calculations were not exclusively about astrology, but the significant role astrology played cannot be overlooked. The question, then, is whether Taqi al-Din's observatory faced opposition and was ultimately dismantled as the result of the widespread adversarial stance against *munajjims* and their astrological expertise. Put another way, was the disdain for Taqi al-Din rooted not in the ʿulamaʾ's resistance to science in its modern understanding but rather in their aversion to *munajjims*?

As stated above, *munajjims* were among the most unwelcome learned men in the medieval and early modern Islamicate world, despite being sought after for their expertise, which was deemed indispensable by both the ruling elite and the general public.[22] The alarming allure of astral expertise demonstrates the ambivalence inherent in *munajjims*' practice. Their unpopularity is most evident in the vast polemical literature targeting them, authored by a diverse array of scholarly figures.[23] Although there was no disciplinary institutional mechanism in medieval Islamicate society akin to the Roman or Spanish Inquisitions, where astrologers could face trial, prominent Muslim religious authorities occasionally issued fatwas calling for the persecution of *munajjims*.[24] For example, the Hanbali jurist Ibn Taymiyya and his prolific student Ibn Qayyim al-Jawziyya wrote extensive denunciations of *munajjims*' practice as being astral determinism and heresy.[25] For Ibn Qayyim al-Jawziyya, *munajjims* dared to lay claim to God's omnipotence and therefore deserve harsh punishment; he suggests that their hands be severed and their "lying tongues" be pulled out to prevent them from "trading on the gullibility of the *umma*."[26] The Shaykh al-Islam Kadızade Şemseddin's reported letter to Sultan Murad III about Taqi al-Din echoed such language, stating that those who dare to reach out their hands to unveil "the curtain of celestial secrets" bring nothing but misfortune. For Kadızade Şemseddin, "there is no kingdom in the world where celestial spheres were observed that did not fall to ruin."[27]

Was the prevalent scholarly disdain toward *munajjims* truly the sole or at least underlying reason for the destruction of the Istanbul observatory in 1580? If so, how were *munajjims* able to continue their activities at the court and elsewhere in the following decades and centuries without facing similar obstacles? The corps of court *munajjims*, for instance, remained intact until the twentieth century, with the number of its members differing from era to era. Sultans and high-ranking officials consistently received annual almanacs replete with astrological prognostications. The ruling elite kept synchronizing their actions according to the auspicious moments regularly reported by court *munajjims*. Biographical dictionaries and chronicles from the seventeenth and eighteenth centuries offer even richer details about various court *munajjims*, noting their presence on military campaigns or their involvement in courtly intrigues.[28] Intriguingly, several *munajjims* cited in seventeenth- and eighteenth-century biographical sources also held judgeships or teaching posts in madrasas. New textbooks outlining general astrological principles, composed by some of the seventeenth-century chief court *munajjims* and incorporating European sources, attained wider circulation.[29] State-of-the-art European astronomical tables, such as those prepared by Giovanni Domenico Cassini (d. 1712) at the Paris observatory, were increasingly referred to in Ottoman *munajjims*' astrological computations. During the reign of Mustafa III (r. 1757–74), whose fascination with the astral sciences surpassed even that of his forefathers, several classics from the Islamic astrological corpus, including ʿAli-Shah Bukhari's *Fruits and Trees*, were translated for the first time into Ottoman Turkish.[30]

The continued and at times heightened presence and activities of *munajjims* after the late sixteenth century thus renders it highly unlikely that the destruction of the Istanbul observatory was occasioned solely by the ʿulamaʾ's categorically critical stance against *munajjims*. Baki Tezcan has recently suggested interpreting this episode through the lens of politics, characterizing it as a political clash between representatives of two opposing political views, each based on different epistemologies: absolutists putting forward a rather experimental and rationalist epistemology versus constitutionalists striving to preserve the supremacy of a traditionalist epistemology.[31] According to Tezcan, just as Murad III's support for Taqi al-Din's ambitious project was a political statement asserting claims of absolutist monarchy, Kadızade Şemseddin's response was a political one, reflecting the ʿulamaʾ's effort to protect their privilege of making law and controlling the realm of legitimate knowledge. While Tezcan's interpretation, highlighting the conflict between differing epistemological traditions and concomitant political claims, aligns well with the framework of expertise and authority discussed in this book, other possible factors, some pertaining to less tangible and more emotional dynamics, should also be considered. Perhaps we should question how the observatory as an edifice appeared to the residents of Istanbul, and how the eccentric, large-scale instruments in it—some of which were around ten meters tall and clearly visible from the outside—were perceived.[32] Did they arouse fear or awe? Were they seen as strange and uncanny (*garib*)—as Taqi al-Din himself depicts them in his account—even irritatingly or menacingly magical and talismanic? How about Taqi al-Din's purportedly distasteful personality and obstinate nature noted in passing by some of his contemporaries? Might that have played a part in the widespread opposition to him? Could there even have been a personal feud between Taqi al-Din and Kadızade Şemseddin, dating back to the late 1550s and very early 1560s, when both were the protégés of Semiz ʿAli Pasha, or when both were frequenting the scholarly gatherings of prominent figures including Ebu's-suʿud Efendi?[33]

These conjectures and possibilities can easily be multiplied, but they do not change the fact that the observatory was demolished and that Taqi al-Din had to leave Istanbul following the ʿulamaʾ's accusing him of being a heretic.[34] A more intriguing detail, and also more pertinent to the prevailing theme of *Forgotten Experts*, is that beginning in the seventeenth century the story of Taqi al-Din and the Istanbul observatory gradually faded away in the memory of Ottoman learned circles. For instance, Evliya Çelebi, who lived and flourished little less than a century after the observatory's destruction, oddly conflates ʿAli Qushji with Taqi al-Din, stating that the observatory had been founded by the former.[35] Katib Çelebi includes the year of the observatory's construction within the chronological scope of his *Takvimu't-tevarih* but pays no attention to the date of its destruction.[36] No explicit discussion or implicit critique of the destruction has been found so far in the early modern Ottoman textual record. Even in the nineteenth century, studies devoted to the history of science in the Ottoman context, particularly the history of astronomy, including Salih Zeki's extensive publications and Süleyman Sudi's *Tabakat-ı Müneccimin*, did not attribute any particular importance to the episode of the observatory's destruction.

New studies may eventually reveal astral experts from the seventeenth- and eighteenth-century Ottoman world eager to conduct new observations and update existing astronomical tables. These discoveries would nicely complement our knowledge of earlier such figures, like Riyazi and Taqi al-Din, as well as translated works from European languages on new astronomy. If no such figures are found, then the widespread conviction among both scholars and lay readers that Ottoman learned people from the late sixteenth century had little enthusiasm for new research may seem to be confirmed. Yet this may also prompt scholars to undertake a deeper

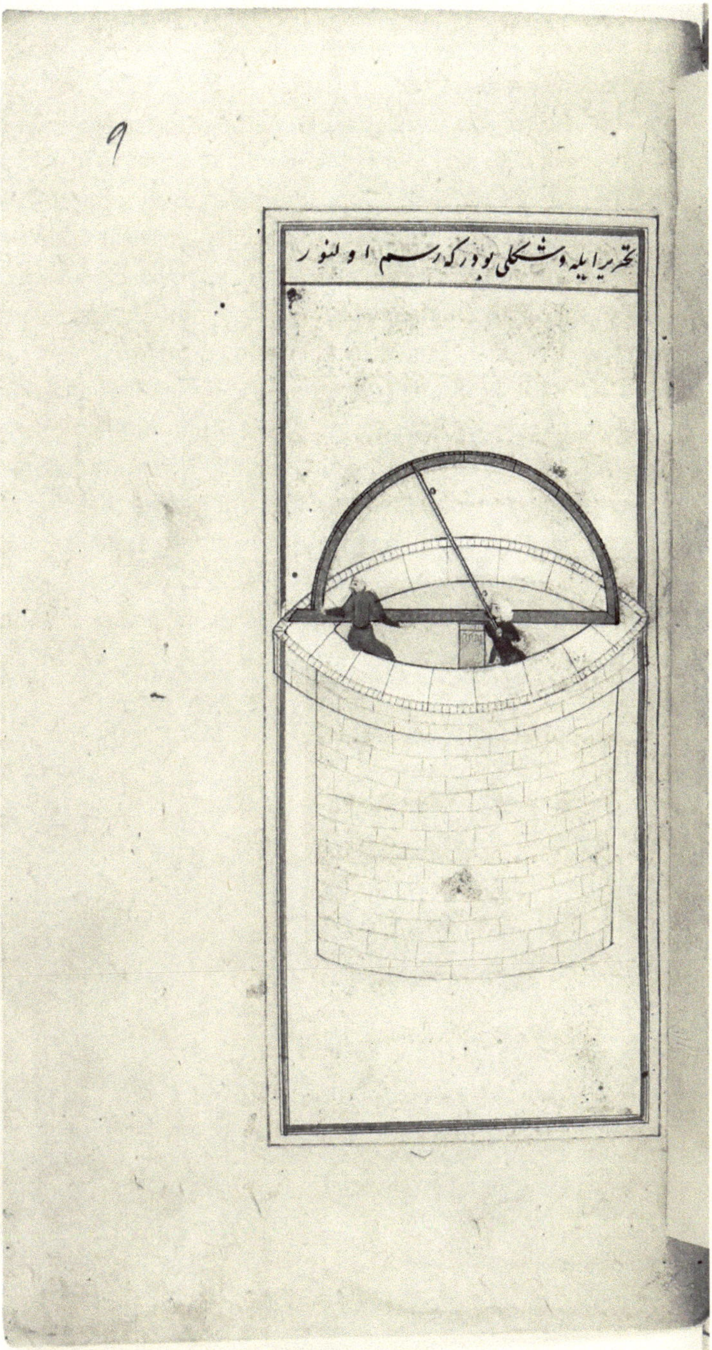

FIGURE 15. Instruments at the Istanbul observatory. From *Al-Alat al-rasadiyya li-Zij al-Shahinshahiyya*. BnF Ms. Suppl. Turc 1126, 9a. Courtesy of the Bibliothèque national de France, Paris.

examination of the material and structural factors underlying this presumed—or real—lack of scientific appetite, perhaps foremost among them the Ottoman system of expert and expertise management, which has been described throughout this book as an administratively efficient system. While it seems true that this patrimonial system effectively administered the relationships between courtly scientific functionaries and political decision-makers, it may also have encouraged individuals to merely pursue positions and titles in the bureaucratic hierarchy rather than demand new research and foster the creation of new scientific circles. Perhaps we should further investigate whether there were observations similar to those of Prospero Alpini (d. 1617), the Venetian physician and botanist visiting Ottoman Egypt in the late sixteenth century, who once noted that the bureaucratization and "concentration of power" in the hands of one person—the so-called *head* practitioner—within the Ottoman realm was a phenomenon different from, and implicitly inferior to, the scientific practices in his own country.[37]

NOTES

Introduction

1. Celālzāde, *Ṭabaḳātü'l-Memālik*, 719: "mehere-i erbāb-ı taḳvīm ve nücūm" (lit. "the experts among the masters of almanacs and stars").

2. ʿAṭā'ī, *Ḥadā'iḳ*, vol. 1, 856: "menbaʿ-ı fitne-i āḫir-zamān."

3. For the story of Taqī al-Dīn and the observatory, see Mordtmann, "Das Observatorium"; Gökmen, "Rasıd Takiyüddin"; Sayılı, *Observatory*, chap. 8; Ünver, *İstanbul Rasathanesi*; Şeşen, "Meşhur Osmanlı Astronomu"; Dizer, *Takiyüddin*; Ben-Zaken, *Cross-cultural Scientific Exchanges*, chap. 1; Ḥasan, *Taqī Al-Dīn*, 33–44.

4. Ḳāḍīzāde Şemseddīn's alleged letter has not survived to date but is cited in ʿAṭā'ī, *Ḥadā'iḳ*, vol. 1, 856.

5. For representative examples tackling the perennial question of why modern science emerged in the West and why the Muslim world did not keep pace—for some from the sixteenth century onward, for others earlier—see Sayılı, "Causes of the Decline," in *Observatory in Islam*, 407–29; Ben-David, "Review Article"; Huff, *Rise of Early Modern Science*; Saliba, *Islamic Science*, 233–55; Livingston, *Rise of Science*, 125; Kuru, *Islam, Authoritarianism, and Underdevelopment*. One should note that the works by Huff and Kuru were the products of political scientists and/or sociologists rather than professional historians of science, which is manifested by their lack of using primary sources in their works.

6. Some scholars define it by "Ottoman Sunnitization," a multilayered process that is considered to have taken place over the course of the sixteenth century. This process was characterized by the Ottomans' increasing awareness of Sharia and a stiffened Sunni perspective, influenced by both top-down state measures within the specific context of the ongoing (Sunni) Ottoman–(Shii) Safavid rivalry and bottom-up concerns from the ʿulamaʾ and different constituents of the society. See Terzioğlu, "How to Conceptualize Ottoman Sunnitization"; Krstić and Terzioğlu, *Historicizing Sunni Islam*.

7. For a succinct treatment of the process and an overview of the secondary literature, see Şahin, "From Frontier Principality."

8. İnalcık, *Ottoman Empire*.

9. Records on Safavid administrative structure mention a similar position for the chief *munajjim*; see *Tadhkirat al-Mulūk*, ed. Minorsky, 57. See also Brentjes, "Mathematical Sciences"; Blake, *Time in Early Modern Islam*, 52–56. For astrologers as regular functionaries at the Mughal court, see Orthmann, "Circular Motions"; Beinorius, "On the Social and Religious Status"; Popp, "Mughal Horoscopes." For Fatimid astrologers, see Rapoport and Savage-Smith, *Lost Maps*, 71–73.

The "Astro-Calendric Bureau" that operated throughout various dynasties in China, from the 2nd century BC to the end of the Qing Dynasty in 1912, also housed many Muslim astral experts that should be considered in this context. See Elman, *On Their Own Terms*, esp. chap. 2.

10. The dearth of Safavid archival sources is likely due to the destruction of the Safavid archives after the Afghan conquest of Isfahan in 1722. Similarly, in the Mughal context, there are no extant central archives, owing to a series of wars and upheavals following British occupation. For a comparative discussion of Ottoman and Mughal archives, see Faroqhi, *Ottoman and Mughal Empires*, chap. 1.

11. For the operation of the corps of architects as organizational structures administering the relevant expertise in that profession, see Orhonlu, "Şehir Mimarları"; Necipoğlu, *Age of Sinan*, chap. 5. For the corps of physicians, see Afyoncu, "Osmanlı Hekimbaşıları." For the corps of artillerymen, see Aydüz, *XV. ve XVI. Yüzyılda Tophane-i Amire*.

12. Stehr and Grundmann, *Experts*, 10–11.

13. Altınay, *Onuncu asr-ı hicride*, 93–94.

14. Ibid.: "ṣanʿatlarında imtiḥān."

15. Göyünç, "XVI. Yüzyılda Ruus."

16. Stehr and Grundmann, *Experts*; Dear, "Mysteries of State"; Ash, "Introduction"; Ash, "By Any Other Name"; Oschema, "Heaven Can Tell . . ."; Oppenheimer et al., *Discerning Experts*, chap. 1. For a discussion on expertise in the nineteenth-century Ottoman context, see Martykánová, "Ottoman Engineers." For an important collection of articles discussing expertise and authority in ancient (Roman) scientific culture, see König and Woolf, *Authority and Expertise*.

17. For a discussion on unified episteme in medieval Islamicate knowledge regimes, see Nasr, *An Introduction to Islamic Cosmological Doctrines*; Bakar, *Classification of Knowledge*. For an alternative view, emphasizing the separation of disciplines, see Dallal, *Islam, Science, and the Challenge of History*, chap. 3.

18. Porter, "Scientific Community"; Elman, *Civil Examinations*, chap. 1.

19. "expert, n." *OED* Online.

20. "expert, adj." *OED* Online.

21. Wehr, *Dictionary of Modern Written Arabic*; Redhouse, *Turkish and English Lexicon*; Steingass, *Comprehensive Persian-English Dictionary*.

22. Mikhail, *Nature and Empire*, 66; White, *Climate of Rebellion*, 244; Husain, *Rivers of the Sultan*, 52. For a discussion of *metis* as a form of practical, experiential knowledge essential for the functioning of societies, in contrast to the more technical knowledge favored by state-led modernist projects, see Scott, *Seeing like a State*, chap. 9.

23. Varlık, *Plague and Empire*, 229–30.

24. Ágoston, *Guns for the Sultan*, 42–46; Casale, *Ottoman Age of Exploration*, 123; Emiralioğlu, *Geographical Knowledge*, 103; Hagen, "Ptolemaeus Triumphans," 242–43.

25. Coşgel and Ergene, *Economics of Ottoman Justice*, 241n38. For the use of experts in settling financial disputes, see Darling, *Revenue-Raising*, 98.

26. Fleischer, *Bureaucrat and Intellectual*, 126; Fleischer, "Preliminaries"; Rhoads Murphey, "Art of Communication," in *Exploring Ottoman Sovereignty*, 256; Şahin, *Empire and Power*, 217; Ferguson, *Proper Order*, 139. For the concomitant bureaucratization and management of expertise in the scholarly establishment of jurists and judges, see Atçıl, *Scholars and Sultans.*

27. Faroqhi, *Artisans of Empire*. For "tacit knowledge," see Polanyi, *Tacit Dimension.*

28. The "first" Ottoman Empire mentioned here is an allusion to Baki Tezcan's *Second Ottoman Empire*. For discussions on the scope and layers of Ottoman (patrimonial) meritocracy, see İnalcık, "Comments on 'Sultanism' "; Barkey, *Bandits and Bureaucrats*, 32–40; Barkey, "Ottoman Empire."

29. M. Genç, *Osmanlı İmparatorluğu*, 55.

30. Fodor, "State and Society"; Howard, "Ottoman Historiography"; Abou-El-Haj, "Ottoman Nasihatname"; Sariyannis, *History of Ottoman Political Thought.*

31. For Muṣṭafā ʿĀlī's critique of a dissolving Ottoman "meritocracy" by his time, see Fleischer, *Bureaucrat and Intellectual*, 96–97; Haarmann, "Plight of the Self-Appointed Genius."

32. Lewis, "Ottoman Observers"; Ergene, "On Ottoman Justice."

33. Goodman, *Power and Penury*; Ash, *Power, Knowledge, and Expertise*; De Renzi, "Medical Expertise"; Rabier, *Fields of Expertise*; Klein and Spary, *Materials and Expertise*; Klein, "Artisanal-Scientific Experts"; Mendelsohn and Kinzelbach, "Common Knowledge"; Bertucci and Courcelle, "Artisanal Knowledge."

34. Blake, "Patrimonial-Bureaucratic Empire"; Duindam, "Royal Courts"; van Berkel and Duindam, *Prince, Pen, and Sword.*

35. In addition to the studies cited in note 16 above, see the following for a more sociological discussion of expertise: Shapin, "Trusting George Cheyne"; Collins and Evans, *Rethinking Expertise*; Eyal, *Crisis of Expertise.*

36. Kütükoğlu, "Berat."

37. Mordtmann, "Das Observatorium," 92: "üsvet-i aṣḥābü'l-heyʾe ve'n-necm, zübde-i erbābü'l-ḥikme ve't-taḳvīm, ʿārif-i maʿārif-i aḥkām-i menāzil ve şems ve ḳamer." Another intriguing example is a *berāt* given by Süleymān to Maṭraḳçı Naṣūḥ in 1530, which established his superiority in the art of using weapons, specifically the mace or cudgel (*maṭraḳ*) from which his expert sobriquet derived. For the transcription of the *berāt*, bound in a copy of Naṣūḥ's treatise on arithmetical calculation, see Yurdaydın, *Matrakçı Nasuh*, 70–71.

38. For some examples of manuals of composition from pre-Ottoman Anatolia that contain copies or samples of appointment letters and/or *berāts* for a wide range of professionals, including physicians and madrasa instructors, see Turan, *Türkiye Selçukluları*, 46–49; Yaḥyā b. Meḥmed el-Kātib, *Menāhicü'l-İnşāʾ*, 86.

39. For some intriguing examples of internal reviews made by men in power before ʿulamaʾ appointments, see Alan and Atçıl, *XVI. Yüzyıl Osmanlı Ulema Defterleri.*

40. The literature on the marginalization of astrology in Western/European culture is too vast to list here. For useful historiographical reviews focusing on the topic, see Rutkin, "How to Accurately Account for Astrology's Marginalization"; Pfeffer, "Reassessing the Marginalization of Astrology." For the conceptual and historical discussion of "pseudoscience," see Gordin, *On the Fringe.*

41. The history of newspaper astrology columns and daily horoscopes does not trace back earlier than the first half of the twentieth century. For an insightful analysis and cultural critique of the popularity of astrology columns, see Adorno, "Stars Down to Earth."

42. Polemics on astrology constitute one of the most extensively examined aspects of the history of astrological practice. See, for instance, Long, "Astrology"; Catani, "Polemics on Astrology"; Akopyan, *Debating the Stars*; Dallal, *Islam, Science and the Challenge of History*, 114–16; Morrison, "Discussions of Astrology."

43. Latour, *Science in Action.*

44. Crombie, *Scientific Change*, 10.

45. There is an inflation of "long centuries" that historians frequently use as a convenient way to distinguish the transformative nature of more than century-long time periods on which their studies focus. Almost every century in human history has a historian's version of a "long Xth century." As for "the long sixteenth century," extending from 1450 to 1640, the term was popularized by Fernand Braudel. For a discussion on the Ottoman adoption of this chronological unit, see Şahin, "Ottoman Empire."

46. The relevant volumes in the Cambridge History of Science series, for instance, or recent compilations on Renaissance and early modern astrology rarely feature the Ottomans. See Park and Daston, *Early Modern Science*; Dooley, *Companion to Astrology.*

47. In his extensive survey of the history of Ottoman-Turkish modernization from the eighteenth century, Niyazi Berkes frequently revisits the question of science; see Berkes, *Development of Secularism.* For an insightful analysis of the debates over the science-religion dichotomy among late Ottoman intellectual circles and their repercussions, see Yalçınkaya, *Learned Patriots*; Yalçınkaya, " 'Science,' 'Religion.' "

48. For a brief history of scholarly trends in modern Ottoman historiography, see Özel and Çetinsaya, "Türkiye'de Osmanlı Tarihçiliğinin." For a succinct analysis of different schools and trends in the historiography of "Ottoman science," see Aydın, "Beyond Culturalism?"; Shefer-Mossensohn, *Science among the Ottomans*, 10–13.

49. Adnan Adıvar's *La Science chez les Turcs Ottomans*, which was translated into Turkish with supplements in 1943 (*Osmanlı Türklerinde İlim*), stands out not only as the first attempt at writing a synoptic book on Ottoman histories of science but also as an account that puts an emphasis on certain absences and downturns in Ottomans' engaging with the sciences from the sixteenth century. Adıvar continued to subscribe to this thesis in his later publications as well. See his following remarks in "Interaction of Islamic and Western Thought" (273): "[After Mehmed II] Turkey once more returned to the obscurantism of the Middle Ages which was to continue until the nineteenth, even the twentieth century."

50. For the admiration on the part of figures such as Machiavelli and Jean Bodin for the structure of the Ottoman Empire, see Hentsch, *Imagining the Middle East*, 49–76.

51. The rhetoric of the "Golden Age of Islam," highlighting the scientific achievements

of Muslims in the first four centuries of Islam and looking down on all that came afterward, a product of the late nineteenth-century colonialist and Orientalist mindset, still influences popular audiences. The rhetoric is reproduced through the nonscholarly views of modern science popularizers, such as Neil DeGrasse Tyson, or by world-touring exhibitions like the *1001 Inventions*. For an example of a popular work circulating the main views of the "Golden Age of Islam" rhetoric, see al-Khalili, *House of Wisdom*. For a sound critique of the *1001 Inventions* and the accompanying events and publications, see Brentjes, Edis, and Richter-Bernburg, *1001 Distortions*; see also Brentjes, "Prison of Categories."

52. The "postclassical" period of science and philosophy in the Islamicate world has finally received deserved attention from scholars in the past two decades. The term is used to refer to the period from the 13th through the 18th century; it takes the 12th century as the time when key institutional and intellectual frameworks of Islamic learning (i.e., madrasas, disciplines, legal schools, etc.) had fully developed and established the groundwork for future progress in these areas. For some studies showing the importance of postclassical Islamic intellectual history, see Brentjes, "Courtly Patronage"; Pfeiffer, *Politics, Patronage*; Griffel, *Formation*; Stearns, *Revealed Sciences*. For a historiographical review of recent trends in the history of science in postclassical Islamicate societies, see Brentjes, "Research Foci."

53. For the Eurocentric historicism inherent in the narrative of the Scientific Revolution, see Elshakry, "When Science Became Western."

54. The shift in scholarly attitudes to the question of "Ottoman decline" is best exemplified in Halil İnalcık's following two publications, published ten years apart: "Heyday and Decline" and "Military and Fiscal Transformation." There is ample literature criticizing the decline paradigm from various perspectives; see, for instance, Faroqhi, "Crisis and Change"; Kafadar, "Question of Ottoman Decline"; Grant, "Rethinking"; Sajdi, "Decline, Its Discontents."

55. El-Rouayheb, *Islamic Intellectual History*, 4.

56. Küçük, *Science without Leisure*, 16.

57. See, for instance, Unat, *Çağını Aşan*. For the continuing appeal to the narrative of "genius Muslim scientists well ahead of their times," which inevitably reproduces the fiction of the Golden Age of Islam, see Starr, *Genius of Their Age*.

58. Ünver, *Ali Kuşci*; Heiderzadeh, "Ali Kuşçu'nun Astronomi Eserleri"; Umut, "Theoretical Astronomy."

59. In addition to the studies cited in note 3 above, see Tekeli, *Osmanlılarda Saat*; Demir, *Takiyüddin'de Matematik ve Astronomi*.

60. The literature on Kātib Çelebi and his wide-ranging intellectual interests is extensive; the following works specifically concern his scientific enterprises: Hagen, *Ein osmanischer Geograph*; Karlığa and Kaçar, *Doğumunun 400. Yıl Dönümünde*; Kātib Çelebi, *Ottoman Cosmography*.

61. Berkes, "İlk Türk Matbaası"; Erginbaş, "Enlightenment"; Küçük, "İbrahim Müteferrika's Copernican Rhetoric"; Küçük, *Science without Leisure*, 183–203.

62. It is no surprise that among the most prolific historians of science in Turkey in the twentieth and twenty-first centuries are physicians (e.g., Süheyl Ünver, Bedi Şehsuvaroğlu, Aykut Kazancıgil), astronomers (e.g., Fatin Gökmen, Muammer Dizer), engineers (e.g., Necati Akgür, Atilla Bir), and physicists (e.g., Erdal İnönü).

63. Halil İnalcık devotes three chapters to learning and sciences in his *The Ottoman Empire: The Classical Age, 1300–1600*, where he recounts the famous episode of the observatory's destruction as "the clear victory of religious fanaticism over the rational sciences." To date, Suraiya Faroqhi has not written anything specifically on the sciences and scientists.

64. For his approach to history of science and conceptualization of "Ottoman science," see İhsanoğlu, *Büyük Cihad'dan*; İhsanoğlu, *Science, Technology, and Learning*; İhsanoğlu, *Studies on Ottoman Science*.

65. IRCICA completed the following published catalogs: *History of Astronomy Literature* (in 1997) in two volumes; *History of Mathematical Literature* (in 1999) in two volumes; *History of Geographical Literature* (in 2000) in two volumes; *History of Music Literature* (in 2003) in one volume; *History of Military Art* (in 2004) in two volumes; *History of Natural and Applied Sciences* (in 2006) in two volumes; *History of Medical Sciences* (in 2008) in four volumes; *History of Astrological Literature* (in 2011) in one volume.

66. İhsanoğlu has later acknowledged the need to incorporate Ottoman non-Muslims' contribution to Ottoman science. See İhsanoğlu, *Ottoman Scientific Heritage*, vol. 1, 8.

67. OASTLT, vol. 1, xxvii.

68. The most emblematic of this is George Sarton's review of a book published in 1949 on Mandaean astrology, which Sarton dismissed as a "wretched collection of omens, debased astrology and miscellaneous nonsense." See Sarton and Siegel, "Seventy-Sixth Critical Bibliography." See also Otto Neugebauer's critique of Sarton's review, "Study of Wretched Subjects."

69. Whewell, *History of the Inductive Sciences*, vol. 1, 300 (emphasis mine).

70. For the famous discussion—the "Renan thesis"—on the incompatibility of science and Islam, put forth by Ernest Renan (d. 1893) in his Sorbonne lecture *L'Islamisme et la science*, see Renan, *Islam and Science*; Cündioğlu, "Ernest Renan ve 'Reddiyeler' "; Murphy, "Islamicate Knowledge Systems," 482–85. For more on nineteenth-century debates regarding Islam's harmony (or lack thereof) with science, see Küçük, "Islam, Christianity"; Yalçınkaya, "Muslims."

71. Pingree, "Indian Influence,"; Kennedy, "Ramifications"; Goldstein, "Book on Eclipses"; Lemay, "Origin and Success." See the Bibliography for the complete list of edited astrological sources in the twentieth century.

72. Ullmann, *Die Natur- und Geheimwissenschaften*; Sezgin, *Geschichte*, vol. 7: *Astrologie*, in *Sezgin Online*.

73. Sarton, *History of Science*.

74. Sarton, *Introduction*, vol. 1, 3–46.

75. See note 68.

76. Sarton, *Six Wings*, 76.

77. Sarton, *Introduction*, vol. 1, 6.

78. In addition to the names cited above in note 71, see the works of Paul Kraus on Jabirian alchemy published in the 1930s and early 1940s. The past three decades have witnessed a boom in edited volumes on the occult sciences in the Islamicate world. Notable works include Lory and Regourd's special journal issue, "Sciences occultes et Islam" (1992); Savage-Smith's *Magic and Divination* (2004); Melvin-Koushki and Gardiner's special journal issue, "Islamicate Occultism" (2017); el-Bizri and Orthmann's *Occult Sci-*

ences (2018); Günther and Pielow's *Magie im Islam* (2018); and *The Islamicate Occult Sciences* edited by Saif et al. (2020). See also the following dissertations—some published—all of which were produced in the last fifteen years: Felek, "(Re-)creating" (2010); Saif, *Arabic Influences* (PhD granted in 2012); Melvin-Koushki, "Quest" (2012); Coulon, *La Magie* (PhD granted in 2013); Artun, "Hearts" (2013); Gardiner, "Esotericism" (2014); Şen, "Astrology" (2016); Lelić, "Ottoman Physiognomy" (2017); Noble, *Philosophising the Occult* (PhD granted in 2017).

79. In the first half of the twentieth century there was already an interest in the occult and magic in medieval and early modern Europe. See, for instance, Lynn Thorndike's *A History of Magic*, published in eight volumes. Despite these earlier studies, a spike in interest occurred after Keith Thomas's *Religion and the Decline of Magic* was published in 1971.

80. The specific reference to the "Messianic Age" is found in Bashir, *Messianic Hopes*, 31. See also Fleischer, "Ancient Wisdom"; Moin, *Millennial Sovereign*; Melvin-Koushki, "Quest"; Mir-Kasimov, *Unity in Diversity*; Binbaş, *Intellectual Networks*; H. Yılmaz, *Caliphate Redefined*.

81. For earlier currents, see Cook, *Studies in Muslim Apocalyptic*; Yücesoy, *Messianic Beliefs*.

82. For the importance of the fifteenth century in terms of political experimentation and ideological innovation, see Woods, *Aqquyunlu*, 1–10; Fazlıoğlu, "Forcing the Boundaries."

83. Melvin-Koushki, "Early Modern Islamicate Empire," 360.

84. Gardiner, "Al-Jafr."

85. On Bistami, see Fazlıoğlu, "İlk Dönem Osmanlı"; Grill, "Ésotérisme"; Fleischer, "Ancient Wisdom"; Coulon, "Building al-Buni's Legend."

86. Moin, *Millennial Sovereign*, 11.

87. For a similar critique of the work, see Anooshahr, "Review."

88. For the sixteenth-century fascination with portents and prophecies, a theme shared across the Mediterranean, see Niccoli, *Prophecy and People*; Fleischer, "Mediterranean Apocalypse"; Green-Mercado, *Visions of Deliverance*; Martin, *Beautiful Ending*.

89. For some notable studies on astrologers in medieval and early modern courts, see Carey, *Courting Disaster*; Grafton, *Cardano's Cosmos*; Ryan, *A Kingdom of Stargazers*; Azzolini, *Duke and the Stars*; Hayton, *Crown and the Cosmos*.

90. Mulsow, *Knowledge Lost*, 1–24, 417–22.

91. See Ripat, "Expelling Misconceptions," for the problems of imposing a hierarchy between elite and street astrologers.

92. The rich textual tradition of the *melḥame* genre in Ottoman Turkish, based on weather prognostications, is a good example of such popular texts transmitting the "metis" about predicting future events from natural phenomena. See Boyraz, *Fal Kitabı*.

93. BL Ms. Add. 5983, 17b.

94. This is particularly evident in extant examples of *rūznāme* and *melhame* genres. In these texts, one frequently encounters remarks stating that their discourse (i.e., *ḳavl-i rūznāme*) is superior to that of *munajjims* (i.e., *ḳavl-i müneccimīn*) or that *munajjims*' almanacs become ineffective each year, whereas *melhame* texts remain relevant for many years.

95. While the term "Persianate" is conceptualized in various ways, *Forgotten Experts* adheres to a basic definition coined by Marshall Hodgson, referring to a cultural and historical phenomenon in which the Persian language and literary-scientific traditions exerted significant influence beyond the geographical boundaries of modern-day Iran.

Chapter 1

1. Wehr, *Dictionary of Modern Written Arabic*, 945.

2. "منجم," Hawramani, *Arabic Lexicon*, last accessed July 2, 2024, http://arabiclexicon.hawramani.com/search/%D9%85%D9%86%D8%AC%D9%85.

3. "منجم," Dehkhoda, *Dehkhoda Dictionary*, last accessed July 2, 2024, https://dehkhoda.ut.ac.ir/fa/dictionary/%D9%85%D9%86%D8%AC%D9%85.

4. Sonja Brentjes proposes "render[ing] the actors' self-designation, *munajjimūn*, as *astral experts*." See Brentjes, "Astral Sciences," 80.

5. Some of Qūshjī's writings, such as his mathematical treatise (*Risāla dar ḥisāb*), as well as the Ulugh Beg tables, include sections intended to be useful for *munajjims* in their astrological calculations. See, for instance, Qūshjī, *Risāla dar ʿilm-i ḥisāb*, 108–23. There are also contemporary anecdotes regarding Ulugh Beg's appeal to geomancers in the presence of ʿAlī Qūshjī. See Barthold, *Four Studies*, vol. 2, 139–40. However, no specific work by Qūshjī on any branch of astrology has been identified.

For a detailed examination of Qūshjī's astronomical project in Samarkand, which questions the Aristotelian causality of astrology and resonates with concurrent developments in early modern Europe, see Saliba, "Al-Qushjī's Reform"; J. Ragep, "Freeing Astronomy," esp. 68–69; J. Ragep, "ʿAlī Qushjī and Regiomontanus."

6. While Fazlıoğlu prefers to transliterate his name as Mīram Çelebi (see Fazlıoğlu, "Mīram Čelebī"), archival registers from the early sixteenth century record it as Mīrīm (میریم). In his own writings, however, he simply writes it as "میرم" without the second long vowel. Throughout the book, I will use the conventional form, "Mīrim." More information about him is provided in Chapter 3.

7. This is reminiscent of the memoirs of Joan Quigley (d. 2014), Ronald and Nancy Reagan's famous astrologer, who derided being referred to as a clairvoyant because, as she said, "clairvoyants are mystics and psychics." Her analyses, however, were based "on the data provided by astronomers and charts calculated by computers." See Quigley, *What Does Joan Say*, 13. I should also add that not all street diviners were lacking the skills for using astronomical instruments or casting horoscopes. For example, the *ghurabā* astrologers in the medieval Islamicate world, who had created a new, large-scale, illustrated book genre known as *bulḥāns*, were reported by some narrative sources as having the ability to "make use of astrological tables (*zīj*)." See Richardson, *Roma*, 84. Since the passage quoted here, taken from a tenth-century source, makes some confusing choices (it uses *jafr*, for instance, for "revolving celestial globes"), I wonder if its use of *zīj* was originally intended to mean something else.

8. For the multiple definitions of "occult" and "esoteric," see the following entries in Hanegraff's *Dictionary of Gnosis and Western Esotericism*: "Esotericism"; "Occult/Occultism." See also Saif, "What Is Esotericism"; Melvin-Koushki, "Is (Islamic) Occult Science."

9. Saliba, "Role of Astrologer"; Michot, "Ibn Taymiyya on Astrology"; Livingston, "Science and the Occult"; Ibn Khaldūn, *Muqaddimah*, vol. 3, 258–67.

10. S. Ökten, "Çağmini." For a critical edition of Jaghmīnī's *Epitome*, see S. Ragep, *Jaghmīnī's Mulakhkhaṣ*. For an astrological treatise by Jaghmīnī, see ibid., 281–82.

11. Van Dalen, "Wābkanawī"; Mozaffari, "Wābkanawī's Prediction"; J. Ragep, "New Light on Shams."

12. Keshaverz, "Horoscope of Iskandar Sultan."

13. Termed *müneccimḫāne*, in Yirmisekiz Çelebi Meḥmed Efendi, *Fransa Sefaretnamesi*, 237; Muṣṭafā Ḫaṭṭī Efendi, *Viyana Sefaretnamesi*, 37.

14. Süleymān Sūdī Efendi, *Ṭabaḳāt-ı Müneccimīn*. An archival document dated 1851 (A. M. 9/30), which includes a map for the expected hours of the solar eclipse according to different cities by the German-American astronomer Christian Heinrich Friedrich Peters, refers to Peters as a "famous *munajjim*."

15. Şemseddīn Sāmī, *Ḳāmūs-ı Türkī*, 1412. In another booklet of his, titled *Gök* (Sky), Şemseddīn Sāmī mentions (10–11) that there are "tricksters" (*dessās*) deceiving "simpletons" (*sāde-dilān*) with the claim that they could draw astrological indications from the stars. By the nineteenth century, there were also indications of unease in using *munajjim* with reference to exact scientific and astronomical endeavors. See, for example, Güneş, "Bir Alimin Gözünden," 369–71.

16. "ἀστρονομίας προγνωστικόν," in Ptolemy, *Tetrabiblos*, xi. See also Burnett, "Certitude of Astrology."

17. Isidore of Seville, *Etymologies*, 66.

18. Saliba, "Islamic Astronomy"; Pines, "Semantic Distinction"; J. Ragep, "Astronomy."

19. In the better-known encyclopedias or books on the classification of the sciences of the sixteenth- and seventeenth-century Ottoman world, including Ṭaşköprülüzāde's (d. 1561) *Miftāḥ al-saʿāda*, Nevʿī Efendi's (d. 1599) *Netāʾicü'l-fünūn*, Ṣadreddīnzāde Şirvānī's (d. 1647) *Al-Fawāʾid al-khaqāniyya*, and Kātib Çelebi's *Kashf al-ẓunūn*, the term "science of the stars" (Ar. *ʿilm al-nujūm*, Tur. *ʿilm-i nücūm*) was deployed separately from *ʿilm al-hayʾa* and in close connection to astrology.

20. J. Ragep, "Astronomy."

21. Nallino, "Sun, Moon, and Stars," 89.

22. Bīrūnī, *Book of Instruction*, 1.

23. Rukn al-Āmulī, *Risāla Panjāh bāb*, SK Ayasofya Ms. 2667, 2a–b.

24. Niẓāmī ʿArūḍī (d. after 1156), in his *Chahār maqāla* (87–88), says, "A man does not merit the title of *munajjim* until he has attained proficiency in four sciences: *handasa*, *ḥisāb*, *hayʾa*, and *aḥkām*." Al-Qabīṣī (d. 967) also mentions in his *Introduction to Astrology* (6–7) that an expert *munajjim* should master Ptolemy's *Almagest*, the mathematical sciences, natural philosophy, and instrumental techniques.

25. Principe, *Scientific Revolution*, chap. 2. For astrologers' subscription to the notion of "as above, so below," see Campion, *What Do Astrologers Believe*.

26. Fārābī, *Iḥsāʾ al-ʿulūm*, 84–86.

27. Druart, "Al-Fārābī's Causation"; Jones, *Method, Structure, and Development*, esp. 44–57.

28. Ibn Sīnā, *Tisʿa rasāʾil*, 105–6.

29. Ibid., 108–11.

30. Al-Ṭūsī, *Nasirean Ethics*, 27.

31. Al-Shīrāzī, *Durrat al-tāj*, 73–75.

32. Al-Rāzī, *Jāmiʿ al-ʿulūm*, 184–86; Vesel, *Les Encyclopédies*, 35–36; Melvin-Koushki, "Powers of One."

33. Al-Fenārī, *Anmūzāj al-ʿulūm*, SK Hüsrev Paşa Ms. 482, 181a–185a.

34. For a transliteration and facsimile of the book, see *Treasures of Knowledge*, eds. Necipoğlu, Kafadar, and Fleischer, vol. 2.

35. See the following articles in the first volume of the *Treasures of Knowledge*: Gardiner, "Occult Sciences"; Şen and Fleischer, "Books on Astrology"; J. Ragep et al., "Astronomical and Other Mathematical Sciences."

36. *Tár*, directed by Todd Field (Focus Features, 2022). Stephen Blake also describes *munajjim* as the "time expert" but adds that "although astrology . . . defined a temporal order dependent on the heavenly bodies, it was a completely unscientific construct." See Blake, *Time in Early Modern Islam*, viii. For a similar characterization of astrologers as "masters of time," see Grafton, *Cardano's Cosmos*, chap. 1.

37. For one of the best treatments of this diversity across different traditions regarding calculations about the beginning of day and night, as well as various calendars, see al-Bīrūnī, *Chronology of Ancient Nations*.

38. The relevant literature is vast. I find the following work particularly useful: Dohrn-van Rossum, *History of the Hour*.

39. This summary on the astrological branches was written based upon the following studies: Burnett, "Astrology"; Nallino, "Sun, Moon, and Stars (Mohammedan)"; Pingree, "Astrology"; Cooper, "Astrology."

40. Pingree, "Kiran."

41. Ibid.; Kennedy and Pingree, *Astrological History of Māshā'allāh*; *Abu Maʿshar on Historical Astrology*; Borrut, "Court Astrologers," esp. 468–69.

42. Borrut, "Court Astrologers"; Borrut, "Histoires astrologiques."

43. In birth horoscopy, for instance, prorogators such as *haylaj* and *kadkhudah* as well as *tasyīr* were to be determined, whereas in electional astrology, the relative motions of the moon were given precedence. For more on these terms, see Burnett, "Astrology"; *Abraham Ibn Ezra on Nativities*; Samsó, *On Both Sides*, 152–323.

44. Casulleras, "Horoscope"; Elwell-Sutton, *Horoscope of Asadullāh Mīrzā*.

45. Franciszek Meniński (d. 1698), the chief translator to the Polish embassy in Istanbul and the author of a multivolume Turkish-Latin dictionary, clearly points out in his *Thesaurus linguarum orientalium* that, as early as the seventeenth century, *ṭāliḥ*, as the transformed version of *ṭāliʿ*, was documented as in use in vernacular Turkish. See *Thesaurus linguarum orientalium*, vol. 3, 597: "*ṭāliʿ*: vulg. *ṭāliḥ*. (1) part. Oriens, prodiens; (2) pl. *ṭawāliʿ*, Primum diluculum; (3) Horoscopus; (4) Sors, fortuna."

46. Saliba, "Role of Astrologer," 68–74; Kheirandish, "Astronomical Poems." For a discussion on the proliferation of the images of astrologers in Timurid painting and manuscripts from the second half of the fifteenth century, see Caiozzo, "Le Prince, l'astrologue et le derviche." Compared to the Timurid, Ilkhanate, or Mughal visual cultures, Ottoman illustrated manuscripts have yielded far fewer examples depicting astrologers. Whether this should be interpreted as the sign of a more pronounced disregard for astrologers in the Ottoman context remains open to discussion.

47. For the use of astronomical tools, specifically of astrolabes, in astrological problems, see Casulleras, "Instruments"; Schmidl, "Using Astrolabes." It is one of the standard themes in the writings of astral experts that conducting a celestial observation with reliable instruments is a costly business, requiring the financial support of a ruler-patron. See, for instance, Chapter 2, note 65, or Chapter 4, note 94.

48. On the dynamics of *zīj* production, see Kennedy, "Survey"; King et al., "Astronomical Handbooks and Tables"; van Dalen, *Islamic Astronomical Tables.*

49. King et al., "Astronomical Handbooks and Tables," 15.

50. For a similar observation in the medieval European context, see Falk, "Learning Medieval Astronomy." Anthropological studies on the methods of modern astrologers also reveal that there are two distinct phases in their operation, one technical/computational and the other interpretive. See Gunzburg, "How Do Astrologers."

51. Van Dalen, "Commentary."

52. Elwell-Sutton, *Horoscope of Asadullāh Mīrzā*, 57–59.

53. Van Dalen, "Commentary." Edward Kennedy also details the mathematical procedures involved in astrological operations by drawing on Jamshīd al-Kāshī's *Zīj-i Khāqānī.* See Kennedy, "On the Contents."

54. Sayılı, *Observatory in Islam*, passim. Ḥasan Beg Rumlu mentions in his *Aḥsan al-tavārīkh* that Shah Ismāʿil decided to build a new observatory after seeing the remnants of the Maragha observatory. Once his *munajjim* told him that it would require at least thirty years of operation to have precise observational results, he abandoned his plans to construct the observatory. See ibid., 166.

55. Tekeli, "Nasirüddin, Takiyüddin."

56. *Zīj-i Ilkhānī*, Medicea Laurenziana Or. 24, 3b.

57. Ibid., 3a.

58. Savadi and Khubravan, "Ḥarakat-i vasaṭ," 455: "tafāvut-i fāhish."

59. Ibid.

60. Sayılı, *Observatory*, 211–18.

61. Savadi and Khubravan, "Ḥarakat-i vasaṭ," 455: "māl" . . . "jāh" . . . "yārān" . . . "ālāt-i ṣaḥīh" . . . "rūzgār-i dirāz."

62. Ibid.

63. Sayılı, *Observatory*, chap. 7.

64. Savadi and Khubravan, "Ḥarakat-i vasaṭ," 462: "tafāvuthā-yi fāhish va ikhtilāfāt-i ẓāhir." See also Mozaffari, "Wābkanawī's Prediction"; van Dalen, "Wābkanawī"; Mozaffari and Zotti, "Observational Instruments."

65. The numbers of manuscript copies available in the early sixteenth-century Ottoman palace library are taken from Şen and Fleischer, "Books on Astrology."

66. Aḥmed-i Dāʿī, *Muḫtaṣar.*

67. Lemay, "Origin and Success." For Ṭūsī's Persian translation of *Al-Thamara*, see al-Ṭūsī, *Sharḥ-i Ṣamarah-yi Baṭlamyūs.*

68. Lemay, "Religion vs. Science," 573.

69. Kūshyar, *Introduction to Astrology.*

70. Al-Bīrūnī, *Book of Instruction.*

71. For Abū Maʿshar's biography and works, see Pingree, "Abu Maʿshar al-Balkhi";

Burnett, "Abū Ma'shar." Many of his extant works have been translated and published as critical editions. See Pingree, *Thousands of Abu Maʿshar*; Abū Maʿshar, *Abbreviation*; Abū Maʿshar, *On Historical Astrology*.

72. For copies of the text, see "3 Astronomy and Astrology," Storey Online. Last accessed online on 20 July 2024. https://referenceworks.brill.com/display/entries/SPLO/COM-20300000.xml.

73. For copies of the text, see ibid. See also the facsimile edition: Rāzī (Shahmardan b. Abi'l-Khayr), *Ravżat al-munajjimīn*.

74. The canon used by al-Rāzī consists of the works of Ptolemy, Vettius Valens (d. circa 175), Abū Maʿshar, Dorothius of Sidon (d. circa 75), ʿUmar b. al-Farrukhān (d. 815/6), al-Sijzī (d. circa 1020), al-Qaṣrānī, al-Ṭabarī, Kūshyār, and Sahl ibn Bishr (d. circa 845). Cited in TSMK Revan Ms. 1705, 2a.

75. SK Bağdatlı Vehbi Ms. 2005, 47b.

76. Hofelich and Varisco, "Taḳwīm"; Hofelich, "The Making of Taqvims"; Aydüz, "Müneccimbaşı Takvimleri"; M. Kurz, *Ein osmanischer Almanach*; Behrens, "Ottoman Calendar"; Şen, "Astrology," 237–304; Peacock, "Two Royal Almanacs"; Danışan, "Osmanlı Takvimlerinin."

77. For earlier references to *taqwīms* or *taqwīm*-like annual compositions (*daftar al-sana*), see al-Bīrūnī, *Book of Instruction*, 186–91; Pingree and Goldstein, "Astrological Almanacs," parts 1 and 2; Pingree and Goldstein, "Additional Astrological Almanacs." The popularity of *taqwīms*, especially from the fifteenth century, can also be seen from numerous "burlesque almanacs" penned by Ottoman literati, which imitated the language of *munajjims*' almanacs to mock various people in society. See Ekinci, "Klasik Türk Edebiyatında Mizahi Takvimler."

78. King and Fahd, "Al-Ṭāliʿ."

79. This was also the case for early modern European almanac-prognostications. See Barnes, "Hope and Despair."

80. For these chronology systems in use, see van Dalen, "Al-Tā'rīkh."

81. Huntington Library Ms. HM71897, 64b.

82. Ibid.: "aḥkām ki samara va natāyij-i īn dalāʾil ast."

83. TSMK Yeni Yazmalar Ms. 830, 200a.

84. See J. Ragep, "Qāḍīzāda al-Rumi"; Fazlıoğlu, "Qūshjī."

85. TSMA E. 10159/145: "ishtighāl bi-nujūmiyyāt siyammā bi-aḥkāmash mustalzim-i qasāvat-i qalb ast."

86. Tzvi Langermann has also demonstrated that the category of experience (*tajriba*) had a significant place in late medieval and early modern epistemological discussions, particularly in the fields of the applied arts and sciences, including astrology. See Langermann, "From My Notebooks: On *Tajriba/Nissayon*."

87. TSMK Revan Ms. 1711/1, 3a*:* "arbāb-i aḥkām rā sahvhā-yi bisyār oftad."

88. Ibid.

89. TSMK Bağdat Ms. 320, 2a. For the limitations of experiential knowledge in Islamicate taxonomies, see McGinnis, "Scientific Methodologies"; Griffel, "Role of Sense Perception."

90. Morrison, "Discussions of Astrology"; Orthmann, "Charm of Suspicious Calculations."

91. The still-authoritative account of observatories in the medieval Islamic world is Aydın Sayılı's *Observatory in Islam.*

92. Yang, "Like Stars in the Sky."

93. Sayılı, *Observatory in Islam*, 235. See also Allsen, *Culture and Conquest*, 161–75 and 203–7.

94. For a brief discussion on possible reasons for the construction and subsequent destruction of observatories in the Islamic realm, see Fazlıoğlu, "Osmanlı Felsefe-Biliminin Arka Planı," esp. 12–14. One exception is the Maragha observatory, which, after its construction in 1259, was able to survive until 1316 thanks mostly to the steadier funding from waqf revenues. See Sayılı, *Observatory in Islam*, 207–23.

95. The most prominent example seems to be the al-Kāshī family, different members of which served different courts in the Timurid realm. *ʿImād al-munajjim* (whose real name was Maḥmūd b. Yaḥyā) was the author of the famous horoscope of Mīrzā Iskandar in 1411. Another horoscope (Huntington Library Ms. HM71897) was produced on 11 February 1419 for Mīrzā Iskandar's brother Mīrzā Rustam by a certain Yaḥyā b. ʿImād b. Yaḥyā al-munajjim, apparently the son of the *munajjim* that prepared Mirza Iskandar's horoscope. I thank Evrim Binbaş for informing me about the text and generously sharing with me the images of the manuscript. The famous astral expert in the Samarkand observatory, Jamshīd al-Kāshī b. Masʿud b. Maḥmūd al-Kāshī, also seems to have been related to the previous two. Fatema Keshaverz argues in her article "The Horoscope of Iskandar Sultan" that Jamshīd al-Kāshī was the grandson of *ʿimād al-munajjim.*

96. Kennedy, "Letter of Jamshīd al-Kāshī"; Sayılı, *Uluğ Bey*; Bagheri, "Newly Found Letter"; Fazlıoğlu, "Osmanlı Felsefe-Biliminin Arka Planı"; Trigg, "From Samarqand to Istanbul," 152. See also Barthold, *Four Studies*, vol. 2, 119–21, 129–32.

97. Bagheri, "Newly Found Letter," 245.

98. Makdisi, *Rise of Colleges.* Even Aydın Sayılı's dissertation ("Institutions of Science"), defended in 1941 as one of the first PhDs in the new academic discipline of history of science, deals primarily with the question of educational institutions in the medieval Islamic world. See also Tibawi's "Origin and Character."

99. Berkey, *Transmission of Knowledge*; Chamberlain, *Knowledge and Social Practice*; Ephrat, *Learned Society.*

100. Berkey, *Transmission of Knowledge*, 23.

101. A similar critique is also raised by Sally Ragep in her dissertation, "Maḥmūd ibn Muḥammad ibn ʿUmar al-Jaghmīnī's *al-Mulakhkhaṣ*," 154.

102. Sabra, "Appropriation and Subsequent Naturalization," 234. In fact, Sabra himself is not affirmative about the institutional nature of scientific instruction, as he tends to hold that an institutional basis had no real significance and that the accomplishments made by Muslim scholars bore no relation to place. He even argues that religious institutions instead operated as obstructions to authentic scientific inquiry. See his "Situating Arabic Science."

103. Morrison, *Islam and Science.*

104. İzgi, *Osmanlı Medreselerinde İlim*, vol. 1, 189–453; Brentjes, "On the Location"; Fazlıoğlu, "Osmanlı Felsefe-Biliminin Arka Planı"; Morrison, *Islam and Science*; S. Ragep, *Jaghmīnī's Mulakhkhaṣ*; Brentjes, *Teaching and Learning.*

105. Cevat İzgi meticulously establishes, based on paratextual evidence from extant

manuscripts, the names of madrasa instructors who copied *Al-Mulakhkhaṣ* or any of its later commentaries. See especially 370–92 in his *Osmanlı Medreselerinde İlim*, vol. 1.

106. Ṭaşköprülüzāde, *Shaqā'iq*, 854. More information on Mīrim Çelebi is provided in Chapter 3.

107. Hasse, *Success and Suppression*, 23–26; Rutkin, "Astrology"; Azzolini, *Duke and the Stars*, 22–64; Lemay, "Teaching of Astronomy."

108. See, for instance, Qāḍī-yi Baghdadī, who compiled and presented to the court a *taqwīm* in 1508 (BL Ms. Or. 6432/2, 26b–52a) with astrological prognostications. More information about him is presented in Chapter 3. See also the University of Michigan Library Islamic Manuscripts Ms. 794, which contains a *taqwīm* with astrological prognostications from the early seventeenth century, compiled by the instructor at the Surgatmishiyya madrasa in Cairo.

109. Öz, "Zwei Stiftungsurkunden"; *Fatih Mehmed II Vakfiyeleri*, 262–63; Ergin, *Fatih İmareti Vakfiyesi*. See also İhsanoğlu, "Fatih Külliyesi Medreseleri." For the problems of different editions of the *waqfiyya* of the Fatih complex, see Hayashi, "Fatih"; Beyatlı, *Fatih*, vii–xiii.

110. *Süleymaniye Vakfiyesi*, 32.

111. D. HMH. SFTH. 21941-B; TSMA D. 9559.

112. King, "On the Role"; King, "Astronomy of the Mamluks."

113. Saliba, *History of Islamic Astronomy*, 98; Charette, "Ibn al-Majdī."

114. Brentjes, "Shams al-Din al-Sakhawi."

115. King, "Astronomy of the Mamluks," 550–51; De Block, "[Re]drawing the Lines."

116. Necmeddīn, *Taḳvīm* [of the year 932 AH], Arkeoloji Müzesi Ms. 1607/2; *Taḳvīm* [of the year 935], TSMK Revan Ms. 1711/16; *Taḳvīm* [of the year 956], TSMK Revan Ms. 1711/18; *Taḳvīm* [of the year 960], BL Ms. Or. 6432/1; *Taḳvīm* [of the year 961], TSMK Revan Ms. 1711/19.

117. KK 1764, 26 (dated March 1527): "ādat-i Necmeddīn munajjim ki dar rūz-i navrūz taqvīm āvurd, 1000." A later register of payments from the 1540s also lists Necmeddīn as a *munajjim* sporadically receiving cash from the imperial court. See Doğan, "Kanuni Sultan Süleyman Döneminde."

118. Mühimme Defteri 25/1231: "ḫaṣṣa müneccimāna şākird." Based on the price index of the time, a person could purchase only four pounds of bread with two *akçes* daily. See Pamuk, *İstanbul ve Diğer Kentlerde*, 106.

119. More information on Muvaḳḳit Muṣṭafā and his *Tuḥfetu'z-zamān* is presented in Chapter 4.

120. His most comprehensive account, *Tuḥfe*, is especially rich in details showing Muṣṭafā's engagement with astrology. Modern scholarship on Muṣṭafā (see Fazlıoğlu, "Mustafa b. Ali el-Muvakkit") states, based on the unverified reference of an early seventeenth-century biographical dictionary, that he later became the chief court *munajjim* in the 1560s. Neither in contemporary archival documents from the 1550s and 1560s nor in the extant accounts of Muṣṭafā have I come across any definitive evidence regarding his employment as the chief.

121. In MC O. 71 (180) is listed a certain Murād, who as a timekeeper (*sā'ātī*) in Edirne presented an almanac. A similar register now housed at TSMA D. 34 that lists the amounts of cash given to various palace personnel in the late sixteenth and early seven-

teenth centuries shows the timekeepers at the Sultan Selim Mosque and Fatih Mosque as recipients of cash for submitting almanacs with prognostications. See Yiğit, "34 Numaralı Filori," 799, 877.

122. KRK Ms. 294, 2b.

123. ʿAlī-Shāh Bukhārī, *Athmar va ashjar*, SK Ayasofya Ms. 2688.

124. Kasımpaşalı ʿOsmān b. Muḥammed, *Cāmiʿ al-aḥkām*, SK İzmir Ms. 470.

125. *Süleymaniye Vakfiyesi*, 34b: "sāʿir ʿilm-i nücūma müteʿallik̩ külliyāt ve cüzʾiyyāt-ı maʿārife vāk̩ıf ve ʿārif." Based on the price index of the time, a person could purchase around four pounds of olive oil with ten *akçes* daily. See Pamuk, *İstanbul ve Diğer Kentlerde*, 103.

126. Ünver, "Osmanlı Türkleri İlim Tarihinde Muvakkithaneler"; Aydüz, "Osmanlı Devleti'nde Küçük Gözlemevleri."

127. Deniz, "Sultan Abdülmecid'in"; Tomar, "Anadolu Muvakkithaneleri." For the transformation of Ottoman temporal culture and practice in the nineteenth century, see Wishnitzer, *Reading Clocks, Alla Turca.*

128. Ünver, "Osmanlı Türkleri İlim Tarihinde Muvakkithaneler," 241–42; Aydüz, "Osmanlı Devleti'nde Küçük Gözlemevleri."

129. Some late fifteenth-century archival documents about the Bayezid Mosque complex in Edirne mention a "newly established clock room" (*saʿātkhāne ki nav-bīnā shud*) close to the hospital. See Barkan, "Edirne ve Civarındaki," 281; Kazancıgil, *Edirne Sultan II*, 77. Evliyā Çelebi mentions (in volume 1 of his *Seyahatname*) that near the southern door of the Süleymaniye Mosque, there were rooms that various servants of the mosque occupied, including the *muwaqqit*, doormen, cleaners, and others.

130. This was also the case in fifteenth- and sixteenth-century Europe, where different institutional structures other than princely courts did exist, such as universities housing chairs in mathematics and astronomy. For the importance of noble patronage in the cultivation of the sciences, including astrology, within the late medieval and early modern European context, which can be viewed, in Pamela Smith's words, as "the first glimmerings of state sponsorship for science," see Westman, "Astronomer's Role," esp. 121–25; Moran, "German Prince-Practitioners"; Moran, *Patronage and Institutions*; Biagioli, *Galileo, Courtier.* For an overview of the literature and the source for the above quotation, see P. Smith, "Science on the Move."

131. Perhaps the most emblematic example is Ṭaşköprülüzāde's comprehensive biography of Ottoman ʿulamaʾ, which does not include a single *munajjim* in its gallery of learned individuals. The remarks frequently noted in the introductory parts of astrological treatises, stating that "their science is the noblest of all disciplines," also seem to echo *munajjims*' reactions against the considerably lower scholarly prestige of their expertise.

132. For an extensive discussion on the immense popularity of printed almanacs, calendars, and annual prognostications across various early modern European countries and the proliferation of printing houses publishing such materials, see Westman, *Copernican Question*, chap. 2; Capp, *Astrology and the Popular Press*; Carlebach, *Palaces of Time*; Green, *Printing and Prophecy.*

133. Faroqhi, *Artisans of Empire*; Cohen, *Guilds of Ottoman Jerusalem.* For an intriguing case of medical guilds, see Shefer-Mossensohn, "Medical and Professional Ethics." It

should also be noted that in many *futuwwa* treatises of late medieval Anatolia, which set the rules for entry-level conditions to brotherhoods, *munajjims* are traditionally not permitted to be part of the community. See, for instance, Köksal, "Fütüvvetnamelere Göre."

134. Brentjes, "Courtly Patronage," 410.

135. Aydüz, "Osmanlı Devleti'nde Müneccimbaşılık ve Müneccimbaşılar"; Aydüz, "Osmanlı Devleti'nde Müneccimbaşılık Müessesesi."

136. For the classical study on the financial problems caused by the discrepancy between solar and lunar years, see Sahillioğlu, "Années sıvış et crises monétaires."

137. For an interesting case, see Sahillioğlu, *Topkapı Sarayı Arşivi H. 951–952*, 60–61. In the late nineteenth century, when a special commission was established to fix the calendar problem, the chief court *munajjim* of the time was one of the members of the commission. See Akyıldız, "19. Yüzyıl Belgelerinde," 51.

138. Al-Şirvānī, *Riyāżu'l-qulūb*, 181.

139. Sayılı, *Observatory in Islam*, 138; Jorati, "Science and Society," 170.

140. The same can also be said for medical experts. For physicians' expertise in keeping time and observing the *kairos*, see Miller, "Doctor on the Clock"; Cooper, "Approaches to the Critical Days."

141. Campagna, *Technic and Magic*, 6.

142. See especially Darin Hayton's *The Crown and the Cosmos* on the uses of astrology for the purposes of imperial propaganda during the reign of the Holy Roman emperor Maximilian I (r. 1493–1519).

143. Reeve, *Philosopher-Kings.*

144. Saliba, "Role of Astrologer"; Saliba, "Astronomy and Astrology"; Lemay, "Religion vs. Science."

145. Ibn Sīnā (Avicenna), *Réfutation de astrologie*; Ibn Sīnā, *Tisʿ rasāʾil*, 105–6, 111.

Chapter 2

1. Aydüz, "Osmanlı Devleti'nde Müneccimbaşılık ve Müneccimbaşılar"; Aydüz, "Osmanlı Devleti'nde Müneccimbaşılık Müessesesi."

2. For a thorough discussion on the circulation of astrological knowledge in early Islam, see Pingree, *From Astral Omens*; Pingree, "Astronomy and Astrology." Many foundational texts of astrological practice in the Islamicate world, including those by Abū Maʿshar, Mashāʾallāh, al-Qabīṣī, and Kūshyār ibn Labbān, have been edited and translated. See Abū Maʿshar, *On Historical Astrology*; Abū Maʿshar, *Abbreviation*; Kennedy and Pingree, *Astrological History of Māshā'allāh*; al-Qabīṣī, *Introduction to Astrology*; Kūshyār, *Introduction to Astrology.*

3. Zilfi, "Diary of a Müderris"; Kafadar, "Self and Others"; Lewis, "First-Person Narrative"; Terzioğlu, "Man in the Image of God." For a recent overview of the state of the field, see Karahasanoğlu, "Ottoman Ego-Documents."

4. A notable exception is the vast literature on the chief architect Sinān from the reign of Süleymān, thanks to the extant (auto)biographies Sinān dictated to his poet friend. See Miʿmār Sinān, *Sinan's Autobiographies.* See also Necipoğlu, *Age of Sinan.* Though he was hardly a "scientific expert," the sixteenth-century silk weaver Kamāl al-Dīn's notebook, recently discovered and edited by Kristina Richardson and Boris Liebrenz, is another welcome contribution.

5. A good example is *Mücerrebnāme* (The book of repeated experiences) of Muṣṭafā Zekī (d. 1735), which contains colorful autobiographical details regarding the professional and family life of a chief court *munajjim* from the eighteenth century. The text is extant in no less than eight copies, cataloged under different names. A reliable manuscript copy of the account can be found in IU Nadir Ms. T 4315. For an insightful study on the autobiographical fragments of a nineteenth-century Ottoman *munajjim*, see Tunalı, "Sadullah el-Ankaravi"; Tunalı, "Ottoman Astrologer."

6. ʿAbdurraḥmān, *Jawhar ḥifẓ al-ṣiḥḥat*, SK Ayasofya Ms. 3635.

7. *OALT*, vol. 1, 70; *Osmanlı Tıbbi Bilimler Literatürü Tarihi*, vol. 1, 84.

8. ʿAbdurraḥmān, *Taqvīm* [of the year 915 AH], SK Esad Efendi Ms. 1978.

9. Çıpa, *Making of Selim*, 29–61.

10. ʿAbdurraḥmān, *Jawḥar*, TSMK III. Ahmed Ms. 2010.

11. SK Ayasofya Ms. 3635, 2b: "takmīl ʿilm al-nujūm wa-taḥqīq daqāʾiqahu."

12. Ibid., 2a.

13. Altınay, "Fatih." Based on the price index of the time, a person could purchase around fifty pounds of flour with ten *akçes* daily. See Pamuk, *İstanbul ve Diğer Kentlerde*, 102.

14. Yazdānbakhsh, *Al-Zīj al-mujmal*, TSMK Revan Ms. 1713.

15. For Persian émigré-scholars and scientists in the Ottoman lands in the later fifteenth and sixteenth centuries, see Heiderzadeh, "Patronage, Networks and Migration"; Sohrewide, "Dichter und Gelehrten"; Nassiri, "Turco-Persian Civilization."

16. De Blois et al., "Zid̲j"; van Dalen, *Islamic Astronomical Tables.*

17. Kennedy, "Survey"; King et al., "Astronomical Handbooks and Tables."

18. For notable examples, see *Zīc-i Uluğ Bey*; Kūshyar, *Ptolemaic Tradition.*

19. Hashemipour, "Buzjani."

20. Savadi and Khubravan, "Ḥarakat."

21. Fazlıoğlu, "Osmanlı Felsefe-Biliminin Arka Planı."

22. Idrīs Bidlīsī, *Hasht Bihisht*, SK Esad Efendi Ms. 2198, 33b–34a; also quoted in Heiderzadeh, "Ali Kuşçu," 15. For an overview of ʿAlī Qūshjī's move to the Ottoman capital and his short tenure there, see Ünver, *Türk Pozitif İlimler Tarihinden Bir Bahis,*; Heiderzadeh, "Ali Kuşçu," esp. 13–17; Umut, "Theoretical Astronomy," chap. 2.

23. For an overview of the transmission to the Ottoman world of texts in the astral sciences produced in the Persianate East in the second half of the fifteenth century, see Şen and Fleischer, "Books on Astrology."

24. For a statistical examination of this transition, see appendix C in Şen, "Astrology."

25. SK Ayasofya Ms. 3635, 2a–b.

26. Ṭaşköprülüzāde, *Al-Shaqāʾiq*, 840.

27. For Mollā Luṭfī, see Gökyay, *Molla Lutfi*; Maraş, "Tokatlı Molla Lutfi."

28. SK Ayasofya Ms. 3635, 2b: "wilāyāt al-ʿAjam."

29. Especially in Shiraz, there was a vibrant community of scholars and scientists in the late fifteenth century with interests in celestial knowledge. See Pourjavady, *Philosophy*; Bdaiwi, "Late Antique Intellectualism."

30. SK Ayasofya Ms. 3635, 3a.

31. Although I browsed several accounts that provide biographical information

about Shirazi scholars, including Kāzirūnī's *Sullam al-samāwāt*, Shūshtarī's *Majālis al-mu'minīn*, and Qummī's *Khulaṣāt al-tawārīkh*, I have not come across any information on Niʿmatullāh Shīrāzī. I'm grateful to Ahab Bdaiwi for his suggested sources.

32. SK Ayasofya Ms. 3635, 3a–4a.

33. Ibid.: "raghbat al-Sulṭān akthar min ʿilm al-nujūm ilā ʿilm al-ṭibb."

34. For Ibn al-Nafīs's commentary on Ibn Sīnā, see Fancy, "Medical Commentaries."

35. MC O. 71, 340.

36. ʿAbdurraḥmān, *Taqvīm* [of the year 915 AH].

37. Yılmazer, "Sahaflar Şeyhizade."

38. A petition by a certain Pīr Muḥammed es-Sivāsī, concerning the preparation of multiple copies of the almanac of the year 1062 AH (1652 CE), documents decisively that multiple copies of almanacs were produced by the staff of chief court *munajjims*. See TSMA E. 1298/25 for the relevant petition. See also Cunbur, "Bir Osmanlı Müneccimbaşısının Nevruz Tebrikleri."

39. Muṣṭafā Zekī, in his *Mücerrebnāme* (IU Nadir Ms. T 4315, 7b), refers to "ʿavām taḳvīmi" (lay almanac), by which he means those compilations lacking detailed, monthly ephemeris tables.

40. ʿAbdurraḥmān, *Taqvīm* [of the year 915 AH], 3a–5a: "Jadval fī bayān-i tavārikh-i āl-i ʿOs̱mān."

41. Ibid., 1b.

42. For a more extensive discussion of chronology sections in fifteenth-century Ottoman almanacs, see Şen, "Science and Politics."

43. In his encyclopedic work on different chronology systems, Bīrūnī often reminds his readers of the difficulty in establishing consistency among dating systems. See Bīrūnī, *Chronology*, 98–129. Kātib Çelebi also makes a reference to the same challenge in establishing pre-Islamic chronology in his *Taḳvīmü't-Tevārīḫ*, 1a.

44. Atsız, *Osmanlı Tarihine ait Takvimler*, 25.

45. Ibid., 69.

46. Atsız, "Fatih Sultan Mehmed'e Sunulmuş Tarihi Bir Takvim," 20; Atsız, "Hicri 858," 262.

47. Al-Bīrūnī, *Book of Instruction*, 188.

48. Tezcan, "Politics," 169. Tezcan contends that the drive toward controlling historical expression began—and ultimately failed—in the second half of the sixteenth century. Given the heightened interest in court-commissioned chronicles and histories in the late fifteenth and early sixteenth centuries, and the emergence of specific genres, such as *Selīmnāmes*, in the first half of the sixteenth century, what Tezcan discusses in the late sixteenth-century context could be extended back in time.

49. The other two extant sixteenth-century almanacs with chronology sections are from 1508 and 1576. The former, compiled by Qāḍī-yi Baghdad, a Persian émigré-scholar fleeing the Safavids in the early sixteenth century, is now available at the British Library (Ms. Or. 6432, 26b–52a). See Chapter 3 for more on Qāḍī-yi Baghdad. The other, a 1576 almanac with a chronology section (now available as BnF Ms. Turc 182), was likely produced—based on its surprising details about the local history of Diyarbakır and adjacent regions—by a *munajjim* named Teberrük, who was based in Diyarbakır at the time. A *mühimme* record from 1573–74, including an order sent to the governor of Diyarbakır,

reveals that a *munajjim* with the name Teberrük was granted 20 *akçes* for regularly sending almanacs to the Imperial Court. See A. DVNSHMHM.d. 25/2984.

50. ʿAbdurraḥmān, *Taqvīm* [of the year 915 AH], 3b: "khiyānat-i lashkar-i Tātār."

51. Ibid., 4a: "fakhr-i āl-i ʿOs̱mān."

52. Ibid., 4b–5a: "sar-surkhiyān-i mulḥidān."

53. Ibid., 8a.

54. Ibid., 6b: "khurūj-i kasāni ki daʿvā-yi haq va pādishāhi kunad."

55. Ibid., 13a: "bar qavl-i ḥukamā-yi Pārsiyān va bar qavl-i ḥukamā-yi Rūmiyān."

56. Ibid., 12a–b.

57. The *OALT* refers to a certain Dellākzāde as if he was the same as Khiṭābī. The confusion arises from the editors' uncritical reliance on the information Ṭaşköprülüzāde provides about a certain Mevlānā Ḥusām, known as Dellākzāde. This Mevlānā Ḥusām, hovewer, was a preacher and known for his expertise in Qur'anic recitation; Ṭaşköprülüzāde does not mention any astral pursuits in regard to him. Thus, it is highly unlikely that Dellākzāde and Khiṭābī were the same person.

The *OALT* also suggests, based on Khiṭābī's regional affiliation (i.e., Lāhījān), that he was the son of Shams al-Dīn al-Lāhījī (d. 1506?), a renowned disciple of Muḥammad Nūrbakhsh (d. 1464), who was an esteemed Sufi shaykh in Iran. See *OALT*, vol. 1, 63–64. Given the strong appeal of astral knowledge within the Nūrbakhshī circles, particularly clear in Muḥammad Nūrbakhsh's *Risālat al-hudā*, where he makes extensive references to Ptolemy and Naṣīr al-Dīn al-Ṭūsī to bolster his messianic claims, it may be tempting to believe that Khiṭābī was Shams al-Dīn al-Lāhījī's son or at least part of the Nūrbakhshī circles. However, conclusive evidence remains elusive in both Khiṭābī's and Shams al-Dīn al-Lāhījī's writings, including the latter's poetry collection. The *OALT* attributes to Shams al-Dīn al-Lāhījī a copy of Ṭūsī's astrological commentary on the pseudo-Ptolemaic *Kitāb al-thamara*. Yet this seemingly promising evidence proves faulty, as the colophon of the extant manuscript (SK Reisülküttap Ms. 572) clearly states that the copy was drafted in 1450 by a different person named Ismāʿil b. Yūsuf Lāhījī.

For the Nūrbakhshī interest in knowledge of the heavens and references to Ptolemy and Ṭūsī in *Risālat al-hudā*, see Bashir, "Risālat al-Hudā."

58. For the publication of the text, based on the sole surviving copy now housed at the Majlis Library in Tehran (catalog number Ms. 6376), see Somi and Bagheri, "Risāla-i tashrīḥ al-ālāt." Somi and Bagheri misattribute the text to another Sayyid Munajjim from the late fourteenth- to early fifteenth-century Timurid realm. See below in the text for discussion on the true identity of the author.

59. For brief references to Rukn al-Āmulī, in addition to Sayılı's *Observatory in Islam* (214–15), see Mozaffari and Zotti, "Observational Instruments," 146.

60. Kennedy, "Letter of Jamshīd al-Kāshī"; Sayılı, *Uluğ Bey*; Bagheri, "Newly Found Letter."

61. "3 Astronomy and Astrology," Storey Online. Last accessed online on 20 July 2024. https://referenceworks.brill.com/display/entries/SPLO/COM-20300000.xml.

62. Al-Āmulī, *Risāla panjāh bāb*. The four extant manuscript copies of the text I was able to locate in Istanbul are SK Ayasofya Ms. 2667; SK Carullah Ms. 1409; SK Ragıp Paşa Ms. 931; SK Lala İsmail Ms. 289. For additional copies listed in Storey Online, see note 61 above.

63. SK Ayasofya Ms. 2667, 2a: "ʿulūm-i ḥikamī."

64. Ibid.: "s̱amara-yi īn ʿulūm baʿd az vasīla ba-maʿrifatullāh ʿilm aḥkām va maʿrifat-i avqāt-ast."

65. Ibid., 3b: "madad va muʿāvanat-i salāṭin va iʿānat-i arbāb-i davlat va dīn."

66. Ibid., 3b–4a.

67. Somi and Bagheri, "Risāla-i tashrīh al-ālāt," 183.

68. Khiṭābī, *Taqvīm* [of the year 894 AH], TSMK Bağdat Ms. 310, 8a.

69. See appendix C in Şen, "Astrology."

70. Kappert, *Die osmanischen Prinzen*, 19–68.

71. Khiṭābī, *Jāmiʿ al-qismayn*, SK Ayasofya Ms. 2414M.

72. Ibid., 21b: "tarjīḥ-i ḥakīm bar ʿalīm."

73. Ibid., 23b.

74. Khiṭābī, *Muwaḍḍiḥ*, SK Ayasofya Ms. 2709.

75. Ibid., 2a–b.

76. For an early Turkish translation of Ṭūsī's *Sī faṣl* presented to Meḥmed I (r. 1413–21), see Aḥmed-i Dāʿī, *Muḫtaṣar.*

77. Khiṭābī, *Ṭāliʿ*, TSMK Yeni Yazmalar Ms. 830. Another copy of his horoscope is located at SK Esad Efendi Ms. 1997.

78. Raby and Tanındı, *Turkish Bookbinding*, 71–73.

79. Mīrzā Rustam Bahadur's nativity is located at Huntington Library Ms. HM 71897. For Iskandar Sultan's more famous nativity, now housed in Wellcome Library, see Elwell-Sutton, "Royal Timurid Nativity Book"; Keshaverz, "Horoscope"; Tourkin, "Iskandar-sultan"; Tourkin, "Medical Astrology"; Caiozzo, "Horoscope"; van Dalen, "Commentary."

80. Khiṭābī, *Ṭāliʿ*, TSMK Yeni Yazmalar Ms. 830, 6a–b.

81. The last major section in both the *Zīj-i Ilkhānī* and *Zīj-i Ulugh Beg* is solely devoted to the elaboration of these procedures. For the technical details of these indicators, see Samsó, *On Both Sides*, 291–300; Elwell-Sutton, *Horoscope of Asadullah Mirza*, 57–98.

82. For the pedagogical uses of horoscope charts in late medieval and Renaissance Europe, see Azzolini, "Refining"; Carvalho, *Astrologer at Work.*

83. Somi and Bagheri, "Risāla-i tashrīh al-ālāt," 184: "ʿārif al-maʿārif al-ʿulwiyya wa-l-sufliyya."

84. Ibid., 192.

85. Ibid.

86. Khiṭābī, *Taqvīm* [of the year 894 AH], 1a: "mustakhrijuhu al-ʿabd Khiṭābī al-Ḥusaynī."

87. Ibid., 4a.

88. Ibid., 9b ("āshūb miyān-i Akrād va Aʿrāb"); 14b ("mukhāṣama va mujādala miyān-i Atrāk va Aʿrāb," "nahb va ghārat dar vilāyat-i ʿArab"); 17b ("angiz-i fitna va āshūb miyān-i Atrāk va Akrād va Aʿrāb").

89. For the Ottoman-Mamluk warfare at the time, see Muslu, *Ottomans and the Mamluks.*

90. Barkan and Ayverdi, *İstanbul Vakıfları*, 155.

91. Barkan, "Edirne ve Civarındaki," 305: "Seyyīd Ḥüsāmeddīn Ḫiṭābī müneccim."

92. TSMA E. 1415/71.

93. Es-Seyyīd el-müneccim el-Tokatī, *Taḳvīm* [of the year 937 AH], TSMK III. Ahmed Ms. 3497, 3a.

94. Sayyid Munajjim (of the Timurid realm), *Laṭāʾif al-kalām*, KRK Ms. 310. For more on Sayyid Munajjim (of the Timurid realm), see Mozaffari, "Effect of Astrological Opinions," 363–64.

95. Sayyid Munajjim (of the Timurid realm), *Risāla al-shakl*, SK Yazma Bağışlar Ms. 1362.

96. MC O. 71.

97. Ibid.

98. TSMA E. 363/41.

99. MC O. 71, 17. The notation reads: "His son deceased; allocation for condolence" (*valadash mutavaffā; tavjīh-i taʿziya*).

100. These references are also extracted from the extensive register of payments and gifts now located at MC O. 71.

101. For this shift in language, see Şen, "Emergence."

102. SK Ayasofya Ms. 2694, 24a.

103. Leiden University Libraries Ms. Or. 8, 1a.

104. Yūsuf b. ʿÖmer, *Taḳvīm* [of the year 967 AH], KRK Ms. 546.

105. Yūsuf b. ʿÖmer, *Taqvīm* [of the year 916 AH], TSMK III. Ahmed Ms. 1960.

106. Ibid., 9b.

107. Yūsuf b. ʿÖmer, *Taqvīm* [of the year 919 AH], TSMK Emanet Hazinesi Ms. 1710, 10b.

108. For the role played by *majlis* in early modern Ottoman scholarly conversations, see Pfeifer, *Empire of Salons.* Pfeifer leaves unaddressed, however, the significance of *majlis* as a space for the transmission of "scientific" knowledge that may not have been conveniently accommodated within the confines of the madrasas.

109. For the figures pertaining to *munajjims*, see appendix A in Şen, "Astrology."

110. Markiewicz, *Crisis of Kingship*; E. Ökten, "Scholars and Mobility"; V. Genç, *Acem'den Rum'a*; Nassiri, "Turco-Persian Civilization."

111. Leālī, a poet from the late fifteenth century, composed the following lines that are emblematic of these concerns over the pro-Persianate patronage dynamics at the time: "Would you be well-liked at court, you must be a Persian or a Kurd [*sic*] . . . Those who come from Persia hope to become *sancaks* [*sic*] or even viziers"; cited in Babinger, *Mehmed the Conqueror*, 472. For more on the competition between Rumi and Persian learned people during the period, see V. Genç, *Acem'den Rum'a*, esp. 203–19. For an extensive discussion on the prestige of learning and writing in Persian in the Ottoman realm during the late fifteenth and early sixteenth centuries, see İnan, "Imperial Ambitions."

Chapter 3

1. In late Ottoman Cairo, not effectively ruled by the Ottoman center, there were several observatories, including the Bulaq observatory established in the 1840s. See Stolz, *Lighthouse.* The Imperial Observatory (Rasathane-i Amire, now the Kandilli Observatory), established in the Ottoman capital in 1868, was founded primarily for carrying out meteorological observations. See Fettahoğlu, *Rasathane.*

2. For an extensive discussion on the immense popularity of printed almanacs and

annual prognostications across various European countries in the early modern period, and the proliferation of printing houses publishing such materials, see Westman, *Copernican Question*, chap. 2; Capp, *Astrology and the Popular Press*; Carlebach, *Palaces of Time*; Green, *Printing and Prophecy*.

3. Géza Dávid's article on Istolni Belgrad (Székesfehérvár in central Hungary) in DİA mentions geomancers' shops in the city in the sixteenth century. A decision issued by the qadi court of Istanbul in 1563 reports the punishment of a certain geomancer (*remmāl*) for committing sodomy in his shop. See İpşirli, "XVI. Asrın İkinci Yarısında," 218. The poet Zātī (d. 1546) was well known for practicing geomancy, bibliomancy, and magic squares in his shop in Istanbul's Beyazıt Mosque square. See Çavuşoğlu, "Zati'nin Letayifi I." The fortune-tellers in the same square were also noted by several European observers visiting the Ottoman capital in the sixteenth century. See, for instance, Gerlach, *Türkiye Günlüğü*, vol. 2, 653.

4. While I have not come across any archival evidence of a sixteenth-century court *munajjim* being dismissed for inaccurate computations and predictions, an eighteenth-century court *munajjim*, Muṣṭafā Zekī (d. 1735), was dismissed on account of his consistent miscalculations of the first day of Ramadan. See Aydüz, "Osmanlı Devleti'nde Müneccimbaşılık ve Müneccimbaşılar," 196.

5. Patricia Blessing also notes the emergence of an "increasingly centralized organization in the creation of art for the Ottoman court" during Bāyezīd II's reign, together with the "larger epistemological project" to systematically organize and catalog knowledge. See Blessing, *Architecture and Material Politics*, 175–77. See also İklil Selçuk and Cihan Yüksel's "Early Modern Reflections" for an overview of studies portraying the growing institutionalization during Bāyezīd II's time.

6. Necipoğlu, *Age of Sinan*, chap. 5; Afyoncu, "Yüzyılda Hassa Mimarları."

7. Afyoncu, "Osmanlı Hekimbaşıları," 85; Aydüz, *XV. ve XVI. Yüzyılda Tophane-i Amire*, 39. Archival records and bureaucratic correspondence related to chief physicians show their evolution from being merely the private servants of the ruler and the royal family to assuming broader responsibilities that included responding to public demands and ensuring best practices within their profession. See Russell, "Physicians"; Sarı, "Medical Organization"; Serin, *Osmanlı Sağlık Sisteminin Yönetimi*.

8. Emecen and Şahin, *Ahkam defterleri*, 2–5; Emecen and Şahin, *906/1501 Tarihli*, xiii–xxiii.

9. Emecen and Şahin, *II. Bayezid Dönemi Ahkam defterleri*; B. Aydın, *XVI. yüzyılda*.

10. Özcan, *Kanunname*. For the use of Meḥmed II's law code by the late sixteenth- and early seventeenth-century Ottoman literati to control rapid social change, see B. Tezcan, " 'Kānūnnāme of Meḥmed II.' " For a recent critical overview of studies on Ottoman *ḳānūns* and *ḳānūnnāmes*, including Meḥmed II's ceremonial codes, see Darling, "Kaṇun and Kanunname."

11. Terzioğlu, "Sufis," 92.

12. İnalcık, *The Ottoman Empire*; Fleischer, *Bureaucrat and Intellectual*; Necipoğlu, "Kanun for the State"; Şahin, *Empire and Power*; Ferguson, *Proper Order*.

13. There is evidence that Bāyezīd II was already known as such during his lifetime. See Raby and Tanındı, *Turkish Bookbinding*, 81. Some of the laudatory poems dedicated

to Bāyezīd II, now housed in the Topkapı Palace archives, also refer to him as *velī*. See TSMA E. 363/13; 363/16.

14. For a similar observation, see Kafadar, "Between Amasya and Istanbul," 80.

15. For examples of extant almanacs from fourteenth-century Anatolia, see SK Nuruosmaniye Ms. 2782 (an almanac dated 1372 and presented to the Eretna court); Turan, *İstanbul'un Fethinden Önce*; Mercier, *Almanac*; Peacock, "Two Royal Almanacs." For a useful overview of narrative sources portraying the presence of *munajjims* during the Seljuqids, see Metin, "Selçuklular Zamanında."

16. BnF Ms. Supp. Pers. 367.

17. Kāşifī, *Ġazānāme-i Rūm*, 14.

18. Beyazıt Ms. 4619.

19. Quoted in Pertusi, *La caduta di Costantinopoli*, vol. 1, 75.

20. Ṭursun Beg, *Tā'rīḫ-i Ebū'l-Feth*, 44: "mehere-yi mühendisīn ve kümmel-i müneccimīn."

21. Tācīzāde Caʿfer Çelebi, *Maḥrūse-i İstanbul Fetḥnāmesi*, 10: "mehere-yi ʿulūm-ı nücūm."

22. Ḳıvāmī, *Fetḥnāme-yi Sulṭān Meḥmed*, 193.

23. The literature on Meḥmed II's intellectual interests, which encompassed Greek learning and Islamic philosophy among other things, is vast. Almost every year, another derivative volume on his intellectual endeavors is published in Turkish. For an extensive study of Meḥmed II's hybrid interests, see Babinger, *Mehmed the Conqueror*, 409–508; Necipoğlu, "Visual Cosmopolitanism"; Mavroudi, "Translations from Greek"; Casale, "Mehmed the Conqueror."

24. For the continuity, rather than rupture, of intellectual interests and cultural hybridity from the time of Meḥmed II to the reign of Bāyezīd II, see Necipoğlu, "Spatial Organization."

25. In 1489, Bāyezīd ordered that a list be drawn up of relics to offer to the king of France in exchange for Cem Sulṭān. See Raby, "El Gran Turco," 98. Necipoğlu, in "Visual Cosmopolitanism" (48) acknowledges that Bāyezīd offered them as gifts in exchange for keeping Cem Sulṭān in captivity.

26. Adnan Adıvar characterizes the execution of Mollā Luṭfī as the "first disaster ever to have occurred in the area of science and [free] thought in Ottoman Turkey" (*Osmanlı Türklerinde İlim*, 54). For a more nuanced assessment of the Mollā Luṭfī episode, see Ocak, *Osmanlı Toplumunda Zındıklar ve Mülhidler*, 239–68; Maraş, "Tokatlı Molla Lutfi."

27. For a review and critique of the studies that contrast the reign of Bāyezīd II with that of his father, see Muslu, "Ottoman-Mamluk Relations."

28. The undated letter sent by Leonardo is now housed in the Topkapı Palace Museum archive (the new catalog number is TSMA E. 754/99). According to the letter, Leonardo also proposed devising several other tools for the sultan, including a new kind of windmill and a sort of pump to empty out the water in the vessels. See Eyice, "II. Bayezid Devrinde."

29. There is no known contemporary evidence documenting any approach by Columbus to the sultan. The earliest implicit criticism of the sultan's alleged rejection of Columbus is from the seventeenth century. The renowned Ottoman traveler Evliyā Çelebi

(d. 1682) recounts in the last volume of his travelogue that Columbus advised the sultan to launch an expedition toward this "virgin" land to seize its riches, Bāyezīd allegedly declined, saying, "Mecca and Medina and this Old Word are enough to conquer. There is no need to go overwhelming distances." See Evliyā Çelebi, *Seyahatname*, vol. 10, 267.

30. Fisher, *Foreign Relations*; Kissling, "Betrachtungen über die Flottenpolitik"; Brummet, *Ottoman Seapower*, 89–121.

31. For a critical examination of the life cycle of the rumors that Bāyezīd II banned printing, see Schwartz, "Did Ottoman Sultans."

32. Pektaş, "Beginnings of Printing," 9.

33. Kafadar, *Between Two Worlds*, 97–98; Özel, "Limits of the Almighty."

34. For Bāyezīd II's support for different Sufi networks, see Karataş, "City as a Historical Actor," chap. 3; Öngören, *Osmanlılar'da Tasavvuf*, 256–59; Le Gall, *Culture of Sufism*, 35–40.

35. For theological debates held in the presence of Meḥmed II, see Balıkçıoğlu, *Verifying the Truth*. For Bāyezīd II's involvement in such debates, see Alper and Apaydın, *Molla Lutfi ve Fütuhat'ı*; Altaş, "Molla Lutfi'nin *'es-Sebʿu'ş-şidad.'* " Sinān Pasha (d. 1486) mentions in his supercommentary on Jaghmīnī's *Epitome* (SK Laleli Ms. 2120) that Meḥmed II often tested those in his presence with questions about the mathematical sciences. See also Fazlıoğlu, "Ali Kuşçu'nun bir Hendese Problemi."

Cem Sulṭān also seems to have developed an interest in studying the astral sciences, as indicated by a certain Yaḥyā, who presented him with an astrological treatise (*Miftāḥ al-nujūm*) in 1470 after hearing Cem Sulṭān's request for a book simple enough to teach him the basics. See TSMK Revan Ms. 1704. For another astrological treatise produced for Cem Sulṭān by a certain Nücūmī, see KRK Ms. 372.

36. For a detailed examination and facsimile publication of the library inventory of the Ottoman palace at the turn of the sixteenth century, see *Treasures of Knowledge*, eds. Necipoğlu, Kafadar, and Fleischer.

37. Idrīs Bidlīsī, *Hasht Bihisht*, SK Esad Efendi Ms. 2198, 33b–34a; also quoted in Heiderzadeh, "Ali Kuşçu'nun Astronomi Eserleri," 15.

38. For a general overview of the contours of scientific activities in the pre-sixteenth-century Ottoman realm, see Fazlıoğlu, "Osmanlılar (İlim ve Kültür)."

39. ʿAbdulwājid b. Muḥammad was born in Khurasan and came to Anatolia in the later fourteenth century. In addition to his treatise on the uses of astrolabe, he also composed a commentary on Ṭūsī's *Sī faṣl*. For a copy of this treatise dated 1412, see SK Carullah Ms. 2108, 1b–38a. See also Fazlıoğlu, "İthaf'tan Enmuzec'e"; J. Ragep, "Astronomy in the Fenari-Circle."

40. Fazlıoğlu, "Kadızade-i Rumi."

41. TSMA E. 970/44.

42. Altınay, "Fatih Devrine ait Vesikalar."

43. For the *müteferriḳa* status, see Uzunçarşılı, *Osmanlı Devletinin Saray Teşkilatı*, 428–31; Gökbilgin, "Müteferrika"; Murphey, "Formation of the Wider Palace Household," in *Exploring Ottoman Sovereignty*.

44. TSMA D. 9587, also available as a transcribed document in Barkan, "H. 933–934 Tarihli," 309.

45. Ibid.: "cemāʿat-i müneccimīn." For the status of *müşāhereḫorāns*, see Darling, "Ottoman Salary Registers."

46. MC O. 71. The full transliteration of the entire register was completed by İlhan Gök as a PhD dissertation. The cited page numbers refer to the original register.

47. TSMA D. 5475, also available as a transcribed document in Barkan, "H. 933–934 Tarihli," 313.

48. TSMA D. 9706; TSMA D. 10141.

49. TSMA D. 7843, also available as a transcribed document in Barkan, "H. 933–934 Tarihli," 323.

50. KK 1764 lists four *munajjims* between the years 1526–27 and 1535–36, but it is unclear whether these *munajjims* were members of the office or if the register only contains the names of experts who presented the court with an almanac. The *munajjims* listed in this register are Luṭfullāh, Yūsuf b. ʿÖmer (noted as Sinān b. ʿÖmer), Muṣliḥiddīn, and Necmeddīn.

MAD 559, a payroll book listing the names and wages of monthly-salaried palace personnel, enumerates two *munajjims* as part of the unit (*cemāʿat-i müneccimān*) for the year 1535–36, namely, İshaḳ and Yūsuf b. ʿÖmer, receiving 14 and 10 *akçes* per diem, respectively. KK 1864, a similar payroll book for the year 1547–48, records only the name of Yūsuf b. ʿÖmer. MAD 192, yet another payroll book, lists Yūsuf as receiving 12 *akçes* per day in January 1548. This information about Yūsuf's receipt of 12 *akçes* per day in 1548 and 1549 is also corroborated by the following registers: D. BRZ.d. 20616 (40, 84, 105), MAD 17888 (18, 75), and MAD 7118 (12, 28, 55, 82). While D. BRZ.d 20616 only mentions Yūsuf, the latter two additionally list [Riyāżī] ʿAlī, who received 6 *akçes* per diem.

51. A. RSK.d. 1459 records Yūsuf as receiving 12 *akçes* per day and 500 *akçes* for his submission of the annual almanac in March 1556. Similarly, MAD 492 indicates that he received 500 *akçes* in April 1555 for the delivery of an almanac.

52. Archival registers from the late sixteenth century also indicate that the amounts court *munajjims* received for their submission of almanacs remained consistent at 500 *akçes*, which was half or even one-third of what court *munajjims* earned during Bāyezīd II's time. For example, KK 1867 lists Muṣliḥiddīn and Ḳurd Çelebi as recipients of 500 *akçes* in March 1568; KK 1769 lists Ḳurd Çelebi and Bālī as receiving 500 *akçes* in March 1574; KK 1771 lists Ḳurd Çelebi and Baḳḳālzāde as receiving 500 *akçes* in March 1582; MAD 12769 records Baḳḳālzāde and Yūsuf as recipients of 500 *akçes* in March 1586; KK 1775 lists Baḳḳālzāde and Yūsuf as receiving 500 *akçes* in March 1595.

53. For solar observations undertaken by Khiṭābī in Istanbul, see Somi and Bagheri, "Risāla-i tashrīh al-ālāt." For alchemical performance in the sultan's presence, see Langermann, "From My Notebooks: A Compendium of Renaissance Science," 311–14.

54. King, "Two Astrolabes," 9. King also notes in his forthcoming article, "Astronomical Instruments," to be published in *Suhayl*, that in 2023, Sergei Maslikov from the Russian Academy of Sciences discovered in the National Library in St. Petersburg the mater of another astrolabe made by Nücūmī el-Rūmī and dedicated to Bāyezīd II (mater is a round metal disc, which serves as a frame that holds all the components of an astrolabe together).

55. Şen and Fleischer, "Books on Astrology."

56. TSMK Ms. III. Ahmed 3495, 88a.

57. Sanuto, *I diarii*, vol. 5, 458.

58. SK Fatih Ms. 4357, 33b.

59. Afṣāḥī, *Ḳaṣāyid-i Afṣāḥi*, 86.

60. Munajjim Bālī, *Tuḥfat al-faqīr*, SK Ayasofya Ms. 2588.

61. Qāḍī-yi Baghdād, *Taqvīm* [of the year 913 AH], BL Ms. Or. 6432/2, 27a: "ʿilm-i nuʿmānī ba-salṭanat-i sulaymānī."

62. His name is recorded on the register of gifts and payments (MC O. 71, 266) as the recipient on 25 April 1508 of 1,500 silver *akçes* for his debut presenting an almanac: "ibtidāʾ-yi ʿādat-i Mavlānā Sinān Qāḍī-yi Baghdād . . . ki taqvīm āvurd."

63. Firdevsī-i Ṭavīl, *Da'vetnāme*.

64. TSMA E. 754/87. The relevant part is written as صنعت الهيه, thus slightly ambiguous without vowel markings.

65. Ibid.

66. Ibid.

67. Janssens, "'Experience' (*tajriba*)"; McGinnis, "Scientific Methodologies"; Langermann, "From My Notebooks: On *Tajriba/Nissayon*"; Griffel, "Role of Sense Perception."

68. Griffel, *Al-Ghazālī's Philosophical Theology*, 208.

69. Langermann, "From My Notebooks: A Compendium of Renaissance Science"; Langermann, "From My Notebooks: Medicine, Mechanics and Magic"; Morrison, "Astronomical Treatise"; Morrison, "Scholarly Intermediary." For the Arabic translation of Zacuto's *Almanach Perpetuum* presented to Müʾeyyedzāde, see Pérez, "Estudio y edición."

70. Langermann, "From My Notebooks: A Compendium of Renaissance Science," 311–14.

71. Artun, "Hearts," 48.

72. Artun, "Bayezid-i Kimyai."

73. Uzunçarşılı, *Osmanlı Devletinin İlmiye Teşkilatı*, 145–49.

74. Fazlıoğlu, "Mīram Čelebī [*sic*]" Fazlıoğlu adds an extra generation between Qāḍīzāda al-Rūmī and Mīrim Çelebi, yet contemporary sources such as Ṭaşköprülüzāde, who personally studied with Mīrim, list his full name as "Maḥmūd b. Muḥammad b. Qāḍīzāda." Fazlıoğlu mentions that Mīrim Çelebi was the son of Quṭb al-dīn, but the information provided by Ṭaşköprülüzāde about Quṭb al-dīn—naming him as Quṭb al-dīn Muḥammad b. Muḥammad b. Qāḍīzāda—suggests that the two were brothers. See Ṭaşköprülüzāde, *Shaqāʾiq*, 522. For more on Mīrim Çelebi, see Morrison, "Cosmography, Cosmology, and Kalām."

75. Ṭaşköprülüzāde, *Shaqāʾiq*, 854.

76. Ibid., 522.

77. Mīrim Çelebi, *Sharḥ-i zīj-i Ulugh Beg*, SK Ayasofya Ms. 2697, 2a. In this autograph copy, Mīrim Çelebi says that he was at the service of a certain ʿAṭāʾullāh while his master was working on the Ulugh Beg tables. Interestingly, this little detail about ʿAṭāʾullāh is not included in other available copies of the text. The copy of the *Zīj-i Ulugh Beg* now housed in Oxford Bodleian (Marsh Ms. 396) was evidently used by Mīrim Çelebi. On several folios of the copy, Mīrim Çelebi has his marginalia, frequently adding his adjusted numerical values according to the latitude of Istanbul.

As for this Persian émigré scholar ʿAṭāʾullāh, who was likely from Kirman in origin, there is only scant information. Ṭaşköprülüzāde describes him as an expert in the use of *zīj* and compilation of almanacs and adds that he personally saw some of ʿAṭāʾullāh's accounts on astronomical instruments. See Ṭaşköprülüzāde, *Shaqāʾiq*, 364. During my research in the manuscript libraries of Istanbul and Europe, I was able to locate a treatise on the quadrant (SK Darülmesnevi Ms. 345), whose colophon registers that it was written by a certain ʿAṭāʾullāh in 1478–79.

78. Barkan, "Ayasofya Camii," 358. This is listed under the status of *zevāʾidḫorān* (lit. "the eaters of the overflow"), which comprised individuals who either possessed useful skills or had privileged familial ties. They were offered modest daily gratuities despite not providing any service within the endowment complex.

79. Ṭaşköprülüzāde, *Shaqāʾiq*, 522.

80. *OALT*, vol. 1, 90–101; *OASTLT*, vol. 1, 14–17.

81. Information is accessible at http://www.islamicmanuscripts.info/reference/books/Sothebys-19941019/Sothebys-19941019-109-128.pdf.

82. TSMA D. 9600.

83. Mīrim Çelebi, *Sharḥ*, 2b: "akmal va aʿqal-i qayāsira-yi ʿālam"; "ḥikam-i Aflāṭūnī"; "Mahdi-yi ākhir-zamān."

84. Ibid., 263b.

85. TSMK Hazine Ms. 1760, 40b: "vāsiṭa-yi ʿaqd-i ṣāḥib-qirānī"; "Mahdi'l-raḥma fī ākhir al-zamān."

86. For a discussion on the vocabulary of sovereignty in the late fifteenth- and early sixteenth-century Ottoman context, see Markiewicz, *Crisis of Kingship*, 151–91.

87. MC O. 71. He is listed at least eighteen different times, but none of these occasions is related to the presentation of an annual almanac.

88. Ibid., 159 (24 November 1505); ibid., 368 (19 February 1510): "barāy-i kharj-i ʿurs-i khud."

89. Idrīs Bidlīsī, *Hasht Bihisht*, SK Esad Efendi Ms. 2198, 370a–b: "bi-ṣoḥbat-i majlis-i humāyūn az sāʾir ʿulamāʾ mumtāz."

90. Uluçay, "Yavuz Sultan Selim Nasıl Padişah Oldu II," 120–21.

91. *OASTLT*, vol. 1, 14–17.

92. Mīrim Çelebi, *Masāʾil*, SK Bağdatlı Vehbi Ms. 2005.

93. Ibid., 47b.

94. Ibid., 10a: "khaṭā dar masāʾil bi-sabab-i khaṭā dar ālāt . . . chun usṭurlāb va rubʿ."

95. MC O. 71, 211.

96. Ibid.: "ibtidāʾ-yi ʿādat-i ʿAli munajjim, ṣākird-i Mīrim Çelebi."

97. ʿĀşıḳ Çelebi, *Meşāʿirü'ş-Şuʿarāʾ*, ed. Kılıç, vol. 2, 1064.

98. Ibid.: "ḫācedaşı ve fenn-i nücūmda pādaşı."

99. Ibid.

100. M. Yılmaz, "Özri ve Şiirleri."

101. SK Ayasofya Ms. 2474.

102. Bursa Türk ve İslam Eserleri Müzesi Ms. 11. I'm grateful to Zeren Tanındı for sharing the information concerning this manuscript.

103. TSMK III. Ahmed Ms. 3328.

104. TSMK III. Ahmed Ms. 3317. For a critical edition and analysis of the *Tadhkira*, see al-Ṭūsī's *Memoir on Astronomy.*

105. SK Ayasofya Ms. 2651.

106. SK Ayasofya Ms. 2702.

107. TSMK III. Ahmed Ms. 2128. For a recent critical edition of Ikhwan al-Safa's treatise on the study of the heavens, see Ikhwān al-Ṣafāʾ, *On Astronomia.*

108. SK Ayasofya Ms. 2618.

109. SK Ayasofya Ms. 2711.

110. TSMK III. Ahmed Ms. 3343.

111. SK Ayasofya Ms. 2432.

112. TSMA E. 10159/145.

113. Ibid.

114. Burns, "Court Astrologers," in *Astrology through History.*

115. Schmidl, "Magic and Medicine."

116. Kennedy, "Ulugh Beg"; "Heritage," in *Astronomy and Astrology in the Medieval Islamic World.*

117. Kennedy, "Letter of Jamshīd al-Kāshī"; Sayılı, *Uluğ Bey*; Bagheri, "Newly Found Letter"; Fazlıoğlu, "Osmanlı Felsefe-Biliminin Arka Planı."

118. Trigg, "From Samarqand," 152.

119. Bayram, *Adli Sultan*; Başpınar, "Sultan II. Bayezid'in Farsça Şiirleri Üzerine."

120. Michael Rogers also refers in his "Centralisation and Timurid Creativity" to the continuity between the Samarkand astronomical tradition and Bāyezīd II's activities. For the impact of Timurid culture on the Ottomans, see Necipoğlu, "From International Timurid to Ottoman"; Birnbaum, "Ottomans and Chagatay Literature"; Bernardini, "Ottoman 'Timuridism' "; İnalcık, *Şair ve Patron.*

121. SK Ayasofya Ms. 2410. The frontispiece of this manuscript comprising twenty-four folios introduces the contents of the volume as such: *Kitāb Aflāṭūn al-musammā bi-Ṭīmāwus fī al-ḥikamiyāt wa-risāla maqāla Aflāṭūn fī al-siyāsa* ("The book of Plato known as *Timaeus* about philosophy and a treatise containing Plato's tract on politics").

122. SK Ayasofya Ms. 2455. Fazlıoğlu, "Forcing the Boundaries." For a detailed examination of the contents of this anonymous treatise, see Treiger, "Fourteenth-Century Arabic Treatise."

123. Reeve, *Philosopher-Kings*, 75–79.

124. Turan, "II. Bayezid."

125. Emecen and Şahin, "Amasya."

126. For Bāyezīd's years in Amasya, see Kappert, *Die osmanischen Prinzen*, 20–95.

127. Vardanyan, *Amirdovlat Amasiatsi.*

128. Erünsal, "Mihri Hatun"; Havlioğlu, *Mihri Hatun.*

129. Sabuncuoğlu, *Cerrāḥiyyetü'l-Ḫāniyye.*

130. Soylu, "Şeyh Hamdullah."

131. Gökbilgin, "Müeyyed-zade"; Kurnaz, "Osmanlı Tarihinde."

132. For the letter to Meḥmed II from the private tutor and provincial treasurer of Prince Bāyezīd in Amasya, informing the sultan of his son's drug use, see TSMA E. 756/93.

133. For Prince Bāyezīd's pardon letter to his father, see TSMA E. 756/92.

134. For Müʾeyyedzāde's time in Shiraz, see Pfeiffer, "Teaching the Learned." For the Shirazi intellectual circles in the fifteenth and sixteenth centuries, see Bdaiwi, "Late Antique Intellectualism"; Anooshahr, "Shirazi Scholars." Specifically for Ghiyāth al-Dīn

Manṣūr al-Dashtakī, see Bdaiwi, "Youth Who Defeated Aristotle."

135. TSMK Revan Ms. 1713, 1a.

136. SK Ayasofya Ms. 2694, 1a.

137. Pfeiffer, "Teaching the Learned."

138. TSMA D. 9291/I–II.

139. Van Gulik et al., *Erasmus and His Books*; Hall, "Galileo's Library Reconsidered."

140. Erünsal, *Osmanlılarda Vakıf Kütüphaneleri*, 123–24; Pfeiffer, "Emerging," 154.

141. Ṭaşköprülüzāde, *Al-Shaqā'iq*, 472.

142. Pfeiffer, "Emerging," 158–60.

143. Ibid., 182–83.

144. Pérez, "Estudio y edición."

145. Pfeiffer, "Emerging," 172–82.

146. TSMA D. 9291/I.

147. Mü'eyyedzāde, *Al-Ḥawāshī ʿalā Sharḥ al-mawāqīf*, SK Ayasofya Ms. 2283.

148. Van Ess, "Jorjani"; Gümüş, *Seyyid Şerif Cürcani*; Sabra, "Science and Philosophy"; Hasan, "Foundations."

149. Hasan, "Foundations," 45–46.

150. Jurjānī, *Sharḥ al-mawāqīf*, vol. 3, 101–3.

151. For example, St. Augustine, in his lengthy critique of astrology in the *City of God*, specifically refers to the twin problem. Cited in Smoller's *History, Prophecy, and the Stars*, 26–27. Martin Luther also resorts to the same argument in his critique of astrology, asking astrologers to explain how "Esau and Jacob . . . born together, of one father and one mother, at one time, and under equal planets," had contrary natures. See Luther, *Table Talk*, 372.

152. Jurjānī, *Sharḥ al-mawāqīf*, vol. 3, 103.

153. SK Ayasofya Ms. 2283, 8a–b.

154. Ibid.

155. Ibid., 9a–b.

156. Ibid., 10b.

157. Antov, *Ottoman "Wild West,"* 88. *Mahdī* is the savior expected to be sent by God to reestablish the rule of justice on Earth before the end times.

158. Oruç Beg, *Oruç Beğ Tarihi*, 165.

159. It is incorrect to assume that Bāyezīd II did not take any measures against Safavid sympathizers in Anatolia and Balkans. See Antov, *Ottoman "Wild West,"* 22; Emecen and Şahin, *II. Bāyezīd dönemine ait 906/1501 tarihli* , xxiv–xxv.

160. Markiewicz, *Crisis of Kingship*, 267.

161. Ibn Kemāl, *Tevārīḫ-i Āl-i ʿOs̱mān*, passim; V. Genç, "Iranian *Shāh-nāma* Writer."

162. Firdevsī-i Ṭavīl, *Ḳuṭb-nāme*.

163. Tansel, "Yeni Vesikalar Karşısında," 208.

164. İnalcık, "Case Study"; Lefort, *Documents grecs*.

165. Eyice, "II. Bayezid Devrinde."

166. TSMA E. 10159/6.

167. TSMA D. 10026. An earlier register dated 1496 likewise lists astronomical instruments. See Necipoğlu, *Architecture, Ceremonial, and Power*, 134–35.

Chapter 4

1. The literature on the efforts to revive Istanbul after 1453 through different policies, including 'forced settlement' (*sürgün*), is vast. See Ergin, "İstanbul'un Fethinden Sonra"; Ayverdi, *Fatih Devri Sonlarında*; İnalcık, "Istanbul"; İnalcık, "Policy of Mehmed II"; Kafesçioğlu, *Constantinopolis/Istanbul*. After defeating the Safavids and Mamluks, Selīm I brought many artists and artisans from Tabriz and Cairo to Constantinople. See Uzunçarşılı, "Osmanlı Sarayı'nda Ehl-i Hıref."

2. Fleischer calls the long fifteenth century down to the 1550s the "Ottomanizing" era, which "represents an intermediate phase in the construction of a new formulation of dynastic legitimacy and its lineage, a new language (Ottoman Turkish), a new genealogy of knowledge particular to the Ottoman lands and their dynastic inheritance." See Fleischer, "Learning and Sovereignty," 155.

3. Flemming, "Der Ǧāmī ül-meknūnāt"; Flemming, "Sāhib-Kirān und Mahdī" (English translations of both articles were published in Flemming, *Essays on Turkish Literature*); Fleischer, "Lawgiver as Messiah"; Fleischer, "Mahdi and Millennium"; Fleischer, "Seer to the Sultan"; Fleischer, "Shadow of Shadows"; H. Yılmaz, *Caliphate Redefined*.

4. TSMA D. 5475: "cemāʿat-i müneccimīn."

5. TSMA D. 10184.

6. Another salary register, TSMA D. 2921/1, documents the list of people accompanying Selīm in his move to Istanbul for the enthronement, but no *munajjim* is cited.

7. TSMA D. 9706; TSMA D. 10141.

8. Qazvīnī, *Tadhkira*, 360: "dar ʿilm-i riyāżī ki mis̱l-i salṭanat-i mirās̱ī-yi u-ast."

9. Keşfī Meḥmed, *Selīmnāme*, SK Esad Efendi Ms. 2147, 133b.

10. Ḫāce Kemāl, *Taḳvīm* [of the year 925 AH], TSMK Emanet Hazinesi Ms. 1695, 6a: "bu yılıñ ibtidāʾsında ḥażret-i pādişāh-ı ʿālem-penāh kendü nefsine sefere çıkmayalar . . . amma sefere çıktıkları vaḳit dördüncü ve beşinci ay ola, vallāhu ʿālem." This Kemāl may be the same individual mentioned in the archival register (MAD 22082, 5) from September 1528.

11. Bidlīsī (Idrīs), *Salīmshāhnāma*, 136: "bi-ittifāq-i arbāb-i aḥkām-i nujūm . . . ān kusūf-i nayyir-i aʿẓam dar mashriq dalīli rūshan . . . bud bar tamam nakbat-i zavāl-i davlat-i khusrav-i mashriq-zamīn."

12. *Taqvīm* [of the year 920 AH], TSMK Revan Ms. 1711/11, 266b: "taʾthīr-i kusūf dar ān diyār [diyār-i ʿAjam] . . . ba-vujūd āmad . . . va dalīl-ast bar nikbat-i shakhṣi-yi ʿaẓīm al-shān dar . . . diyār-i ʿAjam."

13. ʿĀşıḳ Çelebi, *Meşāʿirü'ş-Şuʿarāʾ*, vol. 3, 1396–98; ʿĀlī (Muṣṭafā), *Künhü'l-Aḫbār'ın Tezkire Kısmı*, 146–47. For *kuloğlu*, see Özcan, "Kuloğlu."

14. For a typical *madrasa* student's course of studying the disciplines in an orderly fashion (*ʿulūmu ʿale't-tertīb görmek*), see Ṭaşköprülüzāde's autobiographical recollections in his *Al-Shaqāʾiq*, 850–65. See also Şen, "Sultan's Syllabus Revisited."

15. Riyāżī, *Terceme-i Şāhī*, SK Fatih Ms. 4264, 6a–b: "ʿulūm-ı edebiyye . . . ve fünūn-ı ʿArabiyye."

16. Ibid.

17. Şen, "Sultan's Syllabus Revisited," 207.

18. Riyāżī, *Terceme-i Şāhī*, 6b–7a.

19. ʿĀşıḳ Çelebi, *Meşāʿirü'ş-Şuʿarāʾ*, vol. 3, 1396: "evvelā ṭarīḳ-i ʿulūma sülūk idüp . . .

ṭabᶜı bi'ṭ-ṭabᶜ riyāżī olduġı ḥaseble ᶜulūm-ı ṭabīᶜiyye ve fünūn-ı riyāżiyyeye meyl idüp fenn-i necāmete iştiġāl eyledi."

20. Ṭaşköprülüzāde, *Shaqāʾiq*, 44: "ghalaba ᶜalā ṭabᶜahu al-riyāḍiyyāt."

21. See note 104 in Chapter 1.

22. For an overview of the discussion on Ottoman "Sunnitization," see Terzioğlu, "How to Conceptualize Ottoman Sunnitization." For more on the ongoing discussion regarding the scope and limitations of "Sunnitization," see the recent compilations of volumes: Erginbaş, ed., *Ottoman Sunnism*; Krstić and Terzioğlu, eds., *Historicizing Sunni Islam*.

23. Allouche, *Origins and Development*; Zarinabaf-Shahr, "Qizilbash 'Heresy'"; R. Yıldırım, "Turkomans between Two Empires"; Atçıl, "Safavid Threat"; Karakaya-Stump, *Kizilbash-Alevis*; Baltacıoğlu-Brammer, "Neither Victim nor Accomplice."

24. Pfeifer, "New Hadith Culture?"

25. Kātib Çelebi, *Kashf al-ẓunūn*, vol. 1, 680; also cited in Balıkçıoğlu, "Coherence of Incoherences," 7.

26. Muṣṭafā ᶜĀlī, *Künhü'l-aḫbār*, 152. Intriguingly, ᶜĀlī (Muṣṭafā) notes (3) that learned people enjoyed much more benevolence and recognition during the time of Meḥmed II thanks to the auspicious influences of the stars at the time: "ol devr-i laṭīfiñ nücūmu teʾs̱īrātından."

27. Unan, *Kuruluşundan Günümüze*, 375–78.

28. Fazlıoğlu, "İki Ucu Müphem," 14.

29. Besides his encyclopedia of various branches of knowledge, *Kashf al-ẓunūn* (The removal of doubt), written in Arabic and listing more than 14,500 books in alphabetic order, Kātib Çelebi is known for his comprehensive biographical dictionary of scholars, *Sullam al-wuṣūl* (The ladder of elevation).

30. TSMA E. 968/70: "ᶜulūm-ı ġarībeye ki bu diyārda kimesne meşġūl olmuş değildir heyʾe ve hendese ve ālāt-ı raṣadiye gibi."

31. KK 1764, 25: "ᶜādat-i Luṭfullāh munajjim ki dar rūz-i Navrūz taqvīm dāde 1000." Another archival record from 1534 (KK 1863, 131) also gives similar information about Luṭfullāh.

32. KRK Ms. 325. One specific chapter of Luṭfullāh's horoscope (9a–10b) is titled "Calculating the true longitudes of seven planets at the time of birth by using the information in the Ulugh Beg tables prepared according to the observational enterprise in Samarkand."

33. KRK Ms. 349M. Riyāżī's marginal note is found on page 54a, briefly discussing the difference between a nighttime and daytime ascendant based on Ptolemy.

34. Ibid., 110b.

35. Ibid., 1a.

36. TSMK Hazine Ms. 1760. The four colophons revealing Luṭfullāh b. Mübārek's name are located on pages 15b, 25b, 38a, and 63a.

37. Miroğlu, "İstanbul rasathanesine ait belgeler," 80.

38. Adıvar, *Osmanlı Türklerinde İlim*, 82; İzgi, *Osmanlı Medreselerinde İlim*, vol. 1, 126; Necipoğlu, *Age of Sinan*, 149; B. Tezcan, "Some Thoughts," 145–46.

39. ᶜĀşıḳ Çelebi, *Meşāᶜirü'ş-Şuᶜarāʾ*, vol. 3, 1206–7: "ᶜArab u ᶜAcemden gelen erbāb-ı hünerler . . . meclisinüñ celīsi idi." For more on Kara Bālīzāde, see Velioğlu, "Müneccim Riyazi Ali," 157–65.

40. A. Yıldırım, "16. Yüzyılda"; Emecen, *Kanuni Sultan Süleyman*, 250. For a general discussion on the Ottoman ruling elite's patronage of poets and artists, see İnalcık, *Şair ve Patron*.

41. Velioğlu, "Müneccim Riyazi Ali," 44.

42. Gül, "18. Yüzyılda Yeniçeri Teşkilatı," 59–60; cited in Velioğlu, "Müneccim Riyazi Ali," 10.

43. Velioğlu, "Müneccim Riyazi Ali," 19.

44. Ibid.

45. Diyarbakır Ziya Gökalp Yazma Eserler Kütüphanesi Ms. 1467; also cited in Velioğlu, "Müneccim Riyazi Ali," 20.

46. UPenn Kislak Center for Special Collections, Rare Books and Manuscripts LJS 412, 10b.

47. Velioğlu, "Müneccim Riyazi Ali," 45–54.

48. SK Ragıp Paşa Ms. 913; SK Fatih Ms. 3421. The latter is also cited in Morrison, *Islam and Science*.

49. SK Nuruosmaniye Ms. 2800; SK Hamidiye Ms. 1446. Also cited in Velioğlu, "Müneccim Riyazi Ali," 61.

50. SK Kadızade Mehmed Ms. 339; also cited in Velioğlu, "Müneccim Riyazi Ali," 52.

51. Velioğlu, "Müneccim Riyazi Ali," 48.

52. The copy of the *Zīj-i Ulugh Beg* possessed by Riyāżī is now located in SK Reisülküttab Ms. 571. He also had a copy of Mīrim Çelebi's commentary on it, now found in SK M. Nuri Efendi Ms. 151. For Riyāżī's copies of Abū'l-Fidāʾ and ʿAli b. Abī al-Rijāl, see Velioğlu, "Müneccim Riyazi Ali," 6, 48.

53. Riyāżī, *Terceme-i Şāhī*, SK Fatih Ms. 4264, 265a.

54. ʿĀşıḳ Çelebi, *Meşāʿirü'ş-Şuʿarāʾ*, vol. 3, 1396–97.

55. Ibid., 1397: "ānıñ bir şaṭrı yokdur İbn Şāṭır," "Ebū Maʿşer değüldür ʿöşr-i ʿāşir."

56. Ḳınalızāde (Ḥasan), *Teẕkiretü'ş-Şuʿarāʾ*, 401.

57. MAD 7118. The latest record I was able to find about Riyāżī as a recipient of gifts in the capacity of court *munajjim* is from March 1556. See AE.SSÜL.I., 4/258, 47b.

58. Riyāżī, *Ḳuvvet-i ḳudṣiyye*, SK Nuruosmaniye Ms. 2685, 2a–3a.

59. Riyāżī, *Terceme-i Şāhī*, SK Fatih Ms. 4264, 6a–8a.

60. Ibid., 7b: "dil-i ʿalīl muḥtāc-ı nice ṭabīb-i cāhil . . . olmasun diyü."

61. Ibid., 8b: "Bulurum fażlla gerçi tesellā, ḳanı erbāb-ı fażla ḥürmet ammā."

62. Turan, *Kanuni Süleyman Dönemi*.

63. Riyāżī, *Terceme-i Şāhī*, 5b.

64. Kılıç, *Şehzade Bayezid "Şahi"*; Velioğlu, "Müneccim Riyazi Ali," 172–83.

65. For contemporary European visitors' observations about Selīm II's overindulgence in drinking and courtly pleasures, see K. Şahin, *Peerless among Princes*, 252.

66. For a biography of Semiz ʿAlī Pasha and his architectural patronage, see Afyoncu, "Semiz Ali Paşa"; Necipoğlu, *Age of Sinan*, 384–89.

67. Taqī al-Dīn, *16'ıncı asırda Osmanlılarda saat*.

68. Velioğlu, "Müneccim Riyazi Ali," 34.

69. ʿĀşıḳ Çelebi, *Meşāʿirü'ş-Şuʿarāʾ*, vol. 3, 1397: "ḥālā vilāyet-i Ṭamışvar ve Budun'da zaʿīm olub."

70. Fekete, "XVI. Yüzyılda Taşralı."

71. Ibid.

72. MAD 7118.

73. ʿĀşıḳ Çelebi, *Meşāʾirüʾş-Şuʿarāʾ*, vol. 3, 1397–98.

74. Velioğlu, "Müneccim Riyazi Ali," 140–55.

75. This text has been discussed at length by Şen in "Practicing Astral Magic."

76. The text has already been transcribed and published by Sayın Dalkıran based on the copy now housed at SK Esad Efendi Ms. 3782. See Dalkıran, "İbn-i Kemal'in Tılsımat Risalesi." Menderes Velioğlu has recently identified twelve copies of the text; see Velioğlu, "Müneccim Riyazi Ali," 113–14.

77. Ménage, "Kemāl Pasha-zāde"; Atsız, "Kemalpaşaoğlu'nun Eserleri" I and II; Öçal, *Kışladan Medreseye*.

78. SK Esad Efendi Ms. 3782.

79. Varlık, *Plague and Empire*, 233. This treatise has survived in over forty copies, the oldest of which seems to have come down to us from 1560–61. See Atsız, "Kemalpaşaoğlu'nun Eserleri II," 133–34.

80. SK Hacı Mahmud Efendi Ms. 5584, 42b, 45b.

81. Velioğlu, "Müneccim Riyazi Ali," 117.

82. Riyāżī, *Terceme-i Şāhī*, 132b, 170b.

83. Velioğlu, "Müneccim Riyazi Ali," 116.

84. SK Hacı Mahmud Efendi Ms. 5584, 37b–38a.

85. Varlık, *Plague and Empire*, 179–84.

86. The prime example is the tenth-century compendium on celestial magic, titled *Ghāyat al-ḥakīm* (The aim of the sage), by the Andalusi mathematician Maslama al-Qurṭubī (d. 964). The book is also known in Europe, as the Picatrix, having been translated into Castilian from Arabic around the mid-thirteenth century. For the German translation of the Arabic edition of the Ghāyat al-ḥakīm, see Ritter and Plessner, "*Picatrix*."

87. SK Hacı Mahmud Efendi Ms. 5584, 39b. For Yada stone and its use in Ottoman-Turkish culture, see Öğreten, "Türk Kültüründe 'Yada Taşı.' "

88. Ibid., 40b: "delāʾil ve berāhin . . . tecrübe ve imtiḥān."

89. Ibid., 41b–42a.

90. Ibid., 46a: "ṭılsım beden-i ʿāleme ʿilāc itmekdir."

91. Ibid.

92. Ibid., 46b.

93. Ibid., 48a.

94. Ibid., 46a: "baʿż-ı ālāt ve esbāba muḥtācdır ki mülūk ve selāṭīnden ġayrı kimse āña ḳādir değildir."

95. Ibid., 55a: "fünūn-ı ḥikemiyyenin ḥaḳāyiḳine ve ʿulūm-ı felsefiyyeniñ deḳāyiḳine vuḳūfda küllī kūşişler . . . ve ḫāṣṣaten ʿilm-i nücūma diḳḳatler."

96. SK Esad Efendi Ms. 3782, 89a: "bu aṣl aʿmāl-i ʿaẓīme . . . muʿavenet-i ṣāḥibḳırān-ı zamān ve ḥimāyet-i ḫāḳān-ı Süleymān-mekān olmayınca ẓuhūra gelmek mümkün değildir ammā añlarıñ ʿināyet[i] . . . ẓuhūr bulucak en kemter-i bī-kes . . . Arisṭūṭālīs ve Eflāṭūn olmak muḳarrerdir."

97. Ḳınalızāde (Ḥasan), *Tezkiretüʾş-Şuʿarāʾ*, 401.

98. ʿĀlī (Muṣṭafā), *Tezkire*, 147: "iʿtiḳādāt-ı nāsa ḫalel."

99. Ménage, "Serpent Column"; Yerasimos, *Türk Metinlerinde Konstaniyye ve Ayasofya Efsaneleri*.

100. S. Tezcan, *Topkapı Sarayı Müzesi Koleksiyonundan Tılsımlı Gömlekler*; Felek, "Fears, Hopes, and Dreams"; Ekhtiar and Parikh, "Power and Piety."

101. Contemporary poet Fiġānī (d. 1532) echoed the disapproval in his famous satiric couplet in Persian: "There descended to this world two Ibrahims; one smashed the idols, the other revived them." This verse apparently cost Fiġānī his life after Ibrāhim Pasha, offended, ordered his execution. Cited in Ágoston, *Last Muslim Conquest*, 168.

102. ʿĀşıḳ Çelebi, *Meşāʿirü'ş-Şuʿarā'*, vol. 3, 1396: "ḥiddet-i ṭabʿı . . . āteşden tīz."

103. The earliest available record of him is from the extensive register of gifts and payments housed at MC O. 71; the latest record of him is an almanac of the year 1560 now located at KRK Ms. 546.

104. *OALT*, vol. 1, 137–38.

105. SK Carullah Ms. 1500.

106. MC O. 71, 437: "ba-ṣila khāhad raft."

107. Yūsuf b. ʿÖmer, *Taqvīm* [of the year 916 AH], TSMK III. Ahmed Ms. 1960, 28b: "al-muwaqqit fī al-jāmiʿ al-jadīd al-sulṭān Muḥammad Khān."

108. See, for instance, *Taqvīm* [of the year 919 AH]: "al-munajjim fī bāb al-sulṭān"; TSMK Emanet Hazinesi Ms. 1710, 31b: "al-munajjim li-dargāh-i ʿālī."

109. Necmeddīn signed his *Taḳvīm* [of the year 932 AH] (Arkeoloji Müzesi Ms. 1607/1, 19b) and *Taḳvīm* [of the year 935 AH] (TSMK Revan Ms. 1711/16, 364b) as the timekeeper of the Old Mosque in Edirne (*muwaqqit fī jāmiʿ al-ʿatīq*). He apparently became the timekeeper of the Fatih Mosque complex in Istanbul (*muwaqqit bi-jāmiʿ al-sulṭān Muḥammad*) before he drafted the *taḳvīm* of the year 1535 (BnF Ms. Turc 183, 14b). As two other extant almanacs of his bear the same signature document (TSMA Revan Ms. 1711/18, 398a; BL Ms. Or. 6432/1, 25b), Necmeddīn served in this position at least until 1553.

110. Yūsuf b. ʿÖmer, *Taḳvīm* [of the year 967 AH], KRK Ms. 546.

111. Uludağ, "Ricalü'l-gayb." For the *rijāl al-ghayb* doctrine of Ibn al-ʿArabī, see Chodkiewicz, *Seal of the Saints*, 89–102.

112. For a detailed discussion on the change in almanacs' contents, see Şen, "Astrology," 237–304.

113. KRK Ms. 546, 1b–2a: "şeyḫü'l-muḥaḳḳiḳīn ve ḳuṭbu'l-ʿārifīn."

114. For the circulation of these texts at the court of Süleyman, see Fleischer, "Mediterranean Apocalypse."

115. *OALT*, vol. 1, 161–79.

116. Fazlıoğlu, "ʿAli al-Muwaqqit."

117. Muvaḳḳit Muṣṭafā, *Tuḥfetü'z-zamān.*

118. Fazlıoğlu's entry on Muṣṭafā (n. 116) hints that his first or better-known name was ʿAlī. If that is the case, then the records in the register of gifts and payments (MC O. 71) in the early 1510s mentioning a certain ʿAlī as the student of Mīrim Çelebi can be taken as an evidence.

119. Muvaḳḳit Muṣṭafā, *Tuḥfetü'z-zamān*, 370–74.

120. Ibid., 371: "ḫalḳıñ eks̱eri."

121. Ibid., 372: "evḳāt aḥvāli."

122. Ibid.: "cigerleri ḫūn olmuşdur" (lit. "their lungs bled")

123. A. DVNSHMHM.d. 25/1231.

Chapter 5

1. In his *Futures Past*, Reinhart Koselleck marks the eighteenth century as a transformative moment regarding the changing notions of the "future." For an informative compilation of studies focusing on early modern engagement with the concept of the future within the European context, see Brady and Butterworth, *Uses of the Future*.

2. Pietruska, "Forecasting"; Charmasson, "Divinatory Arts." See also Heiduk, Herbers, and Lehner, *Prognostication*.

3. Wehr's *Dictionary of Modern Written Arabic*, for instance, provides such meanings for the following words (844): *kahāna*: prediction, prophecy; *kihāna*: divination, soothsaying, fortunetelling.

4. How was the potential value of a revenue-generating instrument estimated before the granting of taxing rights remains a question that is yet to be comprehensively examined. For some preliminary observations, based on the distribution of *iltizām* (short-term prerogative of taxing an agricultural land), see Darling, *Revenue-Raising*, 141–45. For the increasing role of finance bureaucrats from the seventeenth century onward as agents of future projections, see Küçük, "Bureaucratic Sense of the Forthcoming."

5. Necipoğlu, "Plans and Models"; Kale, "Intersections"; Kale, "From Measuring."

6. MacDonald and Gardet, "Al-Ghayb."

7. For some useful overviews, see Schmidt, "Occult Sciences"; Sariyannis, "Knowledge"; Sariyannis, *Horizons*.

8. Lelić, "Ottoman Physiognomy." See also Fahd, *La Divination arabe*, 387–88, for its use in the slave trade in Islamic history.

9. Ferhad and Bağcı, "Art of Bibliomancy."

10. The sometimes confusing ways in which such notions as "magic," "occult," "occultism," "esotericism," or "divination" are used have yet to reach a consensus. For a brief discussion on what we may lose by lumping together magic and divination, and how the evidence from classical Greek and Roman sources presents a different view, see Struck, *Divination and Human Nature*, introduction.

11. See note 3 in Chapter 4.

12. Gardiner, "Al-Jafr."

13. At the top of one archival report conveying *munajjims*' calculation of an auspicious hour before an intended military campaign, Sultan ʿAbdulḥamīd I (r. 1774–89) has added a note saying that one should pay attention to the Prophet Muhammed, not to *munajjims*. See HAT 22/1053. Also cited in Aydüz, "Osmanlı Devleti'nde Müneccimbaşılık Müessesesi," 181.

14. TSMA D. 1298.

15. TSMA D. 1298/27.

16. Muslu, *Ottomans and the Mamluks*, 134–55.

17. TSMA D. 1298/27.

18. For a discussion on ʿulamaʾ as "ritual specialists," see Moin, "'Ulama' as Ritual Specialists."

19. For the gradual increase in the use of mechanical timepieces in the Ottoman world, see O. Kurz, *European Clocks*; Gündüz, *Alaturkadan Alafrangaya Zaman*.

20. For an example of a report including a reference to seconds, see TSMA D. 1298/10.3.

21. HAT 14/576.

22. TSMA D. 1298/10.2.
23. Shaw, "Ottoman Archival Materials."
24. AE, I. Abdulhamid 17/1466.
25. Ibid.: "ʿilmlerinde ḳangısı mahāreti ẓāhirse."
26. Riyāżī, *Zāyiçe*, SK Yazma Bağışlar Ms. 4034.
27. Ibid.
28. Ibid., 5b.
29. Ibid., 13a.
30. Ibid., 13b.
31. MAD 7118.
32. ʿĀşıḳ Çelebi, *Meşāʿirü'ş-Şuʿarā*ʾ, vol. 3, 1398.
33. Riyāżī, *Zāyiçe*, 1b–2a.
34. Ibid., 2a.
35. Ibid: "cümle-i müneccimīn ve cumhūr-ı muḳavvimīn."
36. Irani, "Arabic Numeral Forms." For the use of the *abjad* system by astral experts in the Islamicate world, see van Dalen, *Islamic Astronomical Tables*.
37. Riyāżī, *Zāyiçe*, 3a.
38. Ibid.
39. Ibid., 3b.
40. Ibid., 4a.
41. Ibid., 4b: "felāsife-yi İslām."
42. Ibid., 4b–5a. Tankalūshā's book is the *Kitāb fī ṣuwar daraj al-falak* and was translated into Persian before the twelfth century. Riyāżī's lengthy quote in Arabic suggests that he might have had access to the Arabic version of Tankalūshā's account. It is more likely, though, that Riyāżī learned the teachings of Tankalūshā through Fakhr al-Dīn al-Rāzī's book on celestial magic, *Al-Sirr al-maktūm*, which was heavily informed by the teachings of Tankalūshā. As we have already seen, Riyāżī started translating *Al-Sirr al-maktūm*, but he could not complete it. See Riyāżī, *Terceme-i Şāhī*, SK Fatih Ms. 4364, 265a. For the modern Persian edition of Tankalūshā's text, see*Tankalūshā*, ed. Malik. See also Vesel, "Teucros."
43. Riyāżī, *Zāyiçe*, 5a.
44. I thank Barış İlhan, a practicing astrologer in Turkey, for her help in elucidating the difference.
45. Riyāżī, *Zāyiçe*, 5b: "cümlesi bi'l-ittifāḳ semiʿnen ve aṭaʿnen diyüb . . . mustaḫriciñ fażīletine evvelden daḫı ziyāde iʿtirāf ve iʿtiḳād itdiler."
46. Riyāżī, *Terceme-i Şāhī*, SK Fatih Ms. 4264, 114a: "ānlar daḫı erkān-ı devlet ḥużūrunda cemʿiyet idüb faḳīr henüz varmadan bir vaḳt iḫtiyār itmişler ṣoñra faḳīr varub iḫtiyār etdikleri ṭāliʿe naẓar saldum . . . ve bir ġayrı ṭāliʿ daḫı iḫtiyār itdim ki cümle-i erbāb-ı tencīm ve aṣḥāb-ı taḳvīm semiʿnen ve aṭaʿnen didiler."
47. Çulpan, "İstanbul Süleymaniye Camii Kitabesi," 293: "al-bidāya fi awākhir Jumāda'l-ūlā li-sana sabʿa wa-khamsīn wa-tisʿa-miʾa."
48. Miʿmār Sinān, *Sinan's Autobiographies*, 43.
49. Celālzāde (Muṣṭafā), *Ṭabaḳātü'l-Memālik*, 719; Şahin, *Empire and Power*, 208–13.
50. Celālzāde (Muṣṭafā), *Ṭabaḳātü'l-Memālik*, 719: "mehere-i erbāb-ı taḳvīm ve nücūm."
51. Stowasser, *Day Begins at Sunset*; Wishnitzer, "Timekeeping." For a technical dis-

cussion on *munajjims*' computations of lunar crescents, see King, "Lunar Crescent Visibility."

52. (Muṣṭafā) ʿĀlī, *Mustafa Ali's Description of Cairo*, 32.

53. (Muṣṭafā) Sāfī, *Zübdetü't-Tevārīḫ*, vol. 1, 51: "ehl-i tencīm ve istiḫrāc-ı aṣḥāb-ı zīc ü taḳvim." For a similar note on the construction of the Nuruosmaniye Mosque, see Barkan, *Süleymaniye Cami ve İmareti İnşaatı*, vol. 1, 50.

54. Riyāżī, *Zāyiçe*, 6b.

55. Ibid.: "ehl-i zühd ve ʿirfānla ʿArafāt gibi mālāmāl."

56. Ibid., 8b.

57. Ibid., 10a.

58. Ibid., 12b.

59. ʿĀşıḳpaşazāde, *Tevārīḫ-i Āl-i ʿOs̱mān*; quoted in Lindner, *Nomads and Ottomans*, 37.

60. Mottahedeh, *Loyalty and Leadership*, 69–70.

61. Kafadar, *Between Two Worlds*, 132–33.

62. For useful studies on the early modern Ottoman dreamscape, see Niyazioğlu, *Dreams and Lives*; Hagen, "Dreaming Osmans"; Felek "(Re-)creating Image and Identity"; Fleischer, "Secretaries' Dreams"; Kafadar, *Rüya Mektupları*.

63. Şen, "Mirror for Princes," 51.

64. TSMA E. 968/72.

65. C. DH. 142/7081.

66. In his *Mücerrebnāme*, Muṣṭafā Zekī (d. 1735), who held the post of chief court *munajjim*, makes several interesting observations about the uselessness of dream interpretation manuals. For Muṣṭafā Zekī, the subjective nature of dreams necessitates varied interpretations; they cannot be confined to the one-size-fits-all explanatory framework (*taḫt-ı ḳāʿideye gelmez*) that dream interpretation manuals attempt to provide.

67. Felek, "Master of the Master"; Felek, *Kitābü'l-Menāmāt*; Alvan, *Sultan Murad-ı Salis'in Dünyası*.

68. Felek, *Kitābü'l-Menāmāt*.

69. Felek, "(Re)creating Image and Identity."

70. Felek, *Kitābü'l-Menāmāt*, 89, 148, 200, 212.

71. Cited in Kapusuz, "Study on Gelibolulu Mustafa Ali's *Mirror for the Worlds*," 48.

72. Ben-Zaken, *Cross-cultural Scientific Exchanges*, 34–41.

73. Despite his use of Seyyid Loḳman's eye-witness chronicle *Shahīnshāhnāma*, Ben-Zaken argues (38) that the invitation to establish an observatory came to Taqī al-Dīn from the sultan. However, contemporary sources, including the *Shahīnshāhnāma*, clearly document that the idea was developed and proposed by Taqī al-Dīn through different stages and with the assistance of several individuals. See the Conclusion for more details.

74. Felek, "(Re)creating Image and Identity"; Alvan, *Sultan Murad-ı Salis'in Dünyası*.

75. B. Tezcan, "Some Thoughts."

76. Sayılı, "Alauddin Mansur'un," 462–63. Sayılı confuses the copyist of the volume with the author Seyyid Loḳman.

77. Mozaffari and Steele, "Solar and Lunar Observations," 354.

78. For Murād III's letter to his shaykh Şücāʿ Efendi, asking for his opinions about the *jafr* books of İlyās b. ʿĪsā Saruḫānī, see Alvan, *Sultan Murad-ı Salis'in Dünyası*, 735.

79. Ben-Zaken, *Cross-cultural Scientific Exchanges*, 17–18.

80. This view is also shared by Hüseyin Şen, who is completing a dissertation on Taqī al-Dīn. I would like to thank Hüseyin Şen for his personal communication on this matter.

81. See note 4 in the Introduction.

82. Sahillioğlu, *Koca Sinan Paşa'nın Telhisleri*, 216.

83. Ibid.

84. Ö. Yıldırım, *Twin Books.*

85. Ibid., 215.

86. "*şeyṭān-ı ins ve deyyār-ı ḫabīs*"; quoted in Alvan, *Sultan Murad-ı Salis'in Dünyas*, 302. Alvan says that the shaykh in question was ʿAzīz Maḥmūd Hudāʾī. However, Derin Terzioglu has convincingly demonstrated that it must be İbrāhīm-i Ḳırīmī. See Terzioğlu, "Power, Patronage, and Confessionalism."

87. Alvan, *Sultan Murad-ı Salis'in Dünyası*, 837: "raṣadı kaldırmak isteriz."

88. For Gülşenī's life and sociopolitical entanglements, see Emre, *Ibrahim-i Gulshani.*

89. Muḥyī-i Gülşenī, *Menāḳıb*, 121.

90. Ibid.

91. Ibid: "sizin ʿameliniz ḳatʿī olsa."

92. For İlyās b. ʿĪsā Saruḫānī, see Kurnaz and Tatçı, "İbn İsa"; Işık Tamdoğan-Abel, "Le Future dans le *Rumuz-ı Kunuz.*" Hüseyin Yılmaz has also touched upon the significance of the text in the context of apocalyptic-millenarian currents during the reign of Süleyman. See H. Yılmaz, *Caliphate Redefined*, 270–71.

93. İsmail Hakkı Konyalı Kütüphanesi Ms. 153, 12b: "ḳavl-i müneccimīne iʿtibār kılmaya ve remmāl aḥkāmına mültefit olmaya."

94. Ibid.

95. Saruḫānī (İlyās b. ʿĪsā), *Ḳavāʿid-i Tesḫīrāt*; *Fuṣūl-i ʿAşere.*

96. Burckhardt, *Mystical Astrology*; Gardiner, "Stars and Saints."

97. Melvin-Koushki, "Persianate Geomancy."

98. Ibid.

99. Fleischer, "Seer to the Sultan."

100. TSMK Ms. Hazine 1697.

101. Ibid., 1b.

102. Ibid., passim: "ṣāḥib-ḳırān-ı ʿālem"; "ṣāḥibü'z-zamān"; "ḳuṭbu'l-aḳṭāb."

103. Ibid., 2a–7b.

104. Ibid., passim: "delāʾil-i ʿilm-i reml-i Danyal-ı Nebī"; "ʿilm-i cefr."

105. Ibid., 4b.

106. Ibid., 30a–31b.

107. Ibid., 32a–33a.

108. See Kunt, "Sultan Süleyman ve Nikris."

109. TSMK Ms. Hazine 1697, 45b.

110. A. DVNSMHM.2/93, 112; 3/1101; 4/2188; 5/1735.

111. TSMA E. 458/12.

112. Ibid: "nihān ve maḥv idesiz ki bir kimesne aḥvāle muṭṭaliʿ olmaya."

113. Ibid.

114. Ibid: "wa-ttafaqa al-ʿulamāʾ anna ʿilm al-raml muʿjizāt li-sabʿa anbiyāʾ."

115. Ibid.

116. TSMA E. 885/8.

117. Ibid.

118. Ibid: "ḥażret-i resūl ʿilm-i nücūmu ḳabūl buyurmadı ve bizim ʿilmimizi medh eyledi."

Conclusion

1. The list of publications is too long to fit here. The covers of the following works can be taken as representative: al-Khalili, *House of Wisdom*; Freely, *Light from the East*; Aydın et al., *Teoriden Pratiğe Osmanlı'da Bilim*; İhsanoğlu, *Osmanlı Bilim Mirası*, vol. 1.

2. Poskett, *Horizons*, 69.

3. The following remarks by Fatin Gökmen—a student of the last chief court *munajjim*, the director of the Imperial Observatory (*Rasathane-i Amire*) from the 1910s, and an instructor of astronomy, probability, and statistics at *Darülfünun* (the university) from 1909 and 1933—perfectly represent this stance: "If Taqi al-Din's institution in Istanbul had not disintegrated in the midst of ignorance and greed soon after it was established . . . the Orient would have remained at the same level of progress as the Occident, and perhaps would have gone further . . . Then . . . as an advanced nation, we would have been the guardians of a noble civilization." See Gökmen, "Rasıd Takiyüddin."

4. Küçük, *Science without Leisure*, 40.

5. Shafir, "Almighty Akçe"; Küçük, "Response."

6. Although the most comprehensive biographical information about him, provided by Nevʿīzāde ʿAṭāʾī in the early seventeenth century, sets the birthplace of Taqī al-Dīn as Egypt, he was most likely born in Damascus. In the brief report that the grand vizier Koca Sinān Pasha (who held the office of grand vizier for the first time from 1580 to 1582) sent to Murād III around the time or slightly after the observatory was destroyed, Taqī al-Dīn is said to have been born in Damascus ("Şam kendünüñ mevlidi olmağla"). See Sahillioğlu, *Koca Sinan Paşa'nın telhisleri*, 216. His Damascene origin is also confirmed by Aḥmad Yūsuf Ḥasan, who mentions in his introduction to the publication of Taqī al-Dīn's *Al-Ṭuruq al-Sanīyah fī al-Ālāt al-Rūḥānīyah* (Elevated paths about pneumatic/spiritual instruments) that Taqī al-Dīn described himself in his own handwriting on the cover of one of the manuscripts as "Damascene by birth." See Ḥasan, *Taqī al-Dīn*.

7. Baga, "Takıyyüddin Rasıd"; Taqī al-Dīn, *16'ıncı asırda Osmanlılarda saat*; Ḥasan, *Taqī al-Dīn*.

8. Details in this paragraph are taken from the copy of the petition by Taqī al-Dīn found in KRK Ms. 51. A facsimile of this petition was published in Dizer, *Takiyüddin*.

9. Ibid.

10. *Ālāt-ı Raṣadiyye li-Zīc-i Şehinşāhiyye*.

11. This rapport was likely established by Taqī al-Dīn's elder brother Necmeddīn, who was also interested in astrology. See Demir, "Takiyüddin'in Ağabeyi Necmeddin."

12. Sayılı, "Alauddin Mansur'un," 462–63.

13. Despite the absence of any explicit reference to, or a signature by, Taqī al-Dīn, BnF Ms. Arabe 2570 might be his work. This volume contains two almanacs. The first (1b–13b), compiled in Arabic according to the latitude of Cairo, was for the solar year 1573 (up until March 1574) and presented to Sultan Selīm II (r. 1566–74). The compiler of the almanac, whose wording bears affinities to that of Taqī al-Dīn, states on folio 5a

that he made the computations by drawing on the "Samarkand principles." The second almanac (16b–28a), composed in Turkish, was dedicated to Murād III, relaying general prognostications and monthly advice for the solar year 1576. The compiler mentions (21a) the *zīj* of Wābkanawī and the Ulugh Beg tables as his main sources for computation.

14. Al-Qabīṣī, *Introduction to Astrology*, 5–6.

15. Fatin Gökmen says, without citing any source, that Taqī al-Dīn was given 10,000 gold *sultani* coins. According to Şevket Pamuk's *A Monetary History*, the official exchange rate between the silver *akçe* and the gold *sultani* in 1582 was 60:1. Based on this estimation, the initial amount given to Taqī al-Dīn should be around 600,000 *akçes*. Pamuk also notes that in 1582–83, the total revenues of the empire were 313.7 million *akçes*. If we are to believe Fatin Gökmen and Şevket Pamuk, then Taqī al-Dīn's initial support constituted around 0.2 percent of annual revenues. See Pamuk, *Monetary History*, 63.

The contemporary German Lutherian theologian Stephan Gerlach, who was in Constantinople from 1573 to 1578 as the clergyman of the Holy Roman Empire, notes that Taqī al-Dīn was annually receiving 3,000 gold (Venetian) ducats—which was during the period at par against the *sultani* gold coin—and that he would receive an additional 6,000 ducats when the project was completed. Gerlach, *Türkiye Günlüğü*, vol. 2, 678.

The official *berāt* given to Taqī al-Dīn mentions that in addition to his *zeʿāmet* worth 70,000 *akçes* from lands in Rumelia, he was provided an additional *zeʿāmet* worth 46,000 *akçes* from the province of Konya. See Mordtmann, "Das Observatorium," 94.

16. Tekeli, "Nasirüddin, Takiyüddin." Unfortunately, we have no extant report of Taqi al-Din interpreting the 1577 comet.

17. Christianson, *On Tycho's Island*; Christianson, *Tycho Brahe*.

18. R. Smith, "Observatory."

19. For a useful comparison between the observatories founded in the seventeenth century and those in earlier eras, see Špelda, "Collective Empiricism."

20. Bahadır and Danışman, "Late Ottoman and Early Republican Science Periodicals," 286.

21. For a critical analysis of nineteenth- and twentieth-century ʿulamaʾ as "guardians of tradition" see Bein, *Ottoman Ulema, Turkish Republic*.

22. A similar observation can be made regarding medieval and early modern physicians. For the social standing of physicians in medieval Islamicate societies, see Pormann, "Physician and the Other"; Behrens-Abouseif, "Image of the Physician."

23. See note 42 in the Introduction.

24. Baldini, "Roman Inquisition's Condemnation"; Lanuza-Navarro, "Astrology in Court."

25. Michot, "Ibn Taymiyya on Astrology"; Livingston, "Science and the Occult."

26. Livingston, "Science and the Occult," 601. Ibn Qayyim al-Jawziyya even calls Naṣīr al-Dīn al-Ṭūsī "a sorcerer who worshipped idols." See al-Ṭūsī (Naṣīr al-Dīn), *Memoir*, vol. 1, 19.

27. ʿAṭāʾī, *Ḥadāʾiḳ*, vol. 1, 856.

28. See, for instance, the following entries in (Meḥmed) Şeyḫī's (d. 1731) *Veḳāyiʿu'l-Fużalāʾ*: Küfrī Ḥasan Çelebi (1565); Müneccimek Meḥemmed Çelebi (1619); Müneccimbaşı Aḥmed Dede (2229–31); Müneccimbaşı Aḥmed Efendi (2402–3); ʿArabzāde Meḥemmed

Efendi (2483–85); Müneccimbaşı Ḥüseyin (2513–14); Müneccimbaşı Muṣṭafā (3004–5); Müneccim ʿAbdullāh (3181).

29. See, for instance, Müneccimbaşı Meḥemmed's (d. 1630/1) astrological treatise (*Risāle-i Aḥkām-ı Nücūm*) where he says he relied upon Arabic, Persian, and European (*Frengī*) books. KRK Ms. 371, 1b.

30. For the astral interests of Muṣṭafā III, see Kırkoğlu, *Sultan ve Müneccimi*, 77–79.

31. B. Tezcan, "Some Thoughts."

32. *Ālāt-ı Raṣadiyye li-Zīc-i Şehinşāhiyye.*

33. İpşirli, "Kadızade Ahmed Şemseddin."

34. Sahillioğlu, *Koca Sinan Paşa'nın Telhisleri*, 216.

35. Evliyā Çelebi, *Seyahatname*, vol. 1, 184.

36. Kātib Çelebi, *Taḳvīmü't-Tevārīḫ*, 65a: "[987 AH/1579] *Raṣad-ı Taqī al-Dīn der Ġalaṭa.*"

37. Brentjes, "Pride and Prejudice," 234.

BIBLIOGRAPHY

Abbreviations

AKDTYK	Atatürk Kültür, Dil ve Tarih Yüksek Kurumu
AÜDTCF	*Ankara Üniversitesi Dil Tarih Coğrafya Fakültesi*
BEA	*Biographical Encyclopedia of Astronomers*. Electronic resource. Eds. Thomas Hockey et al.
BHM	*Bulletin of the History of Medicine*
BSOAS	*Bulletin of the School of Oriental and African Studies*
DİA	*Türkiye Diyanet Vakfı İslam Ansiklopedisi*
EI^{2}	*Encyclopaedia of Islam*, 2nd ed.
EI^{3}	*Encyclopaedia of Islam*, 3rd ed.
İA	*Milli Eğitim Bakanlığı İslam Ansiklopedisi*
IJMES	*International Journal of Middle East Studies*
IRCICA	Research Centre for Islamic History, Art, and Culture in Istanbul
İÜEF	*İstanbul Üniversitesi Edebiyat Fakültesi*
İÜİFM	*İstanbul Üniversitesi İktisat Fakültesi Mecmuası*
JAOS	*Journal of American Oriental Society*
JESHO	*Journal of the Economic and Social History of the Orient*
JNES	*Journal of Near Eastern Studies*
JOTSA	*Journal of Ottoman and Turkish Studies Association*
JPS	*Journal of Persianate Studies*
JTS	*Journal of Turkish Studies*
OA	*Osmanlı Araştırmaları Dergisi / Journal of Ottoman Studies*
OALT	*Osmanlı Astronomi Literatürü Tarihi = History of Astronomy Literature during the Ottoman Period*. Ed. Ekmeleddin İhsanoğlu (Istanbul: IRCICA, 1997)

OASTLT	*Osmanlı Astroloji Literatürü Tarihi ve Osmanlı Astronomi Literatürü Tarihi Zeyli = History of Ottoman Astrology Literature and Supplement to the History of Ottoman Astronomy Literature* (Istanbul: IRCICA, 2011)
RHS	*Routledge Handbook on Sciences in the Islamicate World: Practices from the 2nd/8th to the 13th/19th Centuries*. Ed. Sonja Brentjes. New York: Routledge, 2022
SI	*Studia Islamica*
TOK	*Treasures of Knowledge: An Inventory of the Ottoman Palace Library (1502/3–1503/4)*. Eds. Gülru Necipoğlu, Cemal Kafadar, and Cornell H. Fleischer. Leiden; Boston: Brill, 2019
TTK	Türk Tarih Kurumu
TÜBA	Türkiye Bilimler Akademisi
TYEK	Türkiye Yazma Eserler Kurumu

Archival Sources

ATATÜRK KITAPLIĞI

Muallim Cevdet (MC) O. 71

DEVLET ARŞIVLERI BAŞKANLIĞI OSMANLI ARŞIVI

Ali Emiri (AE)

Abdülhamid I: 17/1466

Süleyman I: 4/258

Bab-ı Asafi (A.)

Mühimme Defterleri (DVNSHMHM.d.): 2/93, 2/12, 3/1101, 4/2188, 5/1735, 25/1231, 25/2984

Ruus Kalemi Defterleri (RSK.d.): 1459

Bab-ı Defteri (D.)

Büyük Ruznamçe Kalemi Defterleri (BRZ.d): 20616

Fatih Sultan Mehmed Evkafı (HMH. SFTH.): 21941-B

Cevdet (C)

Dahiliye (DH): 142/7081

Hatt-ı Hümayun (HAT): 14/576, 22/1053

Kamil Kepeci (KK): 1764, 1769, 1771, 1775, 1863, 1864, 1867

Maliyeden Müdevver (MAD): 162, 492, 559, 7118, 12769, 17888, 22082

Sadaret (A.)

Müteferrik Evrakı (M): 9/30

TOPKAPI SARAYI MÜZESI ARŞIVI (TSMA)

Defter (D.): 1298, 2921, 5475, 7843, 9291, 9559, 9587, 9600, 9706, 10026, 10141, 10184

Evrak (E.): 363/13, 363/16, 363/41, 458/12, 754/87, 754/99, 756/92, 756/93, 885/8, 968/70, 968/72, 970/44, 1298/25, 1415/71, 10159/6, 10159/145

Unpublished Manuscripts

ARKEOLOJI MÜZESI

Ms. 1607: Necmeddīn. *Taḳvīm*

BEYAZIT KÜTÜPHANESI
Ms. 4619: *Ṭāliʿ-i mavlūd-i Sulṭān Muḥammad b. Murād*

BIBLIOTECA MEDICEA LAURENZIANA
Ms. Or. 24: *Zīj-i Ilkhānī*

BIBLIOTHÈQUE NATIONALE DE FRANCE (BNF)
Arabe Ms. 2570: Taqī al-Dīn. *Taqwīm*
Supp. Pers. Ms. 367: Ibrāhīm b. al-Shaykh. *Taqvīm*
Turc Ms. 182: *Taḳvīm*
Turc Ms. 183: Necmeddīn. *Taḳvīm*

BODLEIAN
Marsh Ms. 396: ʿAlī Qūshjī. *Sharḥ-i Zīj-i Ulugh Beg*

BRITISH LIBRARY (BL)
Ms. Add. 5983: Ḫayreddīn Ḳonevī. *Ṭāliʿ-i mevlūd*
Ms. Or. 6432/1: Necmeddīn. *Taḳvīm*
Ms. Or. 6432/2: Qāḍī-yi Baghdad. *Taqvīm*

BURSA TÜRK VE İSLAM ESERLERI MÜZESI
Ms. 11: *Zīj-i Ilkhānī*

DIYARBAKIR ZIYA GÖKALP YAZMA ESERLER KÜTÜPHANESI
Ms. 1467: *Zīj-i Ilkhānī*

HUNTINGTON LIBRARY
Ms. HM71897: Yaḥyā al-Kāshī. *Ṭāliʿ-i mavlūd*

İSMAIL HAKKI KONYALI KÜTÜPHANESI
Ms. 153: İbn ʿĪsā Saruḫānī. *Rumūzü'l-künūz*

İSTANBUL ÜNIVERSITESI NADIR ESERLER KÜTÜPHANESI (İÜ NADIR)
Ms. T4315: Muṣṭafā Zekī. *Mücerrebnāme*
Ms. FY32: *Taqvīm*

KANDILLI RASATHANESI KÜTÜPHANESI (KRK)
Ms. 294: *Ṭāliʿ-i vilādet-i Aḥmed b. Sulṭān Bāyezīd*
Ms. 310: Sayyid Munajjim. *Laṭāʾif al-kalām*
Ms. 325: Luṭfullāh. *Zāyiçe*
Ms. 349M: Sayyid Munajjim, *Laṭāʾif al-kalām*
Ms. 371: Müneccimbaşı Meḥmed. *Risāle-i Aḥkām-ı Nücūm*
Ms. 372: Nücūmī. *Meẕẕāḳu'l-ʿuşşāḳ*
Ms. 546: Yūsuf b. ʿÖmer. *Taḳvīm*

LEIDEN UNIVERSITY LIBRARIES
Ms. Or. 8: Kūshyār. *Al-Zīj al-Jāmiʿ*

SÜLEYMANIYE KÜTÜPHANESI (SK)
Ayasofya
Ms. 2283: Müʿeyyedzāde. *Al-Ḥawāshī ʿalā Sharḥ al-mawāqīf*

Ms. 2410: *Kitāb Aflāṭūn*
Ms. 2414M: Khiṭābī. *Jāmiʿ al-qismayn*
Ms. 2432: Collection of astronomical-astrological treatises
Ms. 2455: *Al-Muthūl al-Aflāṭūniyya*
Ms. 2474: Collection of treatises including Naṣīr al-Dīn al-Ṭūsī's *Sī faṣl*
Ms. 2588: Munajjim Bālī. *Tuḥfat al-faqīr*
Ms. 2618: *Risāla al-usṭurlāb*
Ms. 2651: Jurjānī. *Sharḥ al-Mulakhkhaṣ fī al-hay'a*
Ms. 2667: Rukn al-Āmulī. *Panjāh bāb*
Ms. 2688: ʿAli-Shāh Bukhārī. *Athmār va ashjār*
Ms. 2694: Wābkanawī. *Al-Zīj al-muhaqqaq*
Ms. 2697: Mīrim Çelebi. *Sharḥ-i Zīj-i Ulugh Beg*
Ms. 2702: Abū Maʿshar. *Al-Madkhal ilā aḥkām al-nujūm*
Ms. 2709: Khiṭābī. *Muwaḍḍiḥ al-rusūm* (aka *Sharḥ-i Sī faṣl*)
Ms. 2711: *Kitāb yawāqit al-mawāqīt*
Ms. 3635: ʿAbdurraḥmān. *Jawhar ḥifẓ al-ṣiḥḥat*

Bağdatlı Vehbi
Ms. 2005: Mīrim Çelebi. *Masā'il*

Carullah
Ms. 1409: Rukn al-Āmulī. *Panjāh bāb*
Ms. 1500: Collection of treatises owned by Riyāżī
Ms. 2108: ʿAbdulwājid b. Muḥammad. *Sharḥ-i Sī faṣl*

Darülmesnevi
Ms. 345: ʿAṭā'ullāh. *Risāla-i Rubʿ*

Esad Efendi
Ms. 1978: ʿAbdurraḥmān. *Taqvīm*
Ms. 1997: Khiṭābī. *Ṭāliʿ-i mavlūd-i Sulṭān Muḥammad b. Murād*
Ms. 2147: Keşfī. *Selīmnāme*
Ms. 2198: Idrīs Bidlīsī. *Hasht Bihisht*
Ms. 3782: Riyāżī. *Risāle-i Ṭılsım*

Fatih
Ms. 3421: Nīsābūrī. *Kashf al-ḥaqā'iq*
Ms. 4264: Riyāżī. *Terceme-i Şāhī*
Ms. 4357: Ibn al-ʿUlayf. *Al-Durr al-manẓūm*

Hacı Mahmud Efendi
Ms. 5584: Riyāżī. *Risāle-i Ṭılsım*

Hamidiye
Ms. 1446: Collection of astronomical and astrological treatises

Hüsrev Paşa
Ms. 482: Meḥmed Şāh al-Fenārī. *Anmūzaj al-ʿulūm*

İzmir
Ms. 470: Kasımpaşalı ʿOs̱mān. *Cāmiʿü'l-aḥkām*

Kadızade Mehmed
Ms. 339: Ibn al-Shāṭir. *Nihāyat al-suʾl*

Lala İsmail
Ms. 289: Rukn al-Āmulī. *Panjāh bāb*

Laleli
Ms. 2120: Sinan Paşa. *Ḥāshiya ʿalā Sharḥ al-Mulakhkhaṣ*

M. Nuri Efendi
Ms. 151: Mīrim Çelebi. *Sharḥ-i Zīj-i Ulugh Beg*

Nuruosmaniye
Ms. 2685: Riyāżī. *Ḳuvvet-i ḳudṣiyye*
Ms. 2782: *Jadāvil al-nujūm*
Ms. 2800: Collection of astrological treatises including *Al-Thamara*

Ragıp Paşa
Ms. 913: Naṣīr al-Dīn al-Ṭūsī. *Taḥrīr al-Majisṭī*
Ms. 931: Rukn al-Āmulī. *Panjāh bāb*

Reisülküttab
Ms. 571: *Zīj-i Ulugh Beg*
Ms. 572: *Kitāb al-Thamara*

Yazma Bağışlar
Ms. 1362: Sayyid Munajjim. *Risāla al-shakl*
Ms. 4034: Riyāżī. *Zāyiçe*

TOPKAPI SARAYI MÜZESI KÜTÜPHANESI (TSMK)
III. Ahmed
Ms. 1960: Yūsuf b. ʿÖmer. *Taqvīm*
Ms. 2010: ʿAbdurraḥmān. *Jawhar ḥifẓ al-ṣiḥḥat*
Ms. 2128: Ikhwān al-Ṣafāʾ's epistle on mathematical sciences
Ms. 3317: Naṣīr al-Dīn al-Ṭūsī. *Al-Tadhkira fī'l-hayʾa*
Ms. 3328: Naṣīr al-Dīn al-Ṭūsī. *Taḥrīr al-Majisṭī*
Ms. 3343: Marrakūshī. *Jāmiʿ al-mabādiʾ*
Ms. 3495: ʿAbdussalām. *Maʿrifat al-mawḍūʿāt*
Ms. 3497: es-Seyyid el-Müneccim el-Tokatī. *Taḳvīm*

Bağdat
Ms. 310: Khiṭābī. *Taqvīm*
Ms. 320: *Taqvīm*

Emanet Hazinesi
Ms. 1695: Ḫāce Kemāl. *Taḳvīm*
Ms. 1710: Yūsuf b. ʿÖmer. *Taqvīm*

Hazine
Ms. 1760: Collection of astronomical treatises including Mīrim Çelebi's epistles
Ms. 1697: Remmāl Ḫaydar. *Risāle-i reml*

Revan
Ms. 1704: Yaḥyā. *Miftāh al-nujūm*
Ms. 1705: Fakhr al-Dīn al-Rāzī. *Ikhtiyārāt al-aḥkām al-ʿalāʾiyya*
Ms. 1711/1: *Taqvīm*
Ms. 1711/11: *Taqvīm*
Ms. 1711/16: Necmeddīn. *Taḳvīm*
Ms. 1711/18: Necmeddīn. *Taḳvīm*
Ms. 1711/19: Necmeddīn. *Taḳvīm*
Ms. 1713: Yazdānbakhsh. *Al-Zīj al-mujmal*

Yeni Yazmalar
Ms. 830: Khiṭābī. *Ṭāliʿ-i mavlūd-i Sulṭān Muḥammad b. Murād*

UNIVERSITY OF MICHIGAN LIBRARY
Islamic Manuscripts Ms. 794: *Taqwīm*

UNIVERSITY OF PENNSYLVANIA KISLAK CENTER
Ms. LJS 412: Four Treatises on astrolabes and astronomy

Published Sources

Abū Maʿshar. *The Abbreviation of the Introduction to Astrology Together with the Medieval Latin Translation of Adelard of Bath*. Eds. Charles Burnett, Keiji Yamamoto, and Michio Yano. Leiden: Brill, 1994.

——. *On Historical Astrology: The Book of Religions and Dynasties*. Eds. Keiji Yamamoto and Charles Burnett. 2 vols. Leiden: Brill, 2000.

Afṣaḥī. *Ḳaṣāyid-i Afṣāḥī dar madh-i Sulṭān Bāyezīd*. Ed. Mustafa Çiçekler. Istanbul: Sakıp Sabancı Müzesi, 2012.

Aḥmed-i Dāʿī. *Muḫtaṣar fī ʿilm el-tencīm ve maʿrifet el-taḳvīm (Risāle-i sī faṣl)*. Eds. T. N. Gencan and M. Dizer. Istanbul: Boğaziçi Üniversitesi Kandilli Rasathanesi, 1984.

Alan, Ercan, and Abdurrahman Atçıl, eds. *XVI. Yüzyıl Osmanlı Ulema Defterleri*. Ankara: TÜBA, 2018.

Ālāt-ı Raṣadiyye li-Zīc-i Şehinşāhiyye. Eds. Mustafa Kaçar, Şinasi Acar, and Atilla Bir. Istanbul: Türkiye İş Bankası, 2011.

ʿĀlī (Muṣṭafā). *Mustafa Ali's Description of Cairo of 1599: Text, Transliteration, Translation, Notes*. Ed. Andreas Tietze. Vienna: Verl. d. Österr. Akad. d. Wiss, 1975.

——. *Mustafa ʿAli's Counsel for Sultans of 1581*. Ed. Andreas Tietze. Vienna: Verlag der österreichischen Akademie der Wissenschaften, 1979.

——. *Künhü'l-Aḫbār*. Vol. 2: *Fatih Sultan Mehmed Devri (1451–1481)*. Ed. Hüdai Şentürk. Ankara: TTK, 2003.

——. *Künhü'l-Aḫbār'ın Tezkire Kısmı*, electronic resource. Ed. Mustafa İsen. Ankara: Kültür ve Turizm Bakanlığı, 2017.

Alper, Ömer Mahir, and Yasin Apaydın, eds. *Molla Lutfi ve Fütuhat'ı*. Ankara: TÜBA, 2021.

Altaş, Eşref. "Molla Lutfi'nin '*es-Sebʿu'ş-şidad*' Adlı Risalesinin Tahkik ve Değerlendirmesi." *Tahkik* 5, no. 1 (2022): 1–50.

Altınay, Ahmed Refik. "Fatih Devrine ait Vesikalar." *Tarih-i Osmani Encümeni Mecmuası* 8–11, no. 49–62 (1335 [1919]): 1–58.

——. *Onuncu asr-ı hicride İstanbul hayatı*. Ankara: Kültür ve Turizm Bakanlığı, 1987.

ʿĀşıḳ Çelebi. *Meşāʿirü'ş-Şuʿarā*. Ed. Filiz Kılıç. 3 vols. Istanbul: İstanbul Araştırmaları Vakfı, 2010.

'Aṭā'ī (Nevʿīzāde). *Ḥadāʾiḳü'l-ḥaḳāʾiḳ fī tekmīleti'ş-şaḳāʾiḳ*. Ed. Suat Donuk. 2 vols. Istanbul: TYEK, 2017.

Atsız, Nihal. "Fatih Sultan Mehmed'e Sunulmuş Tarihi Bir Takvim." *İstanbul Enstitüsü Dergisi* 3 (1957): 17–23.

——. *Osmanlı Tarihine ait Takvimler*. Istanbul: Küçükaydın Matbaası, 1961.

——. "Hicri 858 Yılına ait Takvim." *Selçuklu Araştırmaları Dergisi* 4 (1975): 223–83.

Baga, Elif. "Takıyyüddin Rasıd ve Cebir Risalesi: Matematiksel Değerlendirme, Tercüme ve Editio Princepts." *Nazariyat* 7, no. 2 (2021): 1–52.

Bagheri, Mohammad. "A Newly Found Letter of al-Kāshī on Scientific Life in Samarkand." *Historia Mathematica* 24 (1997): 241–56.

Barkan, Ömer L. "H. 933–934 Tarihli Bütçe Cedveli ve Ekleri." *İÜİFM* 15, no. 1–4 (1953–54): 251–329.

——. "Ayasofya Camii ve Eyüp Türbesinin 1489–1491 yıllarına ait Muhasebe Bilançoları." *İÜİFM* 23, no. 1–2 (1962–3): 342–79.

——. "Edirne ve Civarındaki Bazı İmaret Tesislerinin Yıllık Muhasabe Bilançoları." *Belgeler* 1, no. 2 (1964): 235–377.

Barkan, Ömer L., and Ekrem Hakkı Ayverdi, eds. *İstanbul Vakıfları Tahrir Defteri: 953 (1546) Tarihli*. Istanbul: Baha Matbaası, 1970.

Bashir, Shahzad. "The Risālat al-Hudā of Muhammad Nurbakhsh (d. 869/1464): Critical Edition with Introduction." *Rivista degli studi orientali* 75, no. 1–4 (2001): 87–137.

Bayram, Yavuz, ed. *Adli Sultan İkinci Bayezid Han-ı Veli: Hayatı-Şahsiyeti-Şairliği Divanının Tenkidli Metni*. Amasya: T.C. Amasya Valiliği, 2008.

Beyatlı, Ahmet, ed. *Fatih Sultan Mehmed'in 877/1472 tarihli vakfiyesi*. Ankara: TTK, 2013.

Bidlīsī (Idrīs). *Salīmshāhnāma*. Ed. Hicabi Kırlangıç. PhD, Ankara University, 1995.

al-Bīrūnī. *The Chronology of Ancient Nations*. Ed. C. E. Sachau. London: W. H. Allen, 1879.

——. *The Book of Instruction in the Elements of the Art of Astrology*. Ed. R. Ramsay Wright. London: Luzac, 1934.

Çavuşoğlu, Mehmet. "Zati'nin Letayifi I." *İÜEF Türk Dili ve Edebiyatı Dergisi* 18 (1970): 1–51.

Celālzāde (Muṣṭafā). *Ṭabaḳātü'l-Memālik ve Derecātü'l-Mesālik*. Ed. Funda Demirtaş. PhD, Erciyes University, 2009.

Dalkıran, Sayın. "İbn-i Kemal'in Tılsımat Risalesi." *Türk Dünyası Araştırmaları* 104 (1996): 195–208.

Demir, Remzi. "Takiyüddin'in Ağabeyi Necmeddin ibn Maruf ve Onun 'Cedavilü li-Marifet Ruyet Ehiletti'ş-şuhur [*sic*]' adlı risalesi." *Bilim ve Felsefe Metinleri* 1, no. 1 (1992): 9–28.

——, ed. *Taḳiyüddin'de matematik ve astronomi: Ceridetü'd-dürer ve haridetü'l-fiker üzerine bir inceleme*. Ankara: Atatürk Kültür Merkezi Başkanlığı, 2000.

Doğan, Kaan. "Kanuni Sultan Süleyman Döneminde On Yıl: Bir Filori Defteri Işığında Harcamalar ve Rumeli Güzergahı." MA thesis, Istanbul University, 2017.

Elwell-Sutton, Laurence P. *The Horoscope of Asadullāh Mīrzā: A Specimen of Nineteenth-Century Persian Astrology*. Leiden: Brill, 1977.

Emecen, Feridun, and İlhan Şahin, eds. *II. Bāyezīd dönemine ait 906/1501 tarihli Ahkam Defteri*. Istanbul: Türk Dünyası Araştırmaları Vakfı, 1994.

——. *II. Bayezid Dönemi Ahkam Defterleri*. Ankara: AKDTYK, 2021.

——. "Amasya." *DİA*.

Ergin, Osman Nuri. *Fatih İmareti Vakfiyesi*. Istanbul: Belediye Matbaası, 1945.

Evliyā Çelebi. *Seyahatname*. Eds. Orhan Şaik Gökyay, Robert Dankoff, Seyit Ali Kahraman, and Yücel Dağlı. 10 vols. Istanbul: Yapı Kredi Yayınları, 1996.

al-Fārābī. *Iḥṣāʾ al-ʿulūm*. Beirut: Dār wa Maktabat al-Hilāl, 1996.

Fatih Mehmed II Vakfiyeleri. Ankara: Vakıflar Genel Müdürlüğü, 1938.

Felek, Özgen, ed. *Kitābü'l-Menāmāt: Sultan III. Murad'ın Rüya Mektupları*. Istanbul: Tarih Vakfı Yurt Yayınları, 2014.

Firdevsī-i Ṭavīl. *Ḳuṭb-nāme*. Eds. İbrahim Olgun and İsmet Parmaksızoğlu. Ankara: TTK, 1980.

——. *Da'vetnāme*. Ed. Fatma Büyükkarcı. Cambridge, MA: Harvard University Near Eastern Languages and Civilizations, 1995.

Gerlach (Stephan). *Türkiye Günlüğü*. Tr. Türkis Noyan. 2 vols. Istanbul: Kitap Yayınevi, 2007.

Ḥasan, Aḥmad Yūsuf, ed. *Taqī Al-Dīn wa-al-Handasah al-Mīkānīyah al-ʻArabīyah: Maʻa Kitāb al-Ṭuruq al-Sanīyah fī al-Ālāt al-Rūḥānīyah min al-qarn al-sādis ʻashar*. Aleppo: Jāmiʻat Ḥalab, 1976.

Ibn Kemāl. *Tevārīḫ-i Āl-i ʿOs̱mān*. 8. Defter. Ed. Ahmet Uğur. Ankara: TTK, 1997.

Ibn Khaldūn. *The Muqaddimah: An Introduction to History*. Ed. Franz Rosenthal. 3 vols. Princeton: Princeton University Press, 1958.

Ibn Sīnā (Avicenna). *Tisʿ rasāʾil fī l-ḥikma wa-l-ṭabīʿiyyāt*. Cairo: Maṭbaʿat Hindīyah, 1908.

——. *Réfutation de l'astrologie*. Ed. Yahya Michot. Beirut-Paris: Albouraq, 2006.

Ikhwān al-Ṣafāʾ. *On Astronomia: An Arabic Critical Edition and English Translation of Epistle 3*. Eds. Jamil Ragep and Taro Mimura. Oxford: Oxford University Press, 2015.

Isidore of Seville. *The Etymologies*. Eds. Stephen A. Barney et al. Cambridge; New York: Cambridge University Press, 2006.

al-Jurjānī (al-Sayyid al-Sharīf). *Sharḥ al-mawāqif*. 3 vols. Tr. Ömer Türker. Istanbul: TYEK, 2015.

Kāşifī. *Ġazānāme-i Rūm*. Ed. Ebrahim M. Esmail. MA thesis, Mimar Sinan Fine Arts University, 2005.

Kātib Çelebi. *Kashf al-ẓunūn*. Eds. Şerefettin Yaltkaya and Rifat Belge. 2 vols. Istanbul: Maarif, 1941–43.

——. *Taḳvīmü't-Tevārīḫ*. Ankara: TTK, 2009.

——. *Sullam al-wuṣūl ilā Ṭabaqāt al-guhūl*. Ed. Maḥmūd Arnāʾūṭ. 6 vols. Istanbul: IRCICA, 2010.

——. *An Ottoman Cosmography: Translation of Cihānnümā*. Eds. Gottfried Hagen and Robert Dankoff. Leiden; Boston: Brill, 2022.

Kennedy, Edward, and David Pingree, eds. *The Astrological History of Māshā'allāh*. Cambridge, MA: Harvard University Press, 1971.

Kınalızāde (Ḥasan). *Tez̲kiretü'ş-Şuʿarā*, electronic resource. Ed. Aysun Sungurhan-Eyduran. Ankara: TC Kültür ve Turizm Bakanlığı, 2017.

Ḳıvāmī. *Fetḥnāme-yi Sulṭān Meḥmed*. Ed. Ruşan Türken. MA thesis, Fırat University, 2006.

Kūshyar ibn Labbān. *Introduction to Astrology*. Ed. Michio Yano. Tokyo: Tokyo University of Foreign Studies, 1997.

———. *Ptolemaic Tradition and Islamic Innovation: The Astronomical Tables of Kūshyār ibn Labbān*. Ed. Benno van Dalen. Turnhout: Brepols, 2021.

Lefort, Jacques, ed. *Documents grecs dans les archives de Topkapı Sarayı: Contribution à l'histoire de Cem Sultan*. Ankara: TTK, 1981.

Mercier, Raymond. *An Almanac for Trebizond for the Year 1336*. Louvain-la-Neuve: Academia-Erasme, 1994.

Miʿmār Sinān. *Sinan's Autobiographies: Five Sixteenth-Century Texts*. Eds. Howard Crane, Esra Akın, and Gülru Necipoğlu. Leiden; Boston: Brill, 2006.

Miroğlu, İsmet. "İstanbul Rasathanesine ait Belgeler." *İÜEF Tarih Enstitüsü Dergisi* 3 (1973): 75–82.

Muhyī-i Gülşenī. *Menāḳıb-ı İbrāhim-i Gülşenī*. Eds. Mustafa Koç and Eyyüp Tanrıverdi. Istanbul: TYEK, 2014.

Muṣṭafā Ḥattī Efendi. *Viyana Sefaretnamesi*. Ed. A. İbrahim Savaş. Ankara: TTK, 1999.

Muvaḳḳit Muṣṭafā. *Tuḥfetü'z-zamān*. Ed. Çiğdem Türker. PhD, Çanakkale 18 Mart University, 2016.

Nevʿī Efendi. *Netāʾicü'l-fünūn ve meḥasinü'l-mütūn*. Eds. Gisela Procházka-Eisl and Hülya Çelik. Cambridge, MA: Harvard University, Near Eastern Languages and Civilizations, 2015.

Niẓāmī ʿArūżī. *Chahār maqāla*. Ed. Muḥamad Qazvīnī. Tehran: Kitābfurūshī-yi Zavvār, 1954.

Oruç Beg. *Oruç Beğ Tarihi*. Ed. Necdet Öztürk. Istanbul: Çamlıca, 2007.

Öz, Tahsin. *Zwei Stiftungsurkunden des Sultans Mehmed II. Fatih*. Istanbul: Das Archäologische Institut des Deutschen Reiches, 1935.

Özcan, Abdülkadir, ed. *Kanunname-i Al-i Osman*. Istanbul: Kitabevi, 2003.

Pérez, María José Parra. "Estudio y edición de las traducciones el árabe del Almanach perpetuum de Abraham Zacuto." PhD, Universitat de Barcelona, 2013.

Pingree, David, ed. *The Thousands of Abu Mashar*. London: Warburg Institute, 1968.

Ptolemy. *Tetrabiblos*. Ed. F. E. Robbins. Cambridge, MA: Harvard University Press, 1940.

al-Qabīṣī. *The Introduction to Astrology*. Eds. Charles Burnett, Keiji Yamamoto, and Michio Yano. London: Warburg Institute, 2004.

al-Qazvīnī (Ḥakīm Shāh Muḥammad). *Taz̲kirah-i majālis al-nafāʾis*. Ed. ʿAlī Aṣghar Ḥikmat. Tehran: Kitābfurūshī-i Manūchihrī, 1363 [1984].

al-Qūshjī (ʿAlī). *Risāla dar ʿilm-i ḥisāb*. Ed. Zehra Bilgin. PhD, Istanbul Medeniyet University, 2024.

al-Rāzī (Fakhr al-Dīn). *Jāmiʿ al-ʿulūm, ya Ḥadāyiq al-anwār fī ḥaqāyiq al-asrār*. Ed. Muḥammad Ḥusayn Tasbīḥī. Tehran: Kitābkhānah-i Asadī, 1346 [1967–68].

Rāzī (Shahmardan b. Abi'l-Khayr). *Ravżat al-munajjimīn*. Ed. Jalīl Akhavan Zanjānī. Tehran: Markaz-i Intishār-i Nusakh-i Khaṭṭī, 1989.

Renan, Ernest. *Islam and Science. A Lecture Presented at La Sorbonne 29 March 1883*. Ed. Sally

Ragep. Accessible at: https://www.mcgill.ca/islamicstudies/files/islamicstudies/renan_islamism_cversion.pdf.
Ritter, Hellmut, and M. Plessner, eds. *"Picatrix": Das Ziel des Weisen von Pseudo-Magrīṭī*. London: Warburg Institute, 1962.
Sabuncuoğlu (Şerefeddin). *Cerrāḥiyyetü'l-Ḫāniyye*. Ed. İlter Uzel. Ankara: AKDTYK, 1992.
Sāfī (Muṣṭafā). *Zübdetü't-Tevārīḫ*. Ed. İ. Hakkı Çuhadar. 2 vols. Ankara: TTK, 2003.
Sahillioğlu, Halil, ed. *Topkapı Sarayı Arşivi H. 951–952 Tarihli ve E-12321 Numaralı Mühimme Defteri*. Istanbul: IRCICA, 2002.
——. *Koca Sinan Paşa'nın Telhisleri*. Istanbul: IRCICA, 2004.
Sanuto, Marino. *I diarii*, vol. 5. Ed. Federico Stefani. Venice: F. Visentini, 1881.
Saruḫānī (İlyās b. ʿĪsā). *Rumūzü'l-Künūz*. Ed. Ayhan Özgül. MA thesis, Kırıkkale University, 2004.
——. *Fuṣūl-i ʿAşere*. Ed. Tijen Özbay. MA thesis, Dokuz Eylül University, 2016.
——. *Ḳavāʿid-i Teshīrāt*. Ed. Engin Kaçar. MA thesis, Boğaziçi University, 2019.
Sayılı, Aydın. "Alauddin Mansur'un İstanbul Rasathanesi Hakkındaki Şiirleri." *Belleten* 20, no. 79 (1956): 411–84.
——. *Uluğ Bey ve Semerkanddeki İlim Faaliyeti Hakkında Gıyasüddin-i Kaşi'nin Mektubu*. Ankara: TTK, 1960.
Şemseddīn Sāmī. *Gök*. Istanbul: Mihran, 1296 [1890].
——. *Ḳāmūs-i Türkī*. Istanbul: A. Cevdet, 1318 [1900].
Şeyḫī (Meḥmed). *Veḳāyiʿü'l-Fużalā*. Ed. Ramazan Ekinci. 4 vols. Istanbul: TYEK, 2018.
al-Shīrāzī (Quṭb al-Dīn). *Durrat al-tāj*. Ed. Muḥammad Mushkāt. Tehran: Intishārāt-i Ḥikmat, 1369 [1990].
al-Şirvānī (Şükrullāh). *Riyāżu'l-qulūb*. Ed. Serpil Koç Konuksever. Istanbul: TYEK, 2019.
Şirvānī (Ṣadreddīnzāde Meḥmed). *Al-Fawāʾid al-khāqāniyya*. Ed. Ahmet Kamil Cihan. Istanbul: TYEK, 2019.
Somi, Mortaza, and Mohammad Bagheri. "Risāla-i tashrīḥ al-ālāt fī shaʿn al-imtiḥānāt az Sayyid Munajjim Ḥusaynī." *Mirāth-i ʿIlmī-yi Islām va Īrān* 2, no. 1 (1392 [2013]): 181–205.
Süleymān Sūdī Efendi. *Ṭabaḳāt-ı Müneccimīn*. Ed. Salim Aydüz. Istanbul: Fatih Üniversitesi Yayınları, 2007.
Süleymaniye Vakfiyesi. Ed. Kemal Edib Kürkçüoğlu. Ankara: Vakıflar Genel Müdürlüğü, 1962.
Tācīzāde Caʿfer Çelebi. *Maḥrūse-i İstanbul Fetḥnāmesi*. Istanbul: Ahmed İhsan ve Şurekası Matbaacılık, 1331 [1915–16].
Tadhkirat al-Mulūk: A Manual of Safavid Administration (circa 1137/1725) Persian Text in Facsimile (B.M. Or. 9496). Ed. V. Minorsky. London: Luzac, 1943.
Tankalūshā. Ed. Raḥīm Rižā Zādah Malik. Tehran: Mīrās-i Maktūb, 1384 [2005].
Taqī al-Dīn Muḥammad ibn Maʿrūf. *16'ıncı asırda Osmanlılarda saat ve Takiyüddin'in "Mekanik saat konstrüksüyonuna dair en parlak yıldızlar" adlı eseri*. Ed. Sevim Tekeli. Ankara: Ankara Üniversitesi Basımevi, 1966.
Ṭaşköprülüzāde. *Miftāḥ as-saʿāda wa miṣbāḥ as-siyāda*. 3 vols. Hyderabad: Osmania Oriental Publications Bureau, 1977.
——. *Al-Shaqāʾiq al-nuʿmānīyah fī ʿulamāʾ al-dawlat al-ʿuthmānīyah*. Ed. Muhammet Hekimoğlu. Istanbul: TYEK, 2019.

Turan, Osman. *İstanbul'un Fethinden önce yazılmış Tarihi Takvimler.* Ankara: TTK, 1954.

——. *Türkiye Selçukluları Hakkında Resmi Vesikalar.* Ankara: TTK, 1958.

Ṭursun Beg. *Tā'rīḫ-i Ebū'l-Fetḥ.* Ed. Mertol Tulum. Istanbul: İstanbul Fetih Cemiyeti, 1977.

al-Ṭūsī (Naṣīr al-Dīn). *The Nasirean Ethics.* Tr. G. M. Wickens. London: Allen & Unwin, 1964.

——. *Memoir on Astronomy (al-Tadhkira fī ʿilm al-hay'a).* Ed. Jamil Ragep. 2 vols. New York: Springer-Verlag, 1993.

——. *Sharḥ-i S̱amarah-yi Baṭlamyūs dar aḥkām-i nujūm.* Ed. Khalīl Akhavān Zanjānī. Tehran: Āyīnah-i Mīrās, 1999.

Yaḥyā b. Meḥmed el-Kātib. *Menāhicü'l-İnşā'.* Ed. Fahri Unan. Ankara: TTK, 2014.

Yiğit, Osman. "Topkapı Sarayı Müzesi Arşivi 34 Numaralı Filori Defteri." MA thesis, Marmara University, 2018.

Yirmisekiz Çelebi Meḥmed Efendi. *Fransa Sefaretnamesi.* Ed. Abdullah Uçman. Istanbul: Dergah, 2017.

Zīc-i Uluğ Bey. Eds. Mustafa Kaçar and Atilla Bir. 2 vols. Ankara: Kültür ve Turizm Bakanlığı, 2012.

Studies

Abou-El-Haj, Rifaʿat ʿAli. "The Ottoman Nasihatname as a Discourse over 'Morality'." In *Mélanges Professuer Robert Mantran*, ed. A. Temimi, 17–30. Tunis: Centre d'Etudes et de Recherches Ottomanes, 1988.

Adıvar, Adnan. *La Science chez les Turcs Ottomans.* Paris: Maisonneuve, 1939.

——. *Osmanlı Türklerinde İlim.* Istanbul: Maarif Matbaası, 1943.

——. "Interaction of Islamic and Western Thought in Turkey." In *Near Eastern Culture and Society: A Symposium on the Meeting of East and West*, ed. T. Cuyler Young, 119–29. Princeton: Princeton University Press, 1951.

Adorno, Theodor. "The Stars Down to Earth: The Los Angeles Times Astrology Column." *Telos* 19 (1974): 13–90.

Afyoncu, Erhan. "XVI. Yüzyılda Hassa Mimarları." In *Prof. Dr. İsmail Aka Armağanı*, 207–15. Istanbul: Beta Basım, 1999.

——. "Osmanlı Hekimbaşıları ve Hassa Hekimleri." In *Osmanlılarda Sağlık*, ed. Coşkun Yılmaz and Necdet Yılmaz, vol. 1, 85–97. Istanbul: Biofarma, 2006.

——. "Semiz Ali Paşa." *DİA.*

Ágoston, Gábor. *Guns for the Sultan: Military Power and the Weapons Industry in the Ottoman Empire.* New York: Cambridge University Press, 2005.

——. *The Last Muslim Conquest: The Ottoman Empire and Its Wars in Europe.* Princeton: Princeton University Press, 2021.

Akopyan, Ovanes. *Debating the Stars in the Italian Renaissance: Giovanni Pico Della Mirandola's Disputationes Adversus Astrologiam Divinatricem and Its Reception.* Leiden; Boston: Brill, 2021.

Akyıldız, Ali. "19. Yüzyıl Belgelerinde Hicri Takvim Sorunu: Hicri Takvimler Doğru Zamanı mı Gösterir?" *Toplumsal Tarih* 326 (2021): 50–55.

Allouche, Adel. *The Origins and Development of the Ottoman-Ṣafavid Conflict (906–962/1500–1555).* Berlin: Klaus Schwarz Verlag, 1983.

Allsen, Thomas. *Culture and Conquest in Mongol Eurasia.* New York: Cambridge University Press, 2011.

Alvan, Türkan. *Sultan Murad-ı Salis'in Dünyası: Mektupları ve Rüyaları Işığında Bir Derviş Padişah*. Istanbul: İz, 2021.

Anooshahr, Ali. "Shirazi Scholars and the Political Culture of the Sixteenth-Century Indo-Persian World." *Indian Economic and Social History Review* 51, no. 3 (2014): 331–52.

——. "Review of *The Millennial Sovereign*." *Medieval History Journal* 18, no. 1 (2015): 183–91.

Antov, Nikolay. *The Ottoman "Wild West": The Balkan Frontier in the Fifteenth and Sixteenth Centuries*. New York: Cambridge University Press, 2017.

Artun, Tuna. "Bayezid-i Kimyai: Osmanlı Kimya Metinlerinde Sultan II. Bayezid." *JTS* 39 (2013): 181–86.

——. "Hearts of Gold and Silver: The Production of Alchemical Knowledge in the Early Modern Ottoman World." PhD, Princeton University, 2013.

Ash, Eric H. *Power, Knowledge, and Expertise in Elizabethan England*. Baltimore: Johns Hopkins University Press, 2004.

——. "Introduction: Expertise and the Early Modern State." *Osiris* 25, no. 1 (2010): 1–24.

——. "By Any Other Name: Early Modern Expertise and the Problem of Anachronism." *History and Technology* 35, no. 1 (2019): 3–30.

Atçıl, Abdurrahman. "The Safavid Threat and Juristic Authority in the Ottoman Empire in the Sixteenth Century." *IJMES* 49, no. 2 (2017): 295–304.

——. *Scholars and Sultans in the Early Modern Ottoman Empire*. Cambridge: Cambridge University Press, 2017.

Atsız, Nihal. "Kemalpaşaoğlu'nun Eserleri I." *Şarkiyat Mecmuası* 6 (1966): 71–112.

——. "Kemalpaşaoğlu'nun Eserleri II." *Şarkiyat Mecmuası* 7 (1972): 83–135.

Aydın, Bilgin. *XVI. yüzyılda Divan-ı Hümayun ve Defter Sistemi*. Ankara: TTK, 2017.

Aydın, Cemil. "Beyond Culturalism? An Overview of the Historiography on Ottoman Science in Turkey." In *Multicultural Science in the Ottoman Empire*, eds. E. İhsanoğlu, K. Chatzis, and E. Nikolaidēs, 201–15. Turnhout: Brepols, 2003.

Aydın, Mükerrem Bedizel, et al., eds. *Teoriden Pratiğe Osmanlı'da Bilim*. Çanakkale: Paradigma Akademi, 2022.

Aydüz, Salim. "Osmanlı Devleti'nde Müneccimbaşılık ve Müneccimbaşılar." MA thesis, Istanbul University, 1993.

——. "Osmanlı Devleti'nde Küçük Gözlemevleri: Muvakkithaneler." In *Osmanlı*, vol. 8, ed. Güler Eren, 664–75. Ankara: Yeni Türkiye Yayınları, 1999.

——. "Müneccimbaşı Takvimleri ve Tarihi Kaynak Olarak Değerleri." *Cogito* 22 (2000): 132–44.

——. "Osmanlı Devleti'nde Müneccimbaşılık Müessesesi." *Belleten* 70, no. 257 (2006): 167–264.

——. *XV. ve XVI. Yüzyılda Tophane-i Amire ve Top Döküm Teknolojisi*. Ankara: AKDTYK, 2006.

——. "Ottoman Time Keeping Houses: Muwaqqitkhanas." *Études Balkaniques* 53, no. 2 (2017): 214–29.

Ayverdi, Ekrem Hakkı. *Fatih Devri Sonlarında İstanbul Mahalleleri, Şehrin İskanı ve Nüfusu*. Ankara: Vakıflar Umum Müdürlüğü, 1958.

Azzolini, Monica. "Refining the Astrologer's Art: Astrological Diagrams in Bodleian MS Canon. Misc. 24 and Cardano's *Libelli Quinque* (1547)." *Journal for the History of Astronomy* 42, no. 1 (2011): 1–25.

——. *The Duke and the Stars: Astrology and Politics in Renaissance Milan*. Cambridge, MA: Harvard University Press, 2013.

Babinger, Franz. *Mehmed the Conqueror and His Time.* Tr. Ralph Manheim. Princeton: Princeton University Press, 1978.

Bahadır, O. Recep, and Günhan Danışman. "Late Ottoman and Early Republican Science Periodicals." In *Turkish Studies in the History and Philosophy of Science*, eds. Gürol Irzık and Güven Güzeldere, 285–308. Dordrecht: Springer, 2005.

Bakar, Osman. *Classification of Knowledge in Islam: A Study in Islamic Philosophies of Science*. Cambridge: Islamic Texts Society, 1998.

Baldini, Ugo. "The Roman Inquisition's Condemnation of Astrology: Antecedents, Reasons and Consequences." In *Church, Censorship, and Culture in Early Modern Italy*, ed. G. Fragnito, 79–110. Cambridge: Cambridge University Press, 2001.

Balıkçıoğlu, Efe Murat. "A Coherence of Incoherences: Graeco-Arabic Philosophy and the Fifteenth-Century Ottoman Synthesis of Philosophy with Sharia." PhD, Harvard University, 2019.

——. *Verifying the Truth on Their Own Terms: Ottoman Philosophical Culture and the Court Debate between Zeyrek and Ḫocazāde*. Venice: Edizioni Ca' Foscari, 2023.

Baltacıoğlu-Brammer, Ayşe. "Neither Victim nor Accomplice: The Qizilbash as a Borderland Actor in the Early Modern Ottoman Realm." In *Rethinking Sunnitization in the Ottoman Empire*, eds. T. Krstić and D. Terzioğlu, 423–50. Leiden: Brill, 2020.

Barkan, Ömer L. *Süleymaniye Cami ve İmareti İnşaatı (1550–1557)*. 2 vols. Ankara: TTK, 1972.

Barkey, Karen. *Bandits and Bureaucrats: The Ottoman Route to State Centralization*. Ithaca: Cornell University Press, 1994.

——. "The Ottoman Empire (1299–1923): The Bureaucratization of Patrimonial Authority." In *Empires and Bureaucracy in World History: from late Antiquity to the Twentieth Century*, eds. Peter Crooks and Timothy H. Parsons, 102–26. Cambridge: Cambridge University Press, 2016.

Barnes, Robin Bruce. "Hope and Despair in Sixteenth-Century German Almanacs." In *The Reformation in Germany and Europe: Interpretations and Issues*, eds. Hans R. Guggisberg and Gottfried G. Krodel, 440–61. Sonderband Washington: Gütershloh, 1993.

Barthold, V. V. *Four Studies on the History of Central Asia*, vol. 2. Leiden: Brill, 1963.

Bashir, Shahzad. *Messianic Hopes and Mystical Visions: the Nūrbakhshīya Between Medieval and Modern Islam*. Columbia: University of South Carolina Press, 2003.

Başpınar, Fatih. "Sultan II. Bayezid'in Farsça Şiirleri Üzerine." *Divan Edebiyatı Araştırmaları Dergisi* 8 (2012): 37–66.

Bdaiwi, Ahab. "Late Antique Intellectualism in Medieval Islam: The Shiraz Circle and the Revival of Ancient and Islamic Knowledge." *Journal of Late Antique, Islamic and Byzantine Studies* 2, no. 1–2 (2023): 128–71.

——. "The Youth Who Defeated Aristotle: The Life and Thought of Dashtakī (d. 948/1541)." *Global Intellectual History* (2023), accessible online at DOI: 10.1080/23801883.2022.2163915.

Behrens, Gerhard. "An Ottoman Calendar (takvim) for 1740/41 AD." *Middle East Studies Online Journal* 4, no. 2 (2011): 1–90.

Behrens-Abouseif, Doris. "The Image of the Physician in Arab Biographies of the Post-classical Age." *Der Islam* 66 (1989): 331–43.

Bein, Amit. *Ottoman Ulema, Turkish Republic: Agents of Change and Guardians of Tradition.* Stanford: Stanford University Press, 2011.

Beinorius, Audrius. "On the Social and Religious Status of an Indian Astrologer at the Royal Court." *Acta Orientalia Vilnensia* 9, no. 2 (2008): 39–55.

Ben-David, Joseph. "Review Article, Scientific Growth: A Sociological View." *Minerva* 2, no. 4 (1964): 455–76.

Ben-Zaken, Avner. *Cross-cultural Scientific Exchanges in the Eastern Mediterranean, 1560–1660.* Baltimore: Johns Hopkins University Press, 2010.

Berkes, Niyazi. "İlk Türk Matbaası Kurucusunun Dini ve Fikri Kimliği." *Belleten* 26, no. 104 (1962): 715–37.

——. *The Development of Secularism in Turkey.* Montreal: McGill University Press, 1964.

Berkey, Jonathan. *The Transmission of Knowledge in Medieval Cairo: A Social History of Islamic Education.* Princeton: Princeton University Press, 1992.

Bernardini, Michele. "Ottoman 'Timuridism': Lāmiʿi Çelebi and his *Şehrengiz* of Bursa." In *Irano-Turkic Cultural Contacts in the 11th–17th Centuries*, ed. Éva M. Jeremiás, 1–16. Piliscsaba, Hungary: Avicenna Institute of Middle Eastern Studies, 2003.

Bertucci, Paola, and Olivier Courcelle. "Artisanal Knowledge, Expertise, and Patronage in Early Eighteenth-Century Paris: The 'Société Des Arts' (1728–36)." *Eighteenth-Century Studies* 48, no. 2 (2015): 159–79.

Biagioli, Mario. *Galileo, Courtier: The Practice of Science in the Culture of Absolutism.* Chicago: University of Chicago Press, 1993.

Binbaş, İ. Evrim. *Intellectual Networks in Timurid Iran: Sharaf al-Dīn 'Alī Yazdī and the Islamicate Republic of Letters.* Cambridge: Cambridge University Press, 2016.

Birnbaum, Eleazar. "The Ottomans and Chagatay Literature." *Central Asiatic Journal* 20 (1976): 157–90.

el-Bizri, Nader, and Eva Orthmann, eds. *Occult Sciences in Premodern Islamic Culture.* Beirut: Orient-Institut Beirut, 2018.

Blake, Stephen P. "The Patrimonial-Bureaucratic Empire of the Mughals." *Journal of Asian Studies* 39, no. 1 (1979): 77–94.

——. *Time in Early Modern Islam: Calendar, Ceremony, and Chronology in the Safavid, Mughal, and Ottoman Empires.* New York: Cambridge University Press, 2013.

Blessing, Patricia. *Architecture and Material Politics in the Fifteenth-Century Ottoman Empire.* Cambridge: Cambridge University Press, 2022.

Borrut, Antoine. "Court Astrologers and Historical Writing in Early Abbasid Baghdad: An Appraisal." In *Contexts of Learning in Baghdad from the 8th–10th Centuries*, eds. J. Scheiner and Damien Janos, 455–501. Princeton: Darwin Press, 2014.

——. "Histoires astrologiques et construction du temps culturel dans le débuts de l'Islam." In *Mers et rivages d'Islam: De l'Atlantique à la Méditerranée; Mélanges offerts à Christophe Picard*, eds. Alexandra Bill et al., 255–64. Paris: Éditions de la Sorbonne, 2023.

Boyraz, Şeref. *Fal Kitabı: Melhemeler ve Türk Halk Kültürü.* Istanbul: Kitabevi, 2006.

Brady, Andrea, and Emily Butterworth, eds. *The Uses of the Future in Early Modern Europe.* New York: Routledge, 2010.

Brentjes, Sonja. "On the Location of the Ancient or 'Rational' Sciences in Muslim Education Landscapes (AH 500–1000)." *Bulletin of the Royal Institute for Inter-Faith Studies* 4, no. 1 (2002): 47–71.

———. "Pride and Prejudice: The Invention of a 'Historiography of Science' in the Ottoman and Safavid Empires by European Travelers and Writers in the Sixteenth and Seventeenth Centuries." In *Religious Values and the Rise of Science in Europe*, eds. John H. Brooke and E. İhsanoğlu, 229–54. Istanbul: IRCICA, 2005.

———. "Courtly Patronage of the Ancient Sciences in Post-classical Islamic Societies." *al-Qantara* 29 (2008): 403–36.

———. "Shams al-Din al-Sakhawi on *Muwaqqits, Mu'adhdhins*, and the Teachers of Various Astronomical Disciplines in Mamluk Cities in the Fifteenth Century." In *A Shared Legacy: Islamic Science East and West*, eds. Emilia Calvo et al., 129–50. Barcelona: Universitat de Barcelona, 2008.

———. "The Prison of Categories—'Decline' and Its Company." In *Islamic Philosophy, Science, Culture, and Religion: Studies in Honor of Dimitri Gutas*, eds. Felicitas Opwis and David Reisman, 131–56. Leiden: Brill, 2012.

———. *Teaching and Learning the Sciences in Islamicate Societies (800–1700)*. Turnhout: Brepols, 2018.

———. "The Mathematical Sciences and Medicine in Safavid Iran." In *The Safavid World*, ed. Rudi Matthee, 428–46. New York: Routledge, 2021.

———. "Research Foci in the History of Science in Past Islamicate Societies." *Histories* 2, no. 3 (2022): 270–87.

———. "The Astral Sciences through the 7th/13th Century: Attitudes, Experts and Practices." In *RHS*, 80–95.

Brentjes, Sonja, Taner Edis, and Lutz Richter-Bernburg, eds. *1001 Distortions: How (Not) to Narrate History of Science, Medicine, and Technology in Non-Western Cultures*. Würzburg: Ergon Verlag, 2016.

Brummet, Palmira. *Ottoman Seapower and Levantine Diplomacy in the Age of Discovery.* Albany: State University of New York Press, 1994.

Burckhardt, Titus. *Mystical Astrology According to Ibn ʿArabi*. Tr. Bulent Rauf. Louisville, KY: Fons Vitae, 2001.

Burnett, Charles. "The Certitude of Astrology: The Scientific Methodologies of al-Qabīṣī and Abū Maʿshar." *Early Science and Medicine* 7, no. 3 (2002): 198–213.

———. "Abū Ma'shar (A.D. 787–886) and His Major Texts on Astrology." In *Kayd: Studies in History of Mathematics, Astronomy and Astrology in Memory of David Pingree*, eds. G. Gnoli and A. Panaino, 17–29. Rome: Istituto Italiano per l'Africa e l'Oriente, 2009.

———. "Astrology." *EI*³.

Burns, William, ed. *Astrology through History: Interpreting the Stars from Ancient Mesopotamia to the Present*. Santa Barbara: ABC-CLIO, 2018.

Caiozzo, Anna. "The Horoscope of Iskandar Sultan as a Cosmological Vision in the Islamic World." In *Horoscopes and Public Spheres: Essays on the History of Astrology*, eds. Günther Oestmann, H. Darrel Rutkin, and Kocku von Stuckrad, 115–44. Berlin; New York: Walter de Gruyter, 2005.

———. "Le Prince, l'astrologue et le derviche: Experts et conseillers princiers dans la culture visuelle des Timourides d'Hérat (XVᵉ siècle)." In *Experts et expertise au Moyen Âge*, 313–28. Paris: Publications de la Sorbonne, 2012.

Campagna, Federico. *Technic and Magic: The Reconstruction of Reality*. London: Bloomsbury, 2018.

Campion, Nicholas. *What Do Astrologers Believe?* London: Granta Books, 2006.

Capp, Bernard. *Astrology and the Popular Press: English Almanacs, 1500–1800*. London; Boston: Faber, 1979.

Carey, Hilary M. *Courting Disaster: Astrology at the English Court and University in the Later Middle Ages*. New York: St. Martin's Press, 1992.

Carlebach, Elisheva. *Palaces of Time: Jewish Calendar and Culture in Early Modern Europe*. Cambridge, MA: Belknap Press of Harvard University Press, 2011.

Carvalho, Helena Avelar. *An Astrologer at Work in Late Medieval France: The Notebooks of S. Belle*. Leiden; Boston: Brill, 2021.

Casale, Giancarlo. *The Ottoman Age of Exploration*. New York: Oxford University Press, 2010.

——. "Mehmed the Conqueror between *Sulh-i Kull* and *Prisca Theologia*." *Modern Asian Studies* 56 (2022): 840–69.

Casulleras, Josep. "The Instruments and the Exercise of Astrology in the Medieval Arabic Tradition." *Archives internationales d'histoire des sciences* 63, no. 170–171 (2013): 517–40.

——. "Horoscope." *EI*[3].

Catani, Remo. "The Polemics on Astrology 1489–1524." *Culture and Cosmos* 3, no. 2 (1999): 16–30.

Chamberlain, Michael. *Knowledge and Social Practice in Medieval Damascus, 1190–1350*. Cambridge: Cambridge University Press, 1994.

Charette, François. "Ibn al-Majdī." *BEA*.

Charmasson, Thérèse. "Divinatory Arts." In *Dictionary of Gnosis and Western Esotericism*, ed. Wouter J. Hanegraaff, 313–19. Leiden: Brill, 2006.

Chodkiewicz, Michel. *Seal of the Saints: Prophethood and Sainthood in the Doctrine of Ibn ʿArabī*. Cambridge: Islamic Texts Society, 1993.

Christianson, J. Roberts. *On Tycho's Island: Tycho Brahe and His Assistants, 1570–1601*. New York: Cambridge University Press, 2000.

——. *Tycho Brahe and the Measure of the Heavens*. London: Reaktion Books, 2020.

Çıpa, Erdem. *The Making of Selim: Succession, Legitimacy, and Memory in the Early Modern Ottoman World*. Bloomington: Indiana University Press, 2017.

Cohen, Amnon. *The Guilds of Ottoman Jerusalem*. Leiden: Brill, 2001.

Collins, H. M., and Robert Evans. *Rethinking Expertise*. Chicago: University of Chicago Press, 2007.

Cook, David. *Studies in Muslim Apocalyptic*. Berlin: Gerlach Press, 2021.

Cooper, Glen. "Approaches to the Critical Days in Late Medieval and Renaissance Thinkers." *Early Science and Medicine* 18, no. 6 (2013): 536–65.

——. "Astrology: The Science of Signs in the Heavens." In *The Oxford Handbook to Science and Medicine in the Classical World*, eds. J. Scarborough and P. T. Keyser, 381–407. New York: Oxford University Press, 2018.

Coşgel, Metin, and Boğaç A. Ergene. *The Economics of Ottoman Justice: Settlement and Trial in the Sharia Courts*. Cambridge: Cambridge University Press, 2016.

Coulon, Jean-Charles. "Building al-Buni's Legend: The Figure of al-Buni through Abd al-Rahman al-Bistami's *Shams al-afaq*." *Journal of Sufi Studies* 5, no. 1 (2016): 1–26.

——. *La Magie en terre d'islam au Moyen Âge*. Paris: Éditions du Comité des Travaux Historiques et Scientifiques, 2017.

Crombie, A. C., ed. *Scientific Change: Historical Studies in the Intellectual, Social, and Technical*

Conditions for Scientific Discovery and Technical Invention, from Antiquity to the Present. London: Heinemann, 1963.

Çulpan, Cevdet. "İstanbul Süleymaniye Camii Kitabesi." In *Kanuni Armağanı*, 291–99. Ankara: TTK, 1980.

Cunbur, Müjgan. "Bir Osmanlı Müneccimbaşısının Nevruz Tebrikleri." In *Türk Dünyasında Nevruz İkinci Uluslararası Bilgi Şöleni*, 121–30. Ankara: Atatürk Kültür Merkezi, 1996.

Cündioğlu, Dücane. "Ernest Renan ve 'Reddiyeler' Bağlamında İslam-Bilim Tartışmalarına Bibliyografik bir Katkı." *Divan* 2 (1996): 1–94.

Dallal, Ahmad S. *Islam, Science, and the Challenge of History.* New Haven: Yale University Press, 2010.

Danışan, Gaye. "Osmanlı Takvimlerinin Analizinde Uygulanan Yöntemler ve Karşılaşılan Problemler Üzerine bir Değerlendirme." In *2nd International Prof. Dr. Fuat Sezgin Symposium on History of Science in Islam Proceedings*, ed. Cüneyt Kaya et al., 175–94. Istanbul: Istanbul University Press, 2023.

Darling, Linda T. "Ottoman Salary Registers as a Source for Economic and Social History." *Turkish Studies Association Bulletin* 14, no. 1 (1990): 13–33.

——. *Revenue-Raising and Legitimacy: Tax Collection and Finance Administration in the Ottoman Empire, 1560–1660.* Leiden: Brill, 1996.

——. "Kanun and Kanunname in Ottoman Historiography." *JOTSA* 9, no. 1 (2022): 151–77.

Dávid, Géza. "Istolni Belgrad." *DİA.*

Dear, Peter. "Mysteries of State, Mysteries of Nature: Authority, Knowledge, and Expertise in the Seventeenth Century." In *States of Knowledge: The Co-production of Science and Social Order*, ed. Sheila Jasanoff, 206–24. London; New York: Routledge, 2004.

De Block, Fien. "[Re]drawing the Lines: The Science of the Stars in the Late Fifteenth Century Sultanate of Cairo." PhD, Ghent University, 2020.

De Blois, F. C., David A. King, and Julio Samsó. "Zid̲ j." *EI²*

Deniz, Şefaattin. "Sultan Abdülmecid'in İnşa Ettirdiği Muvakkithaneler." *OA* 49 (2017): 267–91.

De Renzi, Silvia. "Medical Expertise, Bodies, and the Law in Early Modern Courts." *Isis* 98, no. 2 (2007): 315–22.

Dizer, Muammer. *Takiyüddin.* Ankara: Kültür Bakanlığı, 1990.

Dohrn-van Rossum, Gerhard. *History of the Hour: Clocks and Modern Temporal Orders.* Tr. Thomas Dunlap. Chicago: University of Chicago Press, 1996.

Dooley, Brendan Maurice, ed. *A Companion to Astrology in the Renaissance.* Leiden; Boston: Brill, 2014.

Druart, Therese-Anne. "Al-Farabi's Causation of the Heavenly Bodies." In *Islamic Philosophy and Mysticism*, ed. Parwiz Morewedge, 35–45. Delmar, NY: Caravan Books, 1981.

Duindam, Jeroen. "Royal Courts in Dynastic States and Empires." In *Royal Courts in Dynastic States and Empires*, eds. Jeroen Duindam, Tülay Artan, and Metin Kunt, 1–23. Leiden; Boston: Brill, 2011.

Ekhtiar, Maryam, and R. Parikh. "Power and Piety: Islamic Talismans on the Battlefield." In *Islamicate Occult Sciences in Theory and Practice*, eds. Liana Saif et al., 420–53. Leiden: Brill, 2020.

Ekinci, Ramazan. "Klasik Türk Edebiyatında Mizahi Takvimler." *Divan Edebiyatı Araştırmaları Dergisi* 17 (2016): 55–90.

Elman, Benjamin A. *On Their Own Terms: Science in China, 1550–1900*. Cambridge, MA: Harvard University Press, 2005.

——. *Civil Examinations and Meritocracy in Late Imperial China*. Cambridge, MA: Harvard University Press, 2013.

Elshakry, Marwa. "When Science Became Western: Historiographical Reflections." *Isis* 101, no. 1 (2010): 98–109.

Elwell-Sutton, Laurence. "A Royal Timurid Nativity Book." In *Logos Islamikos*, eds. Roger M. Savory and Dionisius A. Agius, 119–34. Toronto: Pontifical Institute of Mediaeval Studies, 1984.

Emecen, Feridun. *Kanuni Sultan Süleyman ve Zamanı*. Ankara: TTK, 2022.

Emiralioğlu, Pınar. *Geographical Knowledge and Imperial Culture in the Early Modern Ottoman Empire*. Burlington, VT: Ashgate, 2014.

Emre, Side. *Ibrahim-i Gulshani and the Khalwati-Gulshani Order: Power Brokers in Ottoman Egypt*. Leiden; Boston: Brill, 2017.

Ephrat, Daphne. *A Learned Society in a Period of Transition: The Sunni 'ulama' of Eleventh-Century Baghdad*. Albany: State University of New York Press, 2000.

Ergene, Boğaç. "On Ottoman Justice: Interpretations in Conflict (1600–1800)." *Islamic Law and Society* 8, no. 1 (2001): 52–87.

Ergin, Osman Nuri. "İstanbul'un Fethinden Sonra Şehir Nasıl İmar ve İskan Edildi." *Resimli Tarih Mecmuası* 41 (1953): 2352–58.

Erginbaş, Vefa. "Enlightenment in the Ottoman Context: İbrahim Müteferrika and His Intellectual Landscape." In *Historical Aspects of Printing and Publishing in Languages of the Middle East*, ed. Geoffrey Roper, 53–99. Leiden: Brill, 2014.

——, ed. *Ottoman Sunnism: New Perspectives*. Edinburgh: Edinburgh University Press, 2019.

Erünsal, İsmail. *Osmanlı Vakıf Kütüphaneleri: Tarihi Gelişimi ve Organizasyonu*. Ankara: TTK, 2008.

——. "Mihri Hatun." *DİA*.

Eyal, Gil. *The Crisis of Expertise*. Cambridge: Polity Press, 2019.

Eyice, Semavi. "II. Bayezid Devrinde Davet Edilen Batılılar." *Belgelerle Türk Tarihi Dergisi* 19 (1969): 23–30.

Fahd, Toufic. *La Divination arabe: Études religieuses, sociologiques et folkloriques sur le milieu natif de l'Islam*. Paris: Sindbad, 1987.

Falk, Seb. "Learning Medieval Astronomy through Tables: The Case of the Equatorie of the Planetis." *Centaurus* 58, no. 1–2 (2016): 6–25.

Fancy, Nahyan. "Medical Commentaries: A Preliminary Examination of Ibn al-Nafis's Shuruh, the Mujaz and Subsequent Commentaries on the Mujaz." *Oriens* 41, no. 3/4 (2013): 525–45.

Faroqhi, Suraiya. "Crisis and Change, 1590–1699." In *An Economic and Social History of the Ottoman Empire*, eds. Halil İnalcık and Donald Quataert, 413–636. Cambridge: Cambridge University Press, 1994.

——. *Artisans of Empire: Crafts and Craftspeople under the Ottomans*. London; New York: I. B. Tauris, 2009.

——. *The Ottoman and Mughal Empires: Social History in the Early Modern World*. London: I. B. Tauris, 2019.

Fazlıoğlu, İhsan. "Ali Kuşçu'nun Bir Hendese Problemi ve Sinan Paşa'ya Nisbet Edilen Cevabı." *Divan* 1, no. 1 (1996): 85–105.

——. "İlk Dönem Osmanlı İlim ve Kültür Hayatında İhvanu's-safa ve Abdurrahman Bistami." *Divan* 1, no. 2 (1996): 229–40.

——. "Osmanlı Felsefe-Biliminin Arka Planı: Semerkand Matematik-Astronomi Okulu." *Divan* 14 (2003): 1–66.

——. "İki Ucu Müphem Bir Köprü: 'Bilim' ile 'Tarih' ya da 'Bilim Tarihi'." *TALİD* 2, no. 4 (2004): 9–27.

——. "Forcing the Boundaries in Religion, Politics and Philosophy-Science in the Fifteenth Century" (Paper presented at the conference "Before the Revolutions: Religions, Sciences and Politics in the Fifteenth Century," Berlin, January 13–15, 2005).

——. "İthaf'tan Enmuzec'e Fetih'ten önce Osmanlı Ülkesi'nde Matematik Bilimler." In *Uluslararası Molla Fenari Sempozyumu (4–6 Aralık 2009 Bursa): Bildiriler*, eds. Tevfik Yücedoğru et al., 131–63. Bursa: Bursa Büyükşehir Belediyesi, 2010.

——. "Kadızade-i Rumi." *DİA*.

——. "Mīram Çelebi." *BEA*.

——. "Mustafa b. Ali el-Muvakkit." *DİA*.

——. "Osmanlılar (İlim ve Kültür. 1. Düşünce Hayatı ve Bilim. Kaynaklar)." *DİA*.

——. "Qushji." *BEA*.

——. "ʿAlī al-Muwaqqit." *BEA*.

Fekete, Lajos. "XVI. Yüzyılda Taşralı Bir Türk Efendi Evi." Tr. Tayyib Gökbilgin. *Belleten* 43, no. 170 (1979): 457–80.

Felek, Özgen. "(Re-)creating Image and Identity: Dreams and Visions as a Means of Murad III's Self-Fashioning." PhD, University of Michigan, 2010.

——. "The Master of the Master: The Twisted Story of an Imperial Master and His Disciple." *Journal of Sufi Studies* 1, no. 2 (2012): 169–92.

——. "Fears, Hopes, and Dreams: The Talismanic Shirts of Murād III." *Arabica* 64, no. 3–4 (2017): 647–72.

Ferguson, Heather L. *The Proper Order of Things: Language, Power, and Law in Ottoman Administrative Discourses*. Stanford: Stanford University Press, 2018.

Ferhad, Massumeh, and Serpil Bağcı. "The Art of Bibliomancy and Falnama in the Sixteenth and Seventeenth Centuries." In *Falnama: The Book of Omens*, eds. Massumeh Farhad and Serpil Bağcı, 20–39. Washington, DC: Smithsonian Institution, 2009.

Fettahoğlu, Kübra. *Rasathane-i Amire (1868–1922)*. Ankara: TTK, 2019.

Fisher, Sydney Nettleton. *The Foreign Relations of Turkey, 1481–1512*. Urbana: University of Illinois Press, 1948.

Fleischer, Cornell H. *Bureaucrat and Intellectual in the Ottoman Empire: The Historian Mustafa Ali (1541–1600)*. Princeton: Princeton University Press, 1986.

——. "Preliminaries to the Study of Ottoman Bureaucracy." *JTS* 10 (1986): 135–41.

——. "The Lawgiver as Messiah: The Making of the Imperial Image in the Reign of Süleyman." In *Soliman le Magnifique et son temps: Actes du colloque de Paris, 7–10 Mars 1990*, ed. Gilles Veinstein, 159–77. Paris: La Documentation Française, 1992.

——. "Secretaries' Dreams: Augury and Angst in Ottoman Scribal Service." In *Armağan, Festschrift für Andreas Tietze*, eds. Ingeborg Baldoruf and Suraiya Faroqhi, 77–88. Prague: Enigma Corporation, 1994.

———. "Mahdi and Millennium: Messianic Dimensions in the Development of Ottoman Imperial Ideology." In *The Great Ottoman-Turkish Civilization*. Vol. 3, *Philosophy, Science and Institutions*, ed. Kemal Çiçek, 42–54. Ankara: Yeni Türkiye, 2000.

———. "Seer to the Sultan: Haydar-i Remmal and Sultan Süleyman." In *Cultural Horizons: A Festschrift in Honor of Talat S. Halman*, vol. 1, ed. Jayne Warner, 290–300. Syracuse, NY: Syracuse University Press, 2001.

———. "Shadow of Shadows: Prophecy and Politics in 1530s Istanbul." *International Journal of Turkish Studies* 13, no. 1–2 (2007): 51–62.

———. "Ancient Wisdom and New Sciences: Prophecies at the Ottoman Court in the Fifteenth and Early Sixteenth Centuries." In *Falnama: The Book of Omens*, eds. Massumeh Farhad and Serpil Bağcı, 231–45. Washington, DC: Smithsonian Institution, 2009.

———. "A Mediterranean Apocalypse: Prophecies of Empire in the Fifteenth and Sixteenth Centuries." *JESHO* 61 (2018): 18–90.

———. "Learning and Sovereignty in the Fifteenth and Sixteenth Centuries." In *TOK*, vol. 1, 155–60.

Flemming, Barbara. "Der Ğāmī ül-meknūnāt: Eine Quelle ʿĀlīs aus der Zeit Sultan Süleymāns." In *Studien zur Geschichte und Kultur des Vorderen Orients: Festschrift für Bertold Spuler zum siebzigsten Geburtstag*, eds. H. R. Römer and A. Noth, 79–92. Leiden: Brill, 1981.

———. "Sāhib-Kirān und Mahdī: Türkische Endzeiterwartungen im ersten Jahrzehnt der Regierung Süleymāns." In *Between the Danube and the Caucasus*, ed. György Kara, 43–62. Budapest: Akadémiai Kiadó, 1987.

———. *Essays on Turkish Literature and History*. Leiden; Boston: Brill, 2018.

Fodor, Pal. "State and Society, Crisis and Reform in the Fifteenth–Seventeenth Century Ottoman Mirror for Princes." *Acta Orientalia Academiae Scientiarum Hungaricae* 40, no. 2–3 (1986): 217–40.

Freely, John. *Light from the East: How the Science of Medieval Islam Helped to Shape the Western World*. London: I. B. Tauris, 2011.

Gardiner, Noah. "Esotericism in a Manuscript Culture: Aḥmad al-Būnī and His Readers through the Mamlūk Period." PhD, University of Michigan, 2014.

———. "Stars and Saints: The Esotericist Astrology of the Sufi Occultist Aḥmad al-Būnī." *Journal of Magic, Ritual, and Witchcraft* 12 (2017): 39–65.

———. "Books on Occult Sciences." In *TOK*, vol. 1, 735–65.

———. "Al-Jafr." *EI*[3].

Genç, Mehmet. *Osmanlı İmparatorluğu'nda Devlet ve Ekonomi*. Istanbul: Ötüken, 2000.

Genç, Vural. *Acem'den Rum'a: Bir Bürokrat ve Tarihçi İdris-i Bidlisi (1457–1520)*. Ankara: TTK, 2019.

———. "An Iranian *Shāh-nāma* Writer at the Court of Bāyezid II: Malekzāda Āhi." *JPS* 13 (2020): 119–45.

Gökbilgin, Tayyib. "Müeyyed-zade." *İA*.

———. "Müteferrika." *İA*.

Gökmen, Fatin. "Rasıd Takiyüddin." *Cumhuriyet Gazetesi*, 14 Nisan–6 Haziran 1341 [1925].

Gökyay, Orhan Şaik. *Molla Lutfi*. Ankara: Kültür ve Turizm Bakanlığı, 1987.

Goldstein, Bernard. "The Book on Eclipses of Mashāʾallāh." *Physis* 6, no. 2 (1964): 205–13.

Goodman, David C. *Power and Penury: Government, Technology, and Science in Philip II's Spain*. New York: Cambridge University Press, 1988.

Gordin, Michael D. *On the Fringe: Where Science Meets Pseudoscience*. New York: Oxford University Press, 2021.

Göyünç, Nejat. "XVI. Yüzyılda Ruus ve Önemi." *İÜEF Tarih Dergisi* 17, no. 22 (1967): 17–33.

Grafton, Anthony. *Cardano's Cosmos: The Worlds and Works of a Renaissance Astrologer*. Cambridge, MA: Harvard University Press, 1999.

Grant, Jonathan. "Rethinking the 'Ottoman Decline': Military Technology Diffusion in the Ottoman Empire, Fifteenth to Eighteenth Centuries." *Journal of World History* 10, no. 1 (1999): 179–201.

Green, Jonathan. *Printing and Prophecy: Prognostication and Media Change, 1450–1550*. Ann Arbor: University of Michigan Press, 2012.

Green-Mercado, Mayte. *Visions of Deliverance: Moriscos and the Politics of Prophecy in the Early Modern Mediterranean*. Ithaca: Cornell University Press, 2019.

Griffel, Frank. *Al-Ghazālī's Philosophical Theology*. New York: Oxford University Press, 2009.

———. *The Formation of Post-classical Philosophy in Islam*. New York: Oxford University Press, 2021.

———. "Role of Sense Perception." In *RHS*, 413–18.

Grill, D. "Esotérisme contre hérésie: 'Abd al-Rahmân al-Bistâmî, un représentant de la science des lettres à Bursa dans la première moitié du XVe siècle." In *Syncrétismes et hérésies dans l'Orient Seljoukide et Ottoman (XIVe–XVIIIe siècles)*, 183–95. Paris: Peeters, 2005.

Gümüş, Sadreddin. *Seyyid Şerif Cürcani*. Istanbul: Fatih Yayınevi, 1984.

Gündüz, Doğan. *Alaturkadan Alafrangaya Zaman: Osmanlı'da Mekanik Saatler*. Istanbul: Efe, 2015.

Güneş, Orhan. "Bir Alimin Gözünden Modern Astronomi: Hayatizade'nin Efkarü'l-Ceberut adlı eseri." PhD, Istanbul Medeniyet University, 2021.

Günther, Sebastian, and Dorothee Anna Maria Pielow, eds. *Die Geheimnisse der Oberen und der Unteren Welt: Magie im Islam Zwischen Glaube und Wissenschaft*. Leiden; Boston: Brill, 2018.

Gunzburg, Darrelyn. "How Do Astrologers Read Charts?" In *Astrologies: Plurality and Diversity*, eds. Nicholas Campion and Liz Greene, 181–200. Ceredigion, Wales: Sophia Centre Press, 2011.

Haarmann, Ulrich. "The Plight of the Self-Appointed Genius: Muṣṭafā ʿĀlī." *Arabica* 38, no. 1 (1991): 73–86.

Hagen, Gottfried. *Ein osmanischer Geograph bei der Arbeit: Entstehung und Gedankenwelt von Katib Čelebis Ǧihānnümā*. Berlin: Schwarz, 2003.

———. "Dreaming Osmans: Of History and Meaning." In *Dreams and Visions in Islamic Societies*, eds. Özlem Felek and Alexander D. Knysh, 99–123. Albany: State University of New York Press, 2012.

———. "Ptolemaeus Triumphans, or: Maps, Knowledge, and Ottoman Patronage." In *Islam on the Margins: Studies in Memory of Michael Bonner*, eds. Robert Haug and Steven Judd, 235–67. Leiden: Brill, 2023.

Hall, Crystal. "Galileo's Library Reconsidered." *Galilaeana* 12 (2015): 29–82.

Hanegraff, Wouter, ed. *The Dictionary of Gnosis and Western Esotericism*. Leiden; Boston: Brill, 2006.

Hasan, Moiz. "Foundations of Science in the Post-classical Islamic Era: The Philosophical, Historical, and Historiographical Significance of Sayyid al-Sharīf al-Jurjānī's (d. 1414) Project." PhD, University of Notre Dame, 2017.

Hashemipour, Behnaz. "Buzjani." *BEA*.

Hasse, Dag Nikolaus. *Success and Suppression: Arabic Sciences and Philosophy in the Renaissance*. Cambridge, MA: Harvard University Press, 2016.

Havlioğlu, Didem. *Mihri Hatun: Performance, Gender-Bending, and Subversion in Ottoman Intellectual History*. Syracuse, NY: Syracuse University Press, 2017.

Hayashi, Kayoko. "Fatih Vakfiyeleri'nin Tanzim Süreci Üzerine." *Belleten* 72, no. 263 (2008): 73–94.

Hayton, Darin. *The Crown and the Cosmos: Astrology and the Politics of Maximilian I*. Pittsburgh: University of Pittsburgh Press, 2015.

Heiderzadeh, Tofigh. "Ali Kuşçu'nun Astronomi Eserleri." MA thesis, Istanbul University, 1997.

——. "Patronage, Networks and Migration: Turco-Persian Scholarly Exchanges in the 15th, 16th and 17th Centuries." *Archives internationales d'histoire des sciences* 55, no. 155 (2005): 419–34.

Heiduk, Matthias, Klaus Herbers, and Hans Christian Lehner, eds. *Prognostication in the Medieval World: A Handbook*. Berlin; Boston: De Gruyter, 2020.

Hentsch, Thierry. *Imagining the Middle East*. Montreal: Black Rose Books, 1992.

Hofelich, M. "The Making of Taqvims in Iran." In *La Science dans le monde Iranien à l'époque islamique*, eds. Ziva Vesel et al., 49–51. Tehran: Institut Français de Recherche en Iran, 1998.

Hofelich, M., and Daniel M. Varisco. "Taḳwīm." *EI*².

Howard, Douglas. "Ottoman Historiography and the Literature of 'Decline' of the Sixteenth and Seventeenth Centuries." *Journal of Asian History* 22 (1988): 52–77.

Huff, Toby. *The Rise of Early Modern Science: Islam, China, and the West*. New York: Cambridge University Press, 1993.

Husain, Faisal. *Rivers of the Sultan: The Tigris and Euphrates in the Ottoman Empire*. New York: Oxford University Press, 2021.

İhsanoğlu, Ekmeleddin. "Fatih Külliyesi Medreseleri Ne Değildi: Tarih yazıcılığı bakımından tenkit ve değerlendirme denemesi." In *İstanbul Armağanı*, vol. 1: *Fetih ve Fatih*, 105–36. Istanbul: İstanbul Büyükşehir Belediyesi Kültür İşleri Daire Başkanlığı, 1995.

——. *Büyük Cihad'dan Frenk Fodulluğuna*. Istanbul: İletişim, 1996.

——. *Science, Technology, and Learning in the Ottoman Empire: Western Influence, Local Institutions, and the Transfer of Knowledge*. Burlington, VT: Ashgate, 2003.

——. *Studies on Ottoman Science and Culture*. New York: Routledge, 2021.

——. *Ottoman Scientific Heritage*. Tr. Maryam Patton. 3 vols. London: Al-Furqan Foundation, 2023.

İnalcık, Halil. "The Policy of Mehmed II toward the Greek Population of Istanbul and the Byzantine Buildings of the City." *Dumbarton Oaks Paper* 23/24 (1969/70): 229–49.

——. "The Heyday and Decline of the Ottoman Empire." In *The Cambridge History of Islam*,

eds. P. M. Holt, B. Lewis, and A. K. S. Lambton, vol. 1, 324–53. Cambridge: Cambridge University Press, 1970.
——. *The Ottoman Empire: The Classical Age, 1300–1600*. New York: Praeger, 1973.
——. "A Case Study in Renaissance Diplomacy: The Agreement between Innocent VIII and Bayezid on the Djem Sultan." *JTS* 3 (1979): 209–30.
——. "Military and Fiscal Transformation in the Ottoman Empire." *Archivum Ottomanicum* 6 (1980): 283–337.
——. "Comments on 'Sultanism': Max Weber's Typification of the Ottoman Polity." *Princeton Papers in Near Eastern Studies* 1 (1992): 49–72.
——. *Şair ve Patron: Patrimonyal Devlet ve Sanat Üzerinde Sosyolojik Bir İnceleme*. Ankara: Doğu Batı, 2003.
——. "Istanbul." *EI*[2].
İnan, Murat. "Imperial Ambitions, Mystical Aspirations: Persian Learning in the Ottoman World." In *The Persianate World: The Frontiers of a Eurasian Lingua Franca*, ed. Nile Green, 75–92. Oakland: University of California Press, 2019.
İpşirli, Mehmet. "XVI. Asrın İkinci Yarısında Kürek Cezası ile İlgili Hükümler." *İÜEF Tarih Enstitüsü Dergisi* 12 (1981–82): 203–48.
——. "Kadızade Ahmed Şemseddin." *DİA*.
Irani, Rida A. K. "Arabic Numeral Forms." *Centaurus* 4, no. 1 (1955): 1–12.
İzgi, Cevat. *Osmanlı Medreselerinde İlim*. 2 vols. Istanbul: İz, 1997.
Janssens, Jules L. "'Experience' (*tajriba*) in Classical Arabic Philosophy (al-Fārābī—Avicenna)." *Quaestio* 4 (2004): 45–62.
Jones, Damien. *Method, Structure, and Development in al-Fārābī's Cosmology*. Leiden: Brill, 2012.
Jorati, Hadi. "Science and Society in Medieval Islam: Nasir al-Din Tusi and the Politics of Patronage." PhD, Yale University, 2014.
Kafadar, Cemal. "Self and Others: The Diary of a Dervish in Seventeenth Century Istanbul and First-Person Narratives in Ottoman Literature." *SI* 69 (1989): 121–50.
——. *Rüya Mektupları*. Istanbul: Oğlak, 1994.
——. *Between Two Worlds*. Berkeley: University of California Press, 1995.
——. "The Question of Ottoman Decline." *Harvard Middle Eastern and Islamic Review* 42, no. 1–2 (1997–98): 30–75.
——. "Between Amasya and Istanbul: Bayezid II, His Librarian, and the Textual Turn of the Late Fifteenth Century." In *TOK*, vol. 1, 79–155.
Kafescioğlu, Çiğdem. *Constantinopolis/Istanbul: Cultural Encounter, Imperial Vision, and the Construction of the Ottoman Capital*. University Park: Pennsylvania State University Press, 2009.
Kale, Gül. "Intersections between the Architect's Cubit, Science of Surveying, and Social Practices in Cafer Efendi's Seventeenth-Century Book on Ottoman Architecture." *Muqarnas* 36 (2019): 131–77.
——. "From Measuring to Estimation: Definitions of Geometry and Architect-Engineer in Early Modern Ottoman Architecture." *Journal of Society of Architectural Historians* 80 (2020): 132–51.
Kappert, Petra. *Die osmanischen Prinzen und ihre Residenz Amasya im 15. und 16. Jahrhundert*. Istanbul: Nederlands Historisch-Archaeologisch Instituut, 1976.

Kapusuz, Müberra. "A Study on Gelibolulu Mustafa Ali's *Mirror for the Worlds (Mir'atü'l-Avalim):* Patronage, Politics, and Millennial Anxieties at the Court of Sultan Murad III (r. 1574–1595)." MA thesis, Central European University, 2021.

Karahasanoğlu, Selim. "Ottoman Ego-Documents: State of the Art." *IJMES* 53, no. 2 (2021): 301–8.

Karakaya-Stump, Ayfer. *The Kizilbash-Alevis in Ottoman Anatolia: Sufism, Politics and Community.* Edinburgh: Edinburgh University Press, 2020.

Karataş, Hasan. "The City as a Historical Actor: The Urbanization and Ottomanization of the Halveiyye Sufi Order by the City of Amasya in the Fifteenth and Sixteenth Centuries." PhD, UC Berkeley, 2011.

Karlığa, Bekir, and Mustafa Kaçar, eds. *Doğumunun 400. yıldönümünde Katip Çelebi.* Ankara: T. C. Kültür ve Turizm Bakanlığı, 2009.

Kazancıgil, Ratip. *Edirne Sultan II: Bayezid Külliyesi.* Edirne: Trakya Üniversitesi, 1997.

Kennedy, Edward. "A Survey of Islamic Astronomical Tables." *Transactions of the American Philosophical Society*, new series, 46, no. 2 (1956): 123–77.

———. "A Letter of Jamshīd al-Kāshī to His Father: Scientific Research and Personalities at a Fifteenth Century Court." *Orientalia* 29 (1960): 191–213.

———. "Ramifications of the World-Year Concept in Islamic Astrology." *Actes du dixième congrès international d'histoire des sciences*, 23–45. Paris, 1964.

———. "On the Contents and Significance of the Khaqani Zij by Jamshid Ghiyath al-Din al-Kashi." In *Islamic Mathematics and Astronomy*, vol. 84, ed. Fuat Sezgin. Frankfurt am Main: Institute for the History of Arabic-Islamic Science, 1998.

———. *Astronomy and Astrology in the Medieval Islamic World.* Brookfield, VT: Ashgate, 1998.

Keshaverz, Fateme. "The Horoscope of Iskandar Sultan." *Journal of Royal Asiatic Society* 2 (1984): 197–208.

al-Khalili, Jim. *The House of Wisdom: How Arabic Science Saved Ancient Knowledge and Gave Us the Renaissance.* New York: Penguin Press, 2011.

Kheirandish, Elaheh. "Astronomical Poems from the Four Corners of Persia." In *Essays in Islamic Philology, History, and Philosophy*, eds. Alireza Korangy et al., 51–90. Berlin; Boston: De Gruyter, 2016.

Kılıç, Filiz. *Şehzade Bayezid "Şahi": Hayatı ve Divanı.* Ankara: Kültür Bakanlığı Yayınları, 2000.

King, David. "The Astronomy of the Mamluks." *Isis* 74 (1983): 531–55.

———. "Lunar Crescent Visibility Predictions in Medieval Islamic Ephemerides." In *Quest for Understanding: Arabic and Islamic Studies in Memory of Malcolm H. Kerr*, eds. S. M. Seikaly, R. Baalbaki, and P. Dodd, vol. 4, 233–52. Beirut: American University of Beirut, 1991.

———. "On the Role of the Muezzin and the *Muwaqqit* in Medieval Islamic Society." In *Tradition, Transmission, Transformation*, eds. F. Jamil Ragep and Sally P. Ragep, 285–346. Brill: Leiden, 1996.

———. "Two Astrolabes for the Ottoman Sultan Bayezit II." In *Essays in Honour of Ekmeleddin İhsanoğlu*, eds. Mustafa Kaçar and Zeynep Durukal, vol. 1, 439–59. Istanbul: IRCICA, 2006.

King, David, and Toufic Fahd. "Al-Taliʿ." *EI*².

King, David, Julio Samsó, and Bernard R. Goldstein, "Astronomical Handbooks and Tables from the Islamic World (750–1900): An Interim Report." *Suhayl* 2 (2001): 9–105.

Kırkoğlu, R. Hakan. *Sultan ve Müneccimi: 18. Yüzyıl Osmanlı Sarayında İlm-i Nücum*. Istanbul: Doğan Kitap, 2017.

Kissling, Hans Joachim. "Betrachtungen über die Flottenpolitik Sultan Bâjezids II (1481–1512)." *Saeculum* 20 (1969): 35–43.

Klein, Ursula. "Artisanal-Scientific Experts in Eighteenth-Century France and Germany." *Annals of Science* 69, no. 3 (2012): 303–6.

Klein, Ursula, and E. C. Spary, eds. *Materials and Expertise in Early Modern Europe: Between Market and Laboratory*. Chicago; London: University of Chicago Press, 2009.

Köksal, Fatih. "Fütüvvetnamelere Göre Cerrahlık ve Tabip Şerefeddin Fütüvvetnamesi." *Divan Edebiyatı Araştırmaları Dergisi* 26 (2021): 426–47.

König, Jason, and Greg Woolf, eds. *Authority and Expertise in Ancient Scientific Culture*. Cambridge; New York: Cambridge University Press, 2017.

Koselleck, Reinhart. *Futures Past: On the Semantics of Historical Time*. Cambridge: MIT Press, 1985.

Krstić, Tijana, and Derin Terzioğlu, eds. *Historicizing Sunni Islam in the Ottoman Empire, c. 1450–c. 1750*. Leiden; Boston: Brill, 2021.

Küçük, Harun. "Islam, Christianity, and the Conflict Thesis." In *Science and Religion: New Historical Perspectives*, eds. Thomas Dixon, Geoffrey Cantore, and Stephen Pumfrey, 111–30. Cambridge: Cambridge University Press, 2010.

———. "İbrahim Müteferrika's Copernican Rhetoric." In *Translating Early Modern Science*, eds. Sietske Fransen, Niall Hodson, and Karl A. E. Enenkel, 258–85. Leiden: Brill, 2017.

———. *Science without Leisure: Practical Naturalism in Istanbul, 1660–1732*. Pittsburgh: University of Pittsburgh Press, 2020.

———. "The Bureaucratic Sense of the Forthcoming in Seventeenth-Century Istanbul." *Journal for the History of Knowledge* 1, no. 1 (2020): 1–16.

———. "Response to Nir Shafir." *OA* 59 (2022): 261–72.

Kunt, Metin. "Sultan Süleyman ve Nikris." In *Muhteşem Süleyman*, ed. Özlem Kumrular, 93–99. Istanbul: Kitap Yayınevi, 2007.

Kurnaz, Cemal. "Osmanlı Tarihinde İz Bırakan Amasyalı bir Aile: Müeyyedzadeler." In *I. Amasya Araştırmaları Sempozyumu Bildirileri*, ed. Yavuz Bayram, vol. 2, 647–66. Amasya: T. C. Amasya Valiliği, 2007.

Kurnaz, Cemal, and Mustafa Tatçı. "İbn İsa." *DİA*.

Kuru, Ahmet. *Islam, Authoritarianism, and Underdevelopment: A Global and Historical Comparison*. Cambridge: Cambridge University Press, 2019.

Kurz, Marlene. *Ein osmanischer Almanach für das Jahr 1239/1240 (1824/25)*. Berlin: Klaus Schwarz, 2007.

Kurz, Otto. *European Clocks and Watches in the Near East*. London: Warburg Institute, 1975.

Kütükoğlu, Mübahat. "Berat." *DİA*.

Langermann, Tzvi. "From My Notebooks: A Compendium of Renaissance Science; *Ta'alumot Hokmah* by Moses Galeano." *Aleph* 7 (2007): 285–318.

———. "From My Notebooks: Medicine, Mechanics and Magic from Moses ben Judah Galeano's *Ta'alumot Ḥokmah*." *Aleph* 9, no. 2 (2009): 353–77.

——. "From My Notebooks: On *Tajriba/Nissayon* ("Experience"); Texts in Hebrew, Judeo-Arabic, and Arabic." *Aleph* 14, no. 2 (2014): 147–76.

Lanuza-Navarro, Tayra. "Astrology in Court: The Spanish Inquisition, Authority, and Expertise." *History of Science* 55, no. 2 (2017): 187–209.

Latour, Bruno. *Science in Action: How to Follow Scientists and Engineers through Society*. Cambridge, MA: Harvard University Press, 1987.

Le Gall, Dina. *A Culture of Sufism: Naqshbandis in the Ottoman World, 1450–1700*. Albany: State University of New York Press, 2005.

Lelić, Emin. "Ottoman Physiognomy (*ʿIlm-i Firaset*): A Window into the Soul of an Empire." PhD, University of Chicago, 2017.

Lemay, Richard. "The Teaching of Astronomy in Medieval Universities, Principally at Paris in the Fourteenth Century." *Manuscripta* 20, no. 3 (1976): 197–217.

——. "Origin and Success of the Kitab Thamara of Abu Jaʿfar Ahmad ibn Yusuf ibn Ibrahim." In *Proceedings of the First International Symposium for the History of Arabic Science: April 5–12, 1976*, eds. Ahmad Y. al-Hassan et al., 91–107. Aleppo: University of Aleppo, 1978.

——. "Religion vs. Science in Islam: The Medieval Debate around Astrology." *Oriente moderno* 19, no. 3 (2000): 557–75.

Lewis, Bernard. "Ottoman Observers of Ottoman Decline." *Islamic Studies* 1 (1962): 71–87.

——. "First-Person Narrative in the Middle East." In *Middle Eastern Lives: The Practice of Biography and Self-Narrative*, ed. Martin Kramer, 20–34. Syracuse, NY: Syracuse University Press, 1991.

Lindner, Rudi Paul. *Nomads and Ottomans in Medieval Anatolia*. Bloomington: Indiana University Press, 1983.

Livingston, John. "Science and the Occult in the Thinking of Ibn Qayyim al-Jawziyya." *JAOS* 112, no. 4 (1992): 598–610.

——. *The Rise of Science in Islam and the West: From Shared Heritage to Parting of the Ways, 8th to 19th Centuries*. New York: Routledge, 2018.

Long, A. A. "Astrology: Arguments pro and contra." In *Science and Speculation: Studies in Hellenistic Theory and Practice*, ed. Jonathan Barnes et al., 165–93. Cambridge: Cambridge University Press, 1982.

MacDonald, D. B., and L. Gardet. "Al-Ghayb." *EI*².

Makdisi, George. *The Rise of Colleges: Institutions of Learning in Islam and the West*. Edinburgh: Edinburgh University Press, 1981.

Maraş, İbrahim. "Tokatlı Molla Lutfi: Hayatı Eserleri ve Felsefesi." *Divan* 14, no. 1 (2003): 119–36.

Markiewicz, Christopher. *The Crisis of Kingship in Late Medieval Islam: Persian Emigres and the Making of Ottoman Sovereignty*. Cambridge: Cambridge University Press, 2019.

Martin, John Jeffries. *A Beautiful Ending: The Apocalyptic Imagination and the Making of the Modern World*. New Haven: Yale University Press, 2022.

Martykánová, Darina. "Ottoman Engineers: The Redefinition of Expert Identities during the Reign of Abdülhamid II and the Early Years of the Second Constitutional Period." *Turcica* 45 (2014): 125–56.

Mavroudi, Maria. "Translations from Greek into Arabic at the Court of Mehmed the Conqueror." In *The Byzantine Court: Source of Power and Culture*, eds. Ayla Ödekan, Nevra Necipoğlu, and Engin Akyürek, 195–207. Istanbul: Koç University Press, 2013.

McGinnis, Jon. "Scientific Methodologies in Medieval Islam." *Journal of the History of Philosophy* 41, no. 3 (2003): 307–27.

Melvin-Koushki, Matthew. "The Quest for a Universal Science: The Occult Philosophy of Ṣā'in al-Dīn Turka Iṣfahānī (1369–1432) and Intellectual Millenarianism in Early Timurid Iran." PhD, Yale University, 2012.

——. "Powers of One: The Mathematicalization of the Occult Sciences in the High Persianate Tradition." *Intellectual History of the Islamicate World* 5, no. 1 (2017): 127–99.

——. "Early Modern Islamicate Empire: New Forms of Religiopolitical Legitimacy." In *The Wiley-Blackwell History of Islam*, eds. Armando Salvatore et al. Hoboken: Wiley-Blackwell, 2018.

——. "Persianate Geomancy from Ṭūsī to the Millennium: A Preliminary Survey." In *Occult Sciences in Premodern Islamic Culture*, eds. Nader el-Bizri and Eva Orthmann, 151–99. Beirut: Orient-Institut Beirut, 2018.

——. "Is (Islamic) Occult Science a Science?" *Theology and Science* 18, no. 2 (2020): 303–24.

Melvin-Koushki, Matthew, and Noah Gardiner. *Islamicate Occultism: New Perspectives*, special double issue of *Arabica* 64, no. 3–4 (2017).

Ménage, V. L. "The Serpent Column in Ottoman Sources." *Anatolian Studies* 14 (1964): 169–73.

——. "Kemāl Pasha-zāde." *EI*[2].

Mendelsohn, Andrew, and Annemarie Kinzelbach. "Common Knowledge: Bodies, Evidence, and Expertise in Early Modern Germany." *Isis* 108, no. 2 (2017): 259–79.

Metin, Tülay. "Selçuklular Zamanında Müneccimliğe Dair Bazı Tespitler." *International Journal of History Studies* 6, no. 3 (2014): 239–52.

Michot, Yahya. "Ibn Taymiyya on Astrology: Annotated Translation of Three Fatwas." *Journal of Islamic Studies* 11, no. 2 (2000): 147–208.

Mikhail, Alan. *Nature and Empire in Ottoman Egypt: An Environmental History*. New York: Cambridge University Press, 2011.

Miller, Kassandra Jackson. "A Doctor on the Clock: Hourly Timekeeping and Galen's Scientific Method." PhD, University of Chicago, 2017.

Mir-Kasimov, Orhan, ed. *Unity in Diversity: Mysticism, Messianism and the Construction of Religious Authority in Islam*. Leiden; Boston: Brill, 2014.

Moin, Azfar. *The Millennial Sovereign: Sacred Kingship and Sainthood in Islam*. New York: Columbia University Press, 2012.

——. "The 'Ulama' as Ritual Specialists: Cosmic Knowledge and Political Rituals." In *The Wiley-Blackwell History of Islam*, eds. Armando Salvatore et al., 377–92. Hoboken: Wiley-Blackwell, 2018.

Moran, Bruce T. "German Prince-Practitioners: Aspects in the Development of Courtly Science, Technology, and Procedures in the Renaissance." *Technology and Culture* 22, no. 2 (1981): 253–74.

——. *Patronage and Institutions: Science, Technology, and Medicine at the European Court, 1500–1750*. Rochester, NY: Boydell Press, 1991.

Mordtmann, Johannes Heinrich. "Das Observatorium des Taqi Ed-Din Zu Pera." *Der Islam* 13, no. 1–2 (1923): 82–96.

Morrison, Robert. *Islam and Science: The Intellectual Career of Nizam al-Din al-Nisaburi*. London: Routledge, 2007.

——. “Discussions of Astrology in Early Tafsīr.” *Journal of Qur'anic Studies* 11, no. 2 (2009): 49–71.
——. “An Astronomical Treatise by Mūsā Jālīnūs alias Moses Galeano.” *Aleph* 11, no. 2 (2011): 385–413.
——. “A Scholarly Intermediary between the Ottoman Empire and Renaissance Europe.” *Isis* 105, no. 1 (2014): 32–57.
——. “Cosmography, Cosmology, and Kalām from Samarqand to Istanbul.” *Intellectual History of the Islamicate World* 9, no. 3 (2020): 308–37.
Mottahedeh, Roy P. *Loyalty and Leadership in an Early Islamic Society*. Princeton: Princeton University Press, 1980.
Mozaffari, Mohammad. “The Effect of Astrological Opinions on Society: A Preliminary View.” *TRAMES* 16, no. 4 (2012): 359–68.
——. “Wabkanawi's Prediction and Calculations of the Annual Solar Eclipse of 30 January 1283.” *Historia Mathematica* 40 (2013): 235–61.
Mozaffari, Mohammad, and John M. Steele. “Solar and Lunar Observations at Istanbul in the 1570s.” *Archive for History of Exact Sciences* 69, no. 4 (2015): 343–62.
Mozaffari, Mohammad, and G. Zotti. “The Observational Instruments at the Maragha Observatory after AD 1300.” *Suhayl* 12 (2013): 45–179.
Mulsow, Martin. *Knowledge Lost: A New View of Early Modern Intellectual History*. Tr. Erik Midelfort. Princeton: Princeton University Press, 2022.
Murphey, Rhoads. *Exploring Ottoman Sovereignty: Tradition, Image and Practice in the Ottoman Imperial Household, 1400–1800*. London; New York: Continuum, 2008.
Murphy, Jane H. “Islamicate Knowledge Systems.” In *The Wiley Blackwell History of Islam*, eds. Armando Salvatore et al., 479–98. Hoboken: Wiley-Blackwell, 2018.
Muslu, Cihan Yüksel. “Ottoman-Mamluk Relations and the Complex Image of Bayezid II.” In *Conquête ottoman de l'Égypte (1517): Arrière-plan, impact, échos*, eds. Benjamin Lellouch and Nicholas Michel, 51–76. Leiden: Brill, 2013.
——. *The Ottomans and the Mamluks: Imperial Diplomacy and Warfare in the Islamic World*. London: I. B. Tauris, 2014.
Nallino, Carlo Alfonso. “Sun, Moon, and Stars (Mohammedan).” In *Encyclopedia of Religion and Ethics*, ed. James Hastings, 88–101. New York: C. Scribner, 1970.
Nasr, Seyyed Hossein. *An Introduction to Islamic Cosmological Doctrines*. Cambridge, MA: Belknap Press of Harvard University Press, 1964.
Nassiri, Giv. “Turco-Persian Civilization and the Role of Scholars' Travel and Migration in Its Elaboration and Continuity.” PhD, University of California, Berkeley, 2002.
Necipoğlu, Gülru. “Plans and Models in 15th- and 16th-Century Ottoman Architectural Practice.” *Journal of the Society of Architectural Historians* 45, no. 3 (1986): 224–43.
——. “From International Timurid to Ottoman: A Change of Taste in Sixteenth-Century Ceramic Tiles.” *Muqarnas* 7 (1990): 136–70.
——. *Architecture, Ceremonial, and Power: The Topkapı Palace in the Fifteenth and Sixteenth Centuries*. Cambridge, MA: MIT Press, 1991.
——. “A Kanun for the State, a Canon for the Arts: Conceptualizing the Classical Synthesis of Ottoman Art and Architecture.” In *Soliman le Magnifique et son temps: Actes du Colloque de Paris*, ed. Gilles Veinstein, 195–216. Paris: Documentation Française, 1992.
——. *The Age of Sinan: Architectural Culture in the Ottoman Empire*. London: Reaktion, 2005.

——. "Visual Cosmopolitanism and Creative Translation: Artistic Conversations with Renaissance Italy in Mehmed II's Constantinople." *Muqarnas* 29 (2012): 1–81.

——. "The Spatial Organization of Knowledge in the Ottoman Palace Library: An Encyclopedic Collection and Its Inventory." In *TOK*, vol. 1, 1–77.

Neugebauer, Otto. "The Study of Wretched Subjects." *Isis* 42, no. 2 (1951): 111.

Niccoli, Ottavia. *Prophecy and People in Renaissance Italy*. Tr. Lydia G. Cochran. Princeton: Princeton University Press, 1990.

Niyazioğlu, Aslı. *Dreams and Lives in Ottoman Istanbul: A Seventeenth Century Biographer's Perspective*. Abingdon, UK: Routledge, 2017.

Noble, Michael. *Philosophising the Occult: Avicennan Psychology and "The Hidden Secret" of Fakhr al-Dīn al-Rāzī*. Berlin: De Gruyter, 2021.

Ocak, Ahmet Yaşar. *Osmanlı Toplumunda Zındıklar ve Mülhidler (15.–17. Yüzyıllar)*. Istanbul: Türkiye Ekonomik ve Toplumsal Tarih Vakfı, 1998.

Öçal, Şamil. *Kışladan Medreseye: Osmanlı Bilgini Kemalpaşazade'nin Düşünce Dünyası*. Istanbul: İz, 2013.

Öğreten, Ahmet. "Türk Kültüründe 'Yada Taşı' ve XVIII. Yüzyıl Sonu Osmanlı-Rus Savaşlarında Kullanılması." *Belleten* 64, no. 241 (2000): 863–900.

Ökten, Ertuğrul. "Scholars and Mobility: A Preliminary Assessment from the Perspective of al-Shaqāyiq al-Nuʿmāniyya." *OA* 41 (2013): 55–70.

Ökten, Sadettin. "Çağmini." *DİA*.

Öngören, Reşat. *Osmanlılar'da Tasavvuf: Anadolu'da Sufiler, Devlet ve Ulema (XVI. Yüzyıl)*. Istanbul: Iz, 2000.

Oppenheimer, Michael, et al. *Discerning Experts: The Practices of Scientific Assessment for Environmental Policy*. Chicago; London: University of Chicago Press, 2019.

Orhonlu, Cengiz. "Şehir Mimarları." *OA* 2 (1981): 1–30.

Orthmann, Eva. "The Charm of Suspicious Calculations: Islamic Astrology in the Debates of Controversial Literature." *Beiruter Blätter* 10–11 (2002–3): 110–18.

——. "Circular Motions, Private Pleasure, and Public Prognostication in the Nativities of the Mughal Emperor Akbar." In *Horoscopes and Public Spheres: Essays on the History of Astrology*, eds. Günther Oestmann, H. Darrel Rutkin, and Kocku von Stuckrad, 101–14. Berlin; New York: Walter de Gruyter, 2005.

Oschema, Klaus. "Heaven Can Tell . . . Late Medieval Astrologers as Experts—and What They Can Teach Us about Contemporary Financial Expertise." In *Making the Medieval Relevant*, eds. Conor Kostick, Chris Jones, and Klaus Oschema, 191–210. Berlin; Boston: De Gruyter, 2019.

Özcan, Abdülkadir. "Kuloğlu." *DİA*.

Özel, Oktay. "Limits of the Almighty: Mehmed II's 'Land Reform' Revisited." *JESHO* 42 (1999): 226–46.

Özel, Oktay, and Gökhan Çetinsaya. "Türkiye'de Osmanlı Tarihçiliğinin Son Çeyrek Yüzyılı: Bir Bilanço Denemesi." *Toplum ve Bilim* 91 (2001–2): 8–38.

Pamuk, Şevket. *A Monetary History of the Ottoman Empire*. Cambridge: Cambridge University Press, 2000.

——. *İstanbul ve diğer kentlerde 500 yıllık fiyatlar ve ücretler, 1469–1998 = 500 years of prices and wages in Istanbul and other cities*. Ankara: T.C. Başbakanlık Devlet İstatistik Enstitüsü, 2000.

Pankenier, David. *Astrology and Cosmology in Early China: Conforming Earth to Heaven.* New York: Cambridge University Press, 2013.
Park, Katharine, and Lorraine Daston, eds. *Early Modern Science.* Volume 3 of *The Cambridge History of Science.* Cambridge: Cambridge University Press, 2006.
Peacock, Andrew C. S. "Two Royal Almanacs from Late Fourteenth-Century Anatolia." In *Fruit of Knowledge, Wheel of Learning: Essays in Honour of Carole Hillenbrand,* ed. Ali Ansari, 122–39. London: Gingko, 2022.
Pektaş, Nil. "The Beginnings of Printing in the Ottoman Capital: Book Production and Circulation in Early Modern Istanbul." *Osmanlı Bilimi Araştırmaları* 16, no. 2–3 (2015): 3–32.
Pertusi, Agostino, ed. *La caduta di Costantinopoli,* vol. 1. Rome: Fondazione Lorenzo Valla, 1976.
Pfeffer, Michelle. "Reassessing the Marginalization of Astrology in the Early Modern World." *The Historical Journal* 66 (2023): 1152–76.
Pfeifer, Helen. "A New Hadith Culture? Arab Scholars and Ottoman Sunnitization in the Sixteenth Century." In *Historicizing Sunni Islam in the Ottoman Empire, c. 1450–c. 1750,* eds. T. Krstić and D. Terzioğlu, 31–61. Leiden: Brill, 2021.
——. *Empire of Salons: Conquest and Community in Early Modern Ottoman Lands.* Princeton: Princeton University Press, 2022.
Pfeiffer, Judith, ed. *Politics, Patronage, and the Transmission of Knowledge in 13th–15th Century Tabriz.* Leiden; Boston: Brill, 2014.
——. "Teaching the Learned: Jalāl al-Dīn al-Dawānī's *Ijāza* to Muʾayyadzāda ʿAbd al-Raḥmān Efendi and the Circulation of Knowledge between Fārs and the Ottoman Empire at the Turn of the Sixteenth Century." In *The Heritage of Arabo-Islamic Learning: Studies Presented to Wadad Kadi,* eds. Maurice A. Pomerantz and Aram A. Shahin, 284–332. Leiden: Brill, 2016.
——. "Emerging from the Copernican Eclipse: The Mathematical and Astronomical Sciences in Müʾeyyedzade ʿAbdurrahman Efendi's Private Library (fl. ca. 1480–1516)." In *1st International Prof. Dr. Fuat Sezgin Symposium on History of Science in Islam Proceedings Books,* eds. M. Kaçar, C. Kaya, and A. Z. Furat, 151–91. Istanbul: Istanbul University Press, 2020.
Pietruska, Jamie. "Forecasting." In *Information: A Historical Companion,* eds. Ann Blair et al., 452–57. Princeton: Princeton University Press, 2021.
Pines, S. "The Semantic Distinction between the Terms Astronomy and Astrology according to al-Bīrūnī." *Isis* 55, no. 3 (1964): 343–49.
Pingree, David. "Astronomy and Astrology in India and Iran." *Isis* 54, no. 2 (1963): 229–46.
——. "Indian Influence on Sassanian and Early Islamic Astronomy and Astrology." *Journal of Oriental Research* 34–35 (1964–1965): 118–26.
——. "Abu Maʿshar al-Balkhi, Jaʿfar ibn Muhammad." In *Dictionary of Scientific Biography,* vol. 1, 32–39. New York: Scribner, 1970.
——. "Astrology." In *Dictionary of the History of Ideas,* vol. 1, ed. Philip P. Wiener, 118–26. New York: Scribner, 1973.
——. *From Astral Omens to Astrology: From Babylon to Bīkāner.* Rome: Istitutio Italiano per l'Africa e l'Oriente, 1997.
——. "Kiran." *EI*².

Pingree, David, and B. Goldstein. "Astrological Almanacs from the Cairo Geniza, Part 1." *JNES* 38, no. 3 (1979): 153–75.

———. "Astrological Almanacs from the Cairo Geniza, Part 2." *JNES* 38, no. 4 (1979): 231–56.

———. "Additional Astrological Almanacs from the Cairo Geniza." *JAOS* 103, no. 4 (1983): 673–90.

Polanyi, Michael. *The Tacit Dimension*. Garden City, NY: Doubleday, 1966.

Popp, Stephan. "Mughal Horoscopes as Propaganda." *JPS* 9 (2016): 45–59.

Pormann, Peter E. "The Physician and the Other: Images of the Charlatan in Medieval Islam." *BHM* 79, no. 2 (2005): 189–227.

Porter, Jonathan. "The Scientific Community in Early Modern China." *Isis* 73, no. 4 (1982): 529–44.

Poskett, James. *Horizons: The Global Origins of Modern Science*. Boston: Mariner Books, 2022.

Pourjavady, Reza. *Philosophy in Early Safavid Iran. Najm al-Din Mahmud al-Nayrizi and His Writings*. Leiden: Brill, 2011.

Principe, Lawrence. *The Scientific Revolution: A Very Short Introduction*. Oxford; New York: Oxford University Press, 2011.

Quigley, Joan. *What Does Joan Say? My Seven Years as White House Astrologer to Nancy and Ronald Reagan*. Secaucus, NJ: Birch Lane Press, 1990.

Rabier, Christelle, ed. *Fields of Expertise: A Comparative History of Expert Procedures in Paris and London, 1600 to Present*. Newcastle, UK: Cambridge Scholars, 2007.

Raby, Julian. "El Gran Turco: Mehmed the Conqueror as a Patron of the Arts of Christendom." PhD, Oxford University, 1980.

Raby, Julian, and Zeren Tanındı. *Turkish Bookbinding in the 15th Century*. London: Azimuth Editions, 1993.

Ragep, Jamil. "Freeing Astronomy from Philosophy: An Aspect of Islamic Influence on Science." *Osiris* 16 (2001): 49–71.

———. "ʿAlī Qushjī and Regiomontanus: Eccentric Transformations and Copernican Revolutions." *Journal for the History of Astronomy* 36, no. 4 (2005): 359–71.

———. "Astronomy in the Fenari-Circle: The Critical Background for Qadizade al-Rumi and the Samarkand School." In *Uluslararası Molla Fenari Sempozyumu (4-6 Aralık 2009 Bursa): Bildiriler = International Symposium on Molla Fanārī (4-6 December 2009 Bursa): Proceedings*, 165–76. Bursa: Bursa Büyükşehir Belediyesi, 2010.

———. "New Light on Shams: The Islamic Side of Σὰμψ Πουχάρης." In *Politics, Patronage and the Transmission of Knowledge in 13th–15th Century Tabriz*, ed. Judith Pfeiffer, 166–80. Leiden: Brill, 2013.

———. "Qadizada al-Rumi." *BEA*.

———. "Astronomy." *EI*[3].

Ragep, Jamil, et al. "Astronomical and Other Mathematical Sciences." In *TOK*, vol. 1, 823–55.

Ragep, Sally. "Maḥmūd ibn Muḥammad ibn ʿUmar al-Jaghmīnī's al-Mulakhkhaṣ fī al-hayʾa al-basīṭa: An Edition, Translation, and Study." PhD, McGill University, 2014.

———. *Jaghmīnī's Mulakhkhaṣ: An Islamic Introduction to Ptolemaic Astronomy*, electronic resource. Cham: Springer, 2016.

Rapoport, Yossef, and Emilie Savage-Smith. *Lost Maps of the Caliphs: Drawing the World in Eleventh-Century Cairo*. Chicago: University of Chicago Press, 2018.

Reeve, C. D. C. *Philosopher-Kings: The Argument of Plato's Republic*. Princeton: Princeton University Press, 1988.

Regourd, A., and Pierre Lory, eds. "Sciences occultes et islam." *Bulletin d'études orientales* 14 (1993).

Richardson, Kristina L. *Roma in the Medieval Islamic World: Literacy, Culture, and Migration*. London; New York: I. B. Tauris, 2022.

Ripat, Pauline. "Expelling Misconceptions: Astrologers at Rome." *Classical Philology* 106, no. 2 (2011): 115–54.

Rogers, J. Michael. "Centralisation and Timurid Creativity." *Oriente moderno* 15, no. 2 (1996): 533–55.

el-Rouayheb, Khaled. *Islamic Intellectual History in the Seventeenth Century: Scholarly Currents in the Ottoman Empire and the Maghreb*. New York: Cambridge University Press, 2015.

Russell, G. A. "Physicians at the Ottoman Court." *Medical History* 34 (1990): 243–67.

Rutkin, Darrel H. "Astrology." In *Cambridge History of Science*. Vol. 3: *Early Modern Science*, 541–61.

——. "How to Accurately Account for Astrology's Marginalization in the History of Science and Culture: The Central Importance of an Interpretive Framework." *Early Science and Medicine* 23 (2018): 217–43.

Ryan, Michael A. *A Kingdom of Stargazers: Astrology and Authority in the Late Medieval Crown of Aragon*. Ithaca: Cornell University Press, 2011.

Sabra, A. I. "The Appropriation and Subsequent Naturalization of Greek Science in Medieval Islam: A Preliminary Statement." *History of Science* 25 (1987): 223–43.

——. "Science and Philosophy in Medieval Islamic Theology: The Evidence of the Fourteenth Century." *Zeitschrift für Geschichte der Arabisch-Islamischen Wissenschaften* 9 (1994): 1–42.

——. "Situating Arabic Science: Locality *Versus* Essence." *Isis* 87, no. 4 (1996): 654–70.

Sahillioğlu, Halil. "Années sıvış et crises monétaires dans l'Empire ottoman." *Annales: Économies, sociétés, civilisations* 24, no. 5 (1969): 1071–91.

Şahin, Kaya. *Empire and Power in the Reign of Süleyman: Narrating the Sixteenth-Century Ottoman World*. Cambridge: Cambridge University Press, 2013.

——. "From Frontier Principality to Early Modern Empire: Limitations and Capabilities of Ottoman Governance." In *Routledge History of the Renaissance*, ed. William Caferro, 321–36. London; New York: Routledge, 2017.

——. "The Ottoman Empire in the Long Sixteenth Century." *Renaissance Quarterly* 70, no. 1 (2017): 220–34.

——. *Peerless among Princes: The Life and Times of Sultan Süleyman*. New York: Oxford University Press, 2023.

Saif, Liana. *The Arabic Influences on Early Modern Occult Philosophy*. New York: Palgrave Macmillan, 2015.

——. "What Is Esotericism." *Correspondences* 7, no. 1 (2019): 1–59.

Saif, Liana, et al., eds. *Islamicate Occult Sciences in Theory and Practice*. Leiden; Boston: Brill, 2021.

Sajdi, Dana. "Decline, Its Discontents, and Ottoman Cultural History: By Way of Introduction." In *Ottoman Tulips, Ottoman Coffee: Leisure and Lifestyle in the Eighteenth Century*, ed. Dana Sajdi, 1–40. London: I. B. Tauris, 2004.

Saliba, George. "The Role of Astrologer in Medieval Islamic Society." *Bulletin d'études orientales* 44 (1992): 45–67.

——. "Al-Qushjī's Reform of the Ptolemaic Model for Mercury." *Arabic Sciences and Philosophy* 3 (1993): 161–203.

——. *A History of Islamic Astronomy: Planetary Theories during the Golden Age of Islam*. New York: New York University Press, 1994.

——. "Astronomy and Astrology in Medieval Arabic Thought." In *Les Doctrines de la science de l'Antiquité à l'Âge Classique*, eds. Roshdi Rashed and Joël Biard, 131–64. Leuven: Éditions Peeters, 1999.

——. "Islamic Astronomy in Context: Attacks on Astrology and the Rise of the *Hayʾa* Tradition." *Bulletin of the Royal Institute for Inter-Faith Studies* 4, no. 1 (2002): 25–46.

——. *Islamic Science and the Making of the European Renaissance*. Cambridge, MA: MIT Press, 2007.

Samsó, Julio. *On Both Sides of the Strait of Gibraltar: Studies in the History of Medieval Astronomy in the Iberian Peninsula and the Maghrib*. Leiden; Boston: Brill, 2020.

Sarı, Nil. "The Medical Organization at the Ottoman Court." *Studies in History of Medicine and Science* 16, no. 1–2 (1999/2000): 37–51.

Sariyannis, Marinos. *A History of Ottoman Political Thought up to the Early Nineteenth Century*. Leiden; Boston: Brill, 2019.

——. "Knowledge and Control of the Future in Ottoman Thought." *Acaʿib: Occasional Papers on the Ottoman Perceptions of the Supernatural* 1 (2020): 49–84.

——. *The Horizons, Limits, and Taxonomies of Ottoman Knowledge*. Berlin: EB Verlag, 2021.

Sarton, George. *Introduction to the History of Science*. Vol. 1: *From Homer to Omar Khayyam*. Baltimore: Williams and Wilkins Company, 1927.

——. *Six Wings: Men of Science in the Renaissance*. Bloomington: Indiana University Press, 1957.

——. *The History of Science and the New Humanism*. New Brunswick: Transaction Books, 1988.

Sarton, George, and Frances Siegel. "Seventy-Sixth Critical Bibliography of the History and Philosophy of Science and of the History of Civilization." *Isis* 41, no. 3–4 (1950): 374.

Savadi, Fateme, and Sajjad Nikfahm Khubravan. "Ḥarakat-i vasaṭ-i kavākib dar Zīj-i Ilkhānī va naqdhā-yi vārid bar ān." In *Ustād-i bashar: Pajuhashhāyi dar zandagī, ruzgār, falsafa va ʿilm-i Khwāja Naṣīr al-Dīn Ṭūsī*, eds. H. Masoumi Hamedani and M. Javad Anwari, 363–470. Tehran: Miras-e Maktoob, 1391 [2012].

Savage-Smith, Emilie, ed. *Magic and Divination in Early Islam*. Burlington, VT: Ashgate/Variorum, 2004.

Sayılı, Aydın. *The Observatory in Islam and Its Place in the General History of the Observatory*. Ankara: TTK, 1960.

Schmidl, Petra G. "Magic and Medicine in a 13th-Century Treatise on the Science of the Stars: The *Kitāb al-Tabṣira fī ʿilm al-nujūm* of the Rasulid Sultan al-Ashraf ʿUmar." In *Herbal Medicine in Yemen: Traditional Knowledge and Practice and Their Value for Today's World*, eds. Ingrid Hehmeyer and Hanne Schonig, 43–68. Leiden: Brill, 2012.

——. "Using Astrolabes for Astrological Purposes: The Earliest Evidence Revisited." In *Heaven and Earth United: Instruments in Astrological Contexts*, eds. Richard Dunn, Silke Ackermann, and Giorgio Strano, 4–23. Leiden: Brill, 2018.

Schmidt, Jan. "The Occult Sciences and Their Importance in Ottoman Culture: Evidence from Turkish Manuscripts in Dutch Public Collections." *OA* 23 (2003): 219–54.

Schwartz, Kathryn. "Did Ottoman Sultans Ban Print?" *Book History* 20 (2017): 1–39.

Scott, James C. *Seeing like a State: How Certain Schemes to Improve the Human Condition Have Failed*. New Haven: Yale University Press, 1998.

Selçuk, İklil, and Cihan Yüksel. "Early Modern Reflections on Bayezid II's Reign." In *Crafting History: Essays on the Ottoman World and Beyond in Honor of Cemal Kafadar*, eds. Rachel Goshgarian, Ilham Khuri-Makdisi, and Ali Yaycıoğlu, 513–32. Boston: Academic Studies Press, 2023.

Şen, A. Tunç. "A Mirror for Princes, A Fiction for Readers: The *Habname* of Veysi and Dream Narratives in Ottoman Turkish Literature." *Journal of Turkish Literature* 8 (2011): 41–65.

——. "Astrology in the Service of the Empire: Knowledge, Prognostication, and Politics at the Ottoman Court, 1450s–1550s." PhD, University of Chicago, 2016.

——. "Practicing Astral Magic in Sixteenth-Century Ottoman Istanbul: A Treatise on Talismans Attributed to Ibn Kemāl (d. 1534)." *Journal of Magic, Ritual, and Witchcraft* 12 (2017): 66–88.

——. "The Sultan's Syllabus Revisited: Sixteenth-Century Ottoman Madrasa Libraries and the Question of Canonization." *SI* 116 (2021): 198–235.

——. "The Emergence of New Scholarly Languages: The Case of Ottoman Turkish (13th–19th Centuries)." In *RHS*, 240–47.

——. "The Science and Politics of Chronology in Late Medieval and Early Modern Ottoman Almanacs." In *Osmanlı'da İlm-i Tarih*, eds. Zahit Atçıl, Ercüment Asil, and Cemal Atabaş, 47–72. Istanbul: İSAR, 2023.

Şen, A. Tunç, and Cornell H. Fleischer. "Books on Astrology, Astronomical Tables, and Almanacs in the Library Inventory of Bayezid II." In *TOK*, vol. 1, 767–821.

Serin, Sinem. *Osmanlı Sağlık Sisteminin Yönetimi: Hekimbaşılık Kurumu*. Istanbul: Kitabevi, 2021.

Şeşen, Ramazan. "Meşhur Osmanlı Astronomu Takiyyuddin El-Rasıd'in Soyu Üzerine." *Erdem* 4, no. 10 (1988): 165–80.

Shafir, Nir. "The Almighty Akçe: Review of Harun Küçük's *Science without Leisure*." *OA* 58 (2021): 251–80.

Shapin, Steven. "Trusting George Cheyne: Scientific Expertise, Common Sense, and Moral Authority in Early Eighteenth-Century Dietetic Medicine." *BHM* 77, no. 2 (2003): 263–97.

Shaw, Stanford. "Ottoman Archival Materials for the Nineteenth and Early Twentieth Centuries: The Archives of Istanbul." *IJMES* 6, no. 1 (1975): 94–114.

Shefer-Mossensohn, Miri. "Medical and Professional Ethics in Sixteenth-Century Istanbul: Towards an Understanding of the Relationships between the Ottoman State and the Medical Guilds." *Med Law* 21, no. 2 (2002): 307–19.

——. *Science among the Ottomans: The Cultural Creation and Exchange of Knowledge*. Austin: University of Texas Press, 2015.

Smith, Pamela. "Science on the Move: Recent Trends in the History of Early Modern Science." *Renaissance Quarterly* 62 (2009): 345–75.

Smith, Robert. "Observatory." In *A Companion to the History of Science*, ed. Bernard V. Lightman, 196–209. Chichester, UK: John Wiley & Sons, 2016.

Smoller, Laura Ackerman. *History, Prophecy, and the Stars: The Christian Astrology of Pierre d'Ailly, 1350–1420*. Princeton: Princeton University Press, 1994.

Sohrewide, Hanna. "Dichter und Gelehrten aus dem Ostem im osmanischen Reich." *Der Islam* 46 (1970): 263–302.

Soylu, Sedef. "Şeyh Hamdullah ve II: Bayezid." *Klasik Sanatlar Yıllığı* (2014): 82–87.

Špelda, Daniel. "Collective Empiricism at the Paris Observatoire in the Late Seventeenth Century." *Seventeenth Century* 37, no. 6 (2022): 977–1007.

Starr, S. Frederick. *The Genius of Their Age: Ibn Sina, Biruni and the Lost Enlightenment*. New York: Oxford University Press, 2023.

Stearns, Justin. *Revealed Sciences: The Natural Sciences in Islam in Seventeenth-Century Morocco*. Cambridge: Cambridge University Press, 2021.

Stehr, Nico, and Reiner Grundmann. *Experts: The Knowledge and Power of Expertise*. New York: Routledge, 2011.

Stolz, Daniel. *The Lighthouse and the Observatory: Islam, Science, and Empire in Late Ottoman Egypt*. Cambridge: Cambridge University Press, 2018.

Stowasser, Barbara Freyer. *The Day Begins at Sunset: Perceptions of Time in the Islamic World*. London; New York: I. B. Tauris, 2014.

Struck, Peter T. *Divination and Human Nature: A Cognitive History of Intuition in Classical Antiquity*. Princeton: Princeton University Press, 2016.

Tamdoğan-Abel, Işık. "Le Future dans le *Rumuz-ı Kunuz* de Mejdeddin Ibn Isa: Une utopie, une prophétie, un livre à mystères." In *Melamis-Bayramis: Études sur trois mouvements mystiques musulmans*, eds. N. Clayer, A. Popovic, and T. Zarcone, 145–52. Istanbul: Isis, 1998.

Tansel, Selahattin. "Yeni Vesikalar Karşısında Sultan İkinci Bayezid Hakkında Bazı Mütalaalar." *Belleten* 27, no. 106 (1963): 185–236.

Tekeli, Sevim. "Nasirüddin, Takiyüddin ve Tycho Brahe'nin Rasad Aletlerinin Mukayeyesi." *AÜDTCF Dergisi* 16, no. 3–4 (1958): 301–93.

Terzioğlu, Derin. "Man in the Image of God in the Image of the Times: Sufi Self-Narratives and the Diary of Niyazi-i Misri (1618–94)." *SI* 94 (2002): 139–65.

———. "Sufis in the Age of State-Building and Confessionalization." In *The Ottoman World*, ed. Christine Woodhead, 86–99. London; New York: Routledge, 2012.

———. "How to Conceptualize Ottoman Sunnitization: A Historiographical Discussion." *Turcica* 44 (2012–2013): 301–38.

———. "Power, Patronage, and Confessionalism: Ottoman Politics through the Eyes of a Crimean Sufi, 1580–1593." In *Political Thought and Practice in the Ottoman Empire: Halcyon Days in Crete IX*, eds. Marinos Sariyannis et al., 149–86. Heraklion: Crete University Press, 2019.

Tezcan, Baki. "The 'Kānūnnāme of Mehmed II': A Different Perspective." In *The Great Ottoman-Turkish Civilization*, ed. Kemal Çiçek, vol. 3, 657–65. Ankara: Yeni Türkiye, 2000.

———. "Politics of Early Modern Ottoman Historiography." In *The Early Modern Ottomans: Remapping the Empire*, eds. Virginia H. Aksan and Daniel Goffman, 167–98. Cambridge: Cambridge University Press, 2007.

———. *The Second Ottoman Empire: Political and Social Transformation in the Early Modern World*. New York: Cambridge University Press, 2010.

——. "Some Thoughts on the Politics of Early Modern Ottoman Science." *OA* 36 (2010): 135–56.

Tezcan, Şirin. *Topkapı Sarayı Müzesi Koleksiyonundan Tılsımlı Gömlekler.* Istanbul: Timaş, 2011.

Thomas, Keith. *Religion and the Decline of Magic: Studies in Popular Beliefs in Sixteenth- and Seventeenth-Century England.* New York: Scribner, 1971.

Thorndike, Lynn. *A History of Magic and Experimental Science.* 8 volumes. New York: Macmillan, 1923–58.

Tibawi, A. L. "Origin and Character of al-Madrasah." *BSOAS* 25 (1962): 225–38.

Tomar, Erdal Zeki. "Anadolu Muvakkithaneleri." MA thesis, Konya Selçuk University, 2011.

Tourkin, Sergei. "Iskandar-sultan und sein Geburtsbuch." In *Orakel. Der Blick in die Zukunft. Ausstellungkatalog Museum Rietberg Zürich*, eds. A. Langer and A. Lutz, 180–89. Zürich: Museum Rietberg, 1999.

——. "Medical Astrology in the Horoscope of Iskandar Sultan." In *Sciences, techniques et instruments dans le monde iranien*, eds. N. Poujarvady and Z. Vesel, 105–9. Tehran: IFRI, 2004.

Treiger, Alexander. "A Fourteenth-Century Arabic Treatise *On the Platonic Intellectual Ideas.*" In *Philosophy and the Abrahamic Religions: Scriptural Hermeneutics and Epistemology*, eds. T. Kirby, R. Acar, and B. Baş, 251–76. Newcastle upon Tyne: Cambridge Scholars, 2013.

Trigg, Scott. "From Samarqand to Istanbul: Astronomy and Scientific Education in the Commentaries of Fatḥallāh al-Shirwānī." PhD, University of Wisconsin–Madison, 2016.

Tunalı, Gülçin. "Sadullah el-Ankaravi: Daily Concerns of an Ottoman Astrologer." MA thesis, Bogazici University, 2002.

——. "An Ottoman Astrologer at Work: Sadullah el-Ankaravi and the Everyday Practice of *İlm-i Nücum.*" In *Les Ottomans et le temps*, eds. F. Georgeon and F. Hitzel, 39–60. Leiden: Brill, 2012.

Turan, Şerafettin. *Kanuni Süleyman Dönemi Taht Kavgaları.* Ankara: Bilgi Yayınevi, 1997.

——. "II. Bayezid." *DİA.*

Uluçay, Çağatay. "Yavuz Sultan Selim Nasıl Padişah Oldu II." *İÜEF Tarih Dergisi* 7, no. 10 (1954): 117–42.

Uludağ, Süleyman. "Ricalü'l-gayb." *DİA.*

Umut, Hasan. "Theoretical Astronomy in the Early Modern Ottoman Empire: ʿAlī al-Qūshjī's *Al-Risāla al-Fatḥiyya.*" PhD, McGill University, 2019.

Unan, Fahri. *Kuruluşundan Günümüze Fatih Külliyesi.* Ankara: TTK, 2003.

Unat, Yavuz. *Ali Kuşçu: Çağını Aşan Bilim İnsanı.* Istanbul: Kaynak, 2009.

Ünver, Süheyl. *Türk Pozitif İlimler Tarihinden Bir Bahis, Ali Kuşci: Hayatı ve Eserleri.* Istanbul: Kenan Matbaası, 1948.

——. *İstanbul Rasathanesi.* Ankara: TTK, 1969.

——. "Osmanlı Türkleri İlim Tarihinde Muvakkithaneler." *Atatürk Konferansları V* (1971–72): 217–57.

Uzunçarşılı, İsmail Hakkı. *Osmanlı Devletinin İlmiye Teşkilatı.* Ankara: TTK, 1965.

——. *Osmanlı Devletinin Saray Teşkilatı.* Ankara: AKDTYK, 1984.

——. “Osmanlı Sarayı’nda Ehl-i Hıref (Sanatkarlar) Defterleri.” *Belgeler* 11, no. 15 (1986): 23–76.

van Berkel, Maaike, and Jeroen Duindam, eds. *Prince, Pen, and Sword: Eurasian Perspectives.* Leiden; Boston: Brill, 2018.

van Dalen, Benno. *Islamic Astronomical Tables: Mathematical Analysis and Historical Investigation.* Farnham, UK: Ashgate, 2013.

——. “Commentary on the Horoscope of Iskandar Sultan: An Introduction to Islamic Mathematical Astrology.” *Zeitschrift für Geschichte der Arabisch-Islamischen Wissenschaften* 23 (2022): 1–104.

——. “Al-Tāʾrīkh 2. Era Chronology in Astronomical Handbooks.” *EI*³.

——. “Wabkanawi.” *BEA.*

van Ess, Josef. “Jorjāni, Zayn al-Dn Abu’l-Ḥasan ʿAlī b. Moḥammad b. ʿAlī al-Ḥosaynī.” *Encyclopædia Iranica*, online edition. https://www.iranicaonline.org/articles/jorjani-zayn-al-din-esmail.

van Gulik, Egbertus, et al. *Erasmus and His Books.* Toronto: University of Toronto Press, 2018.

Vardanyan, S. A. *Amirdovlat Amasiatsi: A Fifteenth-Century Armenian Natural Historian and Physician.* Tr. Michael Yoshpa. Delmar, NY: Caravan Books, 1999.

Varlık, Nükhet. *Plague and Empire in the Early Modern Mediterranean World: The Ottoman Experience, 1347–1600.* New York: Cambridge University Press, 2015.

Velioğlu, Menderes. “Müneccim Riyazi Ali Çelebi’nin Hayatı, Telifatı ve Muhallefatı.” PhD, Istanbul University, 2023.

Vesel, Živa. *Les Encylopédies persanes: Essai de typologie et de classification des sciences.* Paris: Éditions Recherche sur les Civilisations, 1986.

——. “Teucros in Nizami’s Haft Paykar.” In *A Key to the Treasure of the Hakim: Artistic and Humanistic Aspects of Nizami Ganjavi’s Khamsa*, eds. J. C. Bürgel and C. van Ruymbeke, 245–52. Leiden: Leiden University Press, 2011.

Westman, Robert. “The Astronomer’s Role in the Sixteenth Century: A Preliminary Study.” *History of Science* 8 (1980): 105–47.

——. *The Copernican Question: Prognostication, Skepticism, and Celestial Order.* Berkeley: University of California Press, 2011.

Whewell, William. *History of the Inductive Sciences.* London: John W. Parker, 1837.

White, Sam. *The Climate of Rebellion in the Early Modern Ottoman Empire.* New York: Cambridge University Press, 2011.

Wishnitzer, Avner. *Reading Clocks, Alla Turca: Time and Society in the Late Ottoman Empire.* Chicago; London: University of Chicago Press, 2015.

——. “Timekeeping.” *EI*³.

Woods, John E. *The Aqquyunlu: Clan, Confederation, Empire*, revised and expanded edition. Salt Lake City: University of Utah Press, 1999.

Yalçınkaya, M. Alper. *Learned Patriots: Debating Science, State, and Society in the Nineteenth-Century Ottoman Empire.* Chicago; London: University of Chicago Press, 2015.

——. “Muslims.” In *The Warfare between Science and Religion: The Idea That Wouldn’t Die*, eds. J. Hardin, R. L. Numbers, and Ronald A. Binzley, 203–19. Baltimore: Johns Hopkins University Press, 2018.

——. “‘Science,’ ‘Religion,’ and ‘Science-and-Religion’ in the Late Ottoman Empire.” *Zygon: Journal of Religion & Science* 54, no. 4 (2019): 1050–66.

Yang, Qiao. "Like Stars in the Sky: Networks of Astronomers in Mongol Eurasia." *JESHO* 62 (2019): 388–427.

Yerasimos, Stephanos. *Türk Metinlerinde Konstaniyye ve Ayasofya Efsaneleri*. Tr. Şirin Tekeli. Istanbul: İletişim, 1993.

Yıldırım, Ali. "16. Yüzyılda Büyük Bir Devlet Adamı ve Edebiyat Hamisi Defterdar İskender." *Fırat Üniversitesi Sosyal Bilimler Dergisi* 10, no. 1 (2000): 217–32.

Yıldırım, Özge Nur. *The Twin Books for Two Ottoman Princesses: An Examination on the Matali'ü's-Saadet Manuscripts (1582)*. Istanbul: Libra, 2023.

Yıldırım, Rıza. "Turkomans between Two Empires: The Origins of the Qizilbash Identity in Anatolia (1447–1514)." PhD, Bilkent University, 2008.

Yılmaz, Hüseyin. *Caliphate Redefined: The Mystical Turn in Ottoman Political Thought*. Princeton: Princeton University Press, 2018.

Yılmaz, Müslüm. "Özri ve Şiirleri." *Divan Edebiyatı Araştırmaları Dergisi* 27 (2021): 713–30.

Yılmazer, Ziya. "Sahaflar Şeyhizade Esad Efendi." *DİA*.

Yücesoy, Hayrettin. *Messianic Beliefs and Imperial Politics in Medieval Islam: The ʿAbbāsid Caliphate in the Early Ninth Century*. Columbia: University of South Carolina Press, 2009.

Yurdaydın, Hüseyin. *Matrakçı Nasuh*. Ankara: Ankara Üniversitesi Basımevi, 1963.

Zarinabaf-Shahr, Fariba. "Qizilbash 'Heresy' and Rebellion in Ottoman Anatolia in the Sixteenth Century." *Anatolia Moderna* 7 (1997): 1–15.

Zilfi, Madeline. "The Diary of a Müderris: A New Source for Ottoman Biography." *JTS* 1 (1977): 157–74.

Dictionaries and Reference Works

Dehkhoda, Ali Akbar. *Dehkhoda Dictionary*. Tehran: University of Tehran. https://dehkhoda.ut.ac.ir/fa/dictionary.

Hawramani, Ibn Munir. *Arabic Lexicon*. http://arabiclexicon.hawramani.com.

Meniński, Franciszek. *Thesaurus linguarum orientalium Turcicae-Arabicae-Persicae = Lexicon Turcico-Arabico-Persicum*. 6 vols. Istanbul: Simurg, 2000.

OED [Oxford English Dictionary] *Online*. Oxford; New York: Oxford University Press, n.d.

Redhouse, James W. *A Turkish and English Lexicon*. Beirut: Librairie du Liban, 1974.

Sezgin, Fuat. *Sezgin Online: A Bio-bibliography for the Arabic Literary Tradition*. Leiden: Brill, 2021.

Steingass, Francis Joseph. *A Comprehensive Persian-English Dictionary*. 1st ed., London: Routledge, 1892. Digital South Asia Library, University of Chicago, https://dsal.uchicago.edu/dictionaries/steingass.

Storey, C. A. *Persian Literature: A Bio-bibliographical Survey*. Eds. François de Blois, J. Lameer, and J. D. Pearson. Leiden; Boston: Brill, 2021. [Storey Online]

Ullmann, Manfred. *Die Natur- und Geheimwissenschaften im Islam*. Leiden: Brill, 1972.

Wehr, Hans. *A Dictionary of Modern Written Arabic: Arabic-English*. Ed. J. Milton Cowan. Ithaca: Spoken Language Services, 1994.

INDEX

Abbasids, 8, 29, 49, 87–88, 93, 103
ʿAbdülhamid I (sultan), 195, 273n13
ʿAbdülkerim b. Kutbuddin, 230
ʿAbdullah (*munajjim*), 278n28
ʿAbdülmecid I (sultan), 71
ʿAbdurrahman (*munajjim*), 59, 82–86, 88–92, 95–97, 106, 111, 123
ʿAbdüsselam, 126
ʿAbdulwajid b. Muhammad, 122
abjad system, 199
Abu Jaʿfar Ahmad b. Yusuf, 56
Abu Maʿshar, 56–57, 63, 80, 90, 103, 135, 168
Abu Muslim, 93
Abu Saʿid Gurakan, 98
Abu'l-Fida, 167
Abu'l-Qasim Babur Mirza, 99
academies, 226, 233
Adam (prophet), 93
Adıvar, Adnan, 20, 164, 242n49
ʿAdli, 139. *See also* Bayezid II (sultan)
Afsahi, 126
Ahi Çelebi, 129, 133
ahkam al-nujum (*ʿilm al-*), 43–47, 62–63, 68, 70
Ahmed I (sultan), 206
Ahmed (prince, Bayezid II's son), 70, 84–86, 89–92, 96–97, 133, 140
Ahmed Efendi (*munajjim*), 278n28
Ajaʾib al-makhluqat (Wonders of Creation), 171
Albohazen Haly. *See* ʿAli b. Abi al-Rijal
alchemy, 28, 47, 84, 125–27, 129–30, 232
Aleppo, 7, 141–42
Alexander the Great, 50, 132, 173
algebra, 47
ʿAli (Mustafa), 13, 73, 158, 161, 178–9, 206, 211, 269n26
ʿAli b. Abi al-Rijal, 167
ʿAli b. Abi Talib, 29, 141, 220
ʿAli Çelebi (Müʾeyyedzade's father), 140
ʿAli Qushji, 23, 40, 62, 67, 87–88, 122, 130–31, 160–61, 230, 236, 246n5
Almagest, 43, 46, 135, 143, 151, 167. *See also* Ptolemy
almanac (with prognostications): 16–8, 32, 34, 48, 51, 53, 56–60, 62–63, 68–69, 70, 74, 79–80, 82; in the 15th century, 118–19; by Abdurrahman, 85–96, 90–96; in Bayezid II's time, 124–25;

almanac (with prognostications) (*cont.*)
ephemeral nature of, 91, 192; by Khitabi, 99, 107–7; by Mirim Çelebi, 131; by Mirim Çelebi's students, 133–35; vs. other prognostications, 224, 230, 234; by Qadi-yi Baghdad, 127; by Riyazi, 169, 172; in Selim I's time, 156–57; by Yusuf b. ʿÖmer, 180–84
almanac-makers, 87–88, 102, 107, 113, 199, 202. *See also* almanac (with prognostications)
Almanach Perpetuum, 129, 143. *See also* Zacuto, Abraham
Alpini, Prospero, 238
Amasiatsi, Amirdovlat, 140
Amasya, 84–86, 89–92, 96–97, 100, 125, 127, 139–41
amulets, 174, 179
Amuli (Rukn al-), 45, 97–99, 104
apocalypticism: 29–32, 37; in Bayezid II's time, 148–50; in Murad III's time, 211–13; in Süleyman's time, 154, 184, 190–91, 208, 219, 224. *See also* Islamic millennium; messianism
approximation, 207–8
Aqquyunlus, 215
Aquarius, 54, 200
Arabic grammar (science of), 159, 169, 229
ʿArabzade Mehemmed Efendi (*munajjim*), 278n28
architect-engineers, 8, 10, 14, 16, 84, 115, 150, 153, 204
Aries, 34, 58, 96, 110, 193, 219
Aristotelian, 45–47, 77, 144
Aristotle, 92, 135, 168, 178
arithmetic. *See hisab (ʿilm al-)*
Artun, Tuna, 129
ʿArudi (Nizami), 247n24
ascendant, 33, 49–51, 53, 58, 61, 65, 99, 103, 146, 149, 176–77, 183, 199–200, 202, 205, 220
Asharites, 77
astral determinism, 63–64, 78, 205, 234
astral sciences. *See ahkam al-nujum (ʿilm al-); hayʾa (ʿilm al-); nujum (ʿilm al-); tanjim*
astrolabe, 45, 50, 72, 98–99, 125, 129, 133, 135, 138, 151, 167, 171, 176–77, 199, 215–16
astrological history, 49
astrological triplicity, 49, 146
astrologer. *See munajjim*
astrology. *See ahkam al-nujum (ʿilm al-); tanjim*
astronomer. *See munajjim*
astronomy. *See hayʾa (ʿilm al-)*
astronomical handbook with tables, 18, 44, 57, 68, 70, 82, 126, 218; errors in 53–55, 178, 212; in *munajjims*' training, 50–55, 65, 169. *See also Zij-i Ilkhani; Zij-Jamiʿ-i [A]bu Saʿidi; al-Zij al-jamiʿ; al-Zij al-muhaqqaq; al-Zij al-mujmal; Zij-i Sultani*
ʿAşık Çelebi, 133–4, 158, 168, 171–72, 198
ʿAtaʾi (Nevʿizade), 230
ʿAtaʾullah, 131, 264n77
Athmar va ashjar (Fruits and Trees), 53, 57, 70, 143. *See also* Bukhari (ʿAli-Shah)
ʿAtufi (Hayreddin), 47, 126
auspicious hours, 6, 16–18, 48–49, 60, 74, 79–80, 106, 114, 119, 168, 173, 176, 190–200, 202–6, 224, 234
Avicenna. *See* Ibn Sina
Ayasofya endowments, 131

Babylonia, 103, 137, 202
Baghdad, 7, 29, 127, 160, 166
Bahrain, 220–21
Bakkalzade (*munajjim*), 263n52
Bali (*munajjim* in late 15th century), 127
Bali (*munajjim* in late 16th century), 185–86, 263n52
Battle of Ankara, 107
Battle of Chaldiran, 157–58
Bayandur ibn Rustam, 215–16
Bayezid (prince, Süleyman's son), 168, 170, 172, 221

Bayezid II (sultan): 4, 7, 35–36; and bureaucratization, 111–12, 114–18, 120–25, 153–57; intellectual interests of, 82, 125–46; and messianic currents, 148–50; and relations with ʿAbdurrahman, 84–86, 90–91; and relations with Khitabi, 99–100, 104–5, 107–8
Bayezid Mosque Complex in Edirne, 253n129
Bayezid Mosque Complex in Istanbul, 71
Bayrami (Sufi order), 213, 216
Ben-Zaken, Avner, 213
berat (certificate), 14–15
Berlin observatory, 232
bibliomancy, 189
Bidlisi (Idris), 132, 149, 157
birth horoscope: 18, 56–57, 80; by Lutfullah (*munajjim*), 164; of Mehmed II, 102, 111, 119; of Prince Ahmed, 70; Timurid examples of, 61
Biruni, 21, 42, 44, 56, 80, 93–94, 143, 167, 256n43
Bistami (ʿAbd al-Rahman al-), 30, 154, 184, 217
Bodin, Jean, 242n50
Brahe, Tycho, 225, 232
Brentjes, Sonja, 67, 69, 74
Buda, 7, 171, 179
Bukhari (ʿAli-Shah), 53–54, 57, 70, 143, 235. *See also Athmar va Ashjar*
Bulaq observatory, 259n1
burhani (demonstrative), 62, 176
burlesque almanacs, 250n77
Burnett, Charles, 26
Bursa, 7, 122, 131, 159, 181
Buni (Ahmad al-), 143
Buzjani (Abu'l-Wafaʾ al-), 87–88

Cairo, 7, 160, 206, 229
calendar systems, 55, 75
caliph, 21, 29, 65, 93, 105, 137, 199
Campagna, Federico, 76
Cancer, 200
Capricorn, 200
Cardinal Isidore, 119
Cassini, Giovanni Domenico, 235
Catholic Church, 233
Celalzade (Mustafa), 204–5
celestial globes, 72, 99, 135, 151
celestial (or astral) magic, 127, 143, 167, 168, 171, 173
celestial observation, 43–44, 52–54, 62, 65, 76, 87, 104, 122, 125, 128–29, 131, 163, 169, 171, 177–78, 205, 212–13, 229–33, 236, 238. *See also* astronomical handbook with tables; Istanbul observatory; Maragha observatory; Samarkand observatory
Cem Sultan, 116, 120, 150, 262n35
Centiloquium, 55, 143, 167, 207
Central Asia, 98, 123
Chagatay Turkish, 157
Charette, François, 69
chief physician's office, 9–10, 17, 95, 115, 129, 133, 153, 172, 186
China, 10, 65, 137, 223
Chinese Uighur animal calendar, 105
Chinggis, 29
Christian IV, 232
chronology tables, 32, 92–96, 168
chronos, 76
colophons, 36, 67, 102, 119, 126, 133, 155, 164, 173, 198
Columbus, Christopher, 31, 121
comet: in the year 1402, 107; in the year 1577, 73, 232
conjectural (*zanni*), 62–3
conquest of Constantinople, 5, 31, 111, 119, 149
Copernicus, Nicolaus, 225–26
Creation (of the universe), 32, 92–3

Damascus, 7, 229
Daniel (prophet), 29, 220, 222
Dashtaki (Ghiyath al-Din Mansur), 141
Dashtaki (Sadr al-Din), 141
Davani (Jalal al-Din), 141–42
de Block, Fien, 69

descendant, 51
devshirme, 189
Divan (of Bayezid II), 138
Divan (of Hafez), 189
Divan-ı Hümayun. See Imperial Council
divination, 28, 30, 41, 43, 92, 188, 218
divine will, 64
Diyarbakır, 7, 166
Dorothius of Sidon, 250n74
dream interpretation, 6, 31, 47, 92, 171, 189, 190, 208–11, 214
dualists, 145

Ebu's-suʿud Efendi (Shayk al-Islam), 1, 2, 204, 229, 236
eclipses, 53, 60, 76, 177. *See also* lunar eclipse; solar eclipse
Ede Bali, 208–9
Edirne, 7, 69–70, 103, 106, 119, 131, 144, 159, 181, 229
ego-documents, 83–84
Egypt, 30, 161, 178, 213, 217, 229–31, 238
ehl-i hibre. See under expertise
ehl-i vukuf. See under expertise
El-Rouayheb, Khaled, 22
electional astrology, 48–49, 56–7, 99, 133, 196–97
empiricism, 17, 63
End Times, 2, 29, 31, 132, 148, 212
epistolography, 15
Erasmus, 142
Erdeşir (*munajjim*), 124
Eretna, 118
esoteric knowledge, 6, 28–30, 34, 37, 41, 148, 152, 154–55, 174, 180, 191, 208, 211, 213–17, 220, 223–24. *See also* occult sciences
Esʿad Efendi, 92
Euclid, 142, 168
Eurocentrism, 26–27, 32, 235
Evliya Çelebi, 236, 253n129, 261n29
expert. *See* expertise
expertise: as actors' category, 10–12; brokers of, 14; bureaucratic organization of, 8–10, 14; and centralized examinations, 9, 10, 14; in future forecasting, 37, 43, 154–55, 188–92, 208, 219–21; Ottoman administration of, 13, 16–17, 79–80, 104, 151; technical-scientific, 18, 76, 84; vocabulary of, 11, 15. *See also under munajjim*

al-Farabi, 45–6
Faridun, 219
Faroqhi, Suraiya, 24
Fatih Mosque Complex: 68–70, 181–82; *waqfiyya* (endowment deed) of, 68, 71
Fatimids, 8
Fazlıoğlu, İhsan, 67, 139
Fenari (Mehmed Şah), 47
Fenari circle, 47, 122–23
Figani, 272n101
firdarat. See prorogators
Firdevsi-i Tavil, 127, 149
Firuzabadi, 39
Fleischer, Cornell H., 154, 191
Flemming, Barbara, 154, 191
fortune-tellers, 4, 17, 41, 43, 69, 113, 171, 190
furuʾ al-fiqh. See substantive law
Frederick II, 232

Galata, 212
Galeano, Moses, 129, 143
Galen, 88
Galilei, Galileo, 142, 225
Garsuddinzade Muhyiddin (chief physician), 9
generation and corruption, 44
geocentric model, 58
geomancy, 6, 31, 37, 47, 134, 189–91, 208, 217–23
geometry, 44–45, 47, 75, 98, 107, 142, 163, 168–69
Gerlach, Stephan, 278n15
ghayb, 41, 64, 189, 195. *See also rijal al-ghayb*
al-Ghazali, 77, 128
Ghazan Khan, 54, 76
Ghazan Khan observatory, 54
Ghaznawi (Abu'l-Mahamid b. Masʿud al-), 56, 167, 200

gök Tengri (sky God), 65
Gökmen, Fatin, 243n62, 277n3, 278n15
Golden Age of Islam (narrative), 21, 27–28, 80, 225
Goldstein, Bernard, 26
greater Syria, 161, 178, 229
Greek antiquity, 179
Gregorian calendar, 58, 204
Gritti, Andrea, 126
guilds, 74, 113
Gülşeni (İbrahim), 215–16
Gülşeni (Sufi order), 215

Habsburgs, 220
hace-i padişahi (princely tutor), 130
hadith (science of), 64, 126, 159–61, 222, 229
Hafez, 189
Halasi-Kun, Tibor, 171
Halveti (Sufi order), 211, 213–14, 216
Hanbalis, 77, 234
handasa (*ʿilm al-*). *See* geometry
Hasan Çelebi (calligrapher), 204
Hatemi, 144. *See also* Müʾeyyedzade (ʿAbdurrahman)
hayʾa (*ʿilm al-*), 43–47, 56, 67, 98, 126, 144, 160–61, 163, 167, 169, 184
Haydar (geomancer), 155, 184, 218–23
haylajat. *See* prorogators
Herat, 7, 87, 98
Hellenistic (astral tradition), 80
heresy, 64, 78, 88, 164, 213, 233–34
Hermes, 103, 133
hermetic, 45
Hezarfen Ahmed, 233
Hijri calendar, 60, 74, 204
hikmat. *See* natural philosophy
Hippocrates, 88
Hippodrome, 179
hisab (*ʿilm al-*), 44–45, 47, 98, 169
history of science (as an academic discipline), 20–27, 55, 84, 179, 225–26, 236
History of Science Society, 27
Hoca Saʿdeddin, 212, 230
Hocazade, 131, 161
Hodgson, Marshall, 246n95
Holy Roman Empire, 232
horary astrology, 48–49, 57, 70, 133
Hülegü Khan, 54, 76
huruf (*ʿilm al-*). *See* science of letters
Hüseyin (*munajjim*), 278n28
Hüseyin (Riyazi's father), 158, 166

Ibn ʿAbbas, 222
Ibn al-ʿArabi, 29, 183–84, 217, 220
Ibn al-Haytham, 21
Ibn Kathir, 159, 165, 168–70, 174, 176, 203
Ibn Kemal. *See* Kemalpaşazade
Ibn Khaldun, 77
Ibn Manzur, 39
Ibn al-Nafis, 90
Ibn al-Shatir, 89, 167–68
Ibn Sina, 21, 45–47, 77, 90, 126, 141
Ibn Taymiyya, 77, 234
Ibn al-ʿUlayf, 126
İbrahim (*munajjim*, son of Khitabi), 108, 124, 156
İbrahim (*munajjim* in mid-15th century), 118
İbrahim Müteferrika, 23
İbrahim Pasha (grand vizier), 179
İbrahim-i Kırımi, 214
Idris (Hermes), 222
İhsanoğlu, Ekmeleddin, 24–26
ijaza (teaching certificate), 65, 67, 138, 141
al-Iji, 141, 144
ikhtiyarat. *See* electional astrology
Ikhwan al-Safaʿ, 135
Ilkhanids, 54, 118
Imperial Council, 9, 186
İnalcık, Halil, 24
India, 99, 223,
Inquisition, 41, 234
intellectual networks, 29–30
Iranian lands, 35, 53, 55, 82, 85, 89, 97, 100, 106, 123, 148, 158
IRCICA (Research Center for Islamic History, Art, and Culture), 24–25
Isfahani (Ghiyath al-Din al-mujallid al-), 102

Ishak (*munajjim*), 124, 156
Ishak (physician), 88–89
Isidore of Seville, 43. *See also Etymologiae*
Isis (journal), 27
İskender Çelebi, 165
İskendername (Alexander legend), 173
Islamic millennium, 4–5, 27, 29, 31, 146, 148–49, 211, 224. *See also* apocalypticism; messianism
Istanbul observatory, 2, 5, 14, 23, 37, 38, 164, 171, 178, 212–14, 225–8, 232–34, 236–37. *See also* Taqi al-Din; Murad III
Istanbul University, 24
istikhraj, 51
İzgi, Cevat, 67, 164

jafr (*ʿilm al-*). *See* lettrist divination
Jaghmini (Mahmud ibn Muhammad ibn ʿUmar al-), 42, 67, 122, 135–36, 142–44
Jalali calendar, 60
Jalinus, Musa. *See* Galeano, Moses
Janissaries, 132, 158, 166
Jawziyya (Ibn Qayyim al-), 77, 234
Jewish natural philosophers, 116, 129
Julian calendar, 58
Jupiter, 49, 54, 104, 176, 193, 200, 219
jurisprudence (science of), 34, 41, 126–27, 134, 144, 159, 163, 229
Jurjani (Sayyid Sharif al-), 135, 141, 144–46, 160

Kadızade Şemseddin (Shaykh al-Islam), 2, 213, 234–36
kadkhuda. *See* prorogators
kairos, 76
kalam (*ʿilm al-*). *See* theology
Kandilli Observatory, 164
kanunname, 115
Kara Balizade, 165
karama, 217
Kaşifi, 118
Kasımpaşalı ʿOsman b. Muhammed,70
Kashi (Jamshid al-), 42, 65, 67, 89, 138, 168, 198, 230
Kashi (Mahmud b. Yahya al-), 42, 61, 102
Katib Çelebi, 23, 161–2, 174, 236, 247n19
Kaysuni (Seydi Ali b.), 172
Kaʿba, 44, 206, 207
Kemal (*munajjim*), 157
Kemalpaşazade, 149, 161, 173–4
Kennedy, Edward, 26, 51
Keşfi (Mehmed Çelebi), 157
Khitabi (Husam b. Shams al-Din al-Lahijani al-Gilani), 82, 97, 99–111, 119, 124, 127, 156, 181
Khurasan, 99
Kifayat al-taʿlim fi sinaʾat al-tanjim (Sufficiency of learning in the art of astrology), 56, 167, 200. *See also* Ghaznawi
kihana, 31
Kınalızade Hasan, 168, 178
King, David, 51, 69
Kıvami, 119
kızıl elma (red apple), 149–50. *See also* apocalypticism; messianism
Koca Sinan Pasha, 213–14
Korkud (prince), 85
Koselleck, Reinhard, 188
Küçük, Harun, 22–23, 228
Küfri Hasan Çelebi (*munajjim*), 278n28
kuloğlu, 158
Kurd Çelebi (*munajjim*), 263n52
Kushyar ibn Labban, 56, 108–9, 143, 197, 200, 206, 250n74

land grant, 171, 221
lay astrology, 33–34
Leaʾli, 259n111
learned societies, 226, 233
legal theory (science of), 159–60
Leiden observatory, 232
Lemay, Richard, 26
Leo, 199–200, 202–3, 205
Leonardo da Vinci, 121, 150
letter divination, 16, 29–30, 37, 148, 155, 191, 208, 213, 216–19
linguistic turn, 20
logic (science of), 22, 34, 41, 88, 134, 144, 159, 163, 168
London observatory, 232

long sixteenth century, 4–6, 13, 19, 31, 115, 224
lower Mid-Heaven (*Imum Coeli*), 51
lunar crescent, 205
lunar eclipse, 104, 158
lunar mansions, 30, 217
Lutfullah (*munajjim*), 156, 164–65, 198, 263n50
Luther, Martin, 31, 267n151

Machiavelli, Nicolo, 242n50
madrasa, 6, 22–23, 34, 41, 64–69, 72, 83, 86, 110, 113, 127, 130–31, 137, 144, 159, 160–63, 169, 173, 214, 229–30, 235
magic, 26–29, 47. *See also* celestial magic
magic squares, 168, 174
magicians, 40, 78, 190
maharet. *See under* expertise
Mahdi, 148, 211
Mahmud II (sultan), 71
majlis (social gathering), 110, 129, 165, 189, 229
majmuʿa, 136, 180
Makdisi, George, 66
Malekzade Ahi, 149
Mamluks, 7, 46, 69, 89, 105, 193, 206
al-Maʾmun (caliph), 65
Maragha observatory, 52, 54, 76, 87, 122
marginalia, 36, 91, 108, 134, 139–40, 155, 164
Marrakushi (Sharaf al-Din Abu ʿAli al-Hasan ibn ʿAli ibn ʿUmar al-), 135
Mars, 104, 193,
masaʾil. *See* horary astrology
Mashaʾallah, 80, 133
mathematical sciences, 6, 23, 25, 36, 39–40, 44–47, 64–67, 74–75, 88, 98–99, 116, 123, 130–31, 139–42, 155, 157–60, 162–63, 167, 169–70, 187, 199, 229–31
mawalid. *See* natal astrology
Mawlana Kuchak. *See* Yazdanbakhsh
Maybudi (Mir Husayn), 141
Mecca, 7, 126, 199
mechanical timepieces, 48, 71–2, 151, 171, 194, 213
medicine (science of), 9, 44, 46–7, 62, 88, 90, 100, 127, 129, 167, 169, 176
Mehmed I (sultan), 118
Mehmed II (sultan), 13, 61, 85, 87, 100–104, 108, 111, 115–16, 119–24, 139–40, 151, 153
Mehmed Suʿudi Efendi, 214
melhame, 33
Meniński, Franciszek, 248n45
meritocracy, 12–13, 15, 114, 197
messianism, 29–30, 37, 132, 146, 148–50, 154, 191, 211–13, 219. *See also* apocalypticism; Islamic millennium
metis, 11
Michalengelo, 150
Mihri Hatun, 140
miqat (*ʿilm al-*). *See* timekeeping
miqati. *See* timekeeper
Miʿmar Sinan, 164, 172, 204, 207
Mirim Çelebi, 35, 40, 57, 67, 116, 130–34, 139, 148, 160–61, 164–65, 185, 230, 246n6
Mirza Iskandar, 61, 102
Mirza Rustam b. ʿUmar-Shaykh, 61, 102
Mitterrand, François, 137
modernization, 20, 233
Moin, Azfar, 30
Molla Fenari, 122–23, 161
Molla Lutfi, 88, 120, 164–65, 233
Mongols, 29, 93
Moon, 47, 193, 200, 206, 219
Morrison, Robert, 67
mudabbir, 145. *See also* astral determinism
Mughals, 8, 29–30
mujaddid, 149
al-Mulakhkhas fi al-hayʾa al-basita (Epitome of plain astronomical theory), 42, 67, 122, 135–6, 142–4. *See also* Jaghmini
Mulsow, Martin, 32
munajjim: as astral experts, 8, 16, 32, 40–45, 106; as charlatans, 4, 18, 40, 190; in classification of sciences, 42, 45–47; courtly patronage of, 5, 7–8, 17, 35, 71–72, 74–77, 79–82, 91, 109–110, 113–14; dictionary definition of, 39,

munajjim (cont.)
42; marginalization of, 18–19, 25–26; as masters of time, 47–49, 76, 94; and *muwaqqits*, 69–71; vs. other diviners, 4, 30–34, 154–55, 213, 220; polemics against, 18, 77–78, 144–46, 215–19, 223–24; as practitioners of an ambivalent science, 4, 6, 178, 234;requests of for new astronomical research, 53–54; as ritual specialists, 193–94, 205; self-reflections of, 62–64, 78, 208, 224; technical expertise of, 13, 16–18, 40–41, 44–45, 50–52, 55–58, 61, 191–92; training of, 64–69; the unit of at the Ottoman court, 7, 114–115, 118–20, 123–25, 151–53, 156–57, 164, 173–74, 180–81, 198, 230
müneccim. *See munajjim*
müneccimbaşı. *See under munajjim*
Müneccimbaşı Ahmed Dede (*munajjim*), 278n28
Müneccimek Mehemmed Çelebi (*munajjim*), 278n28
Murad II (sultan), 118
Murad III (sultan), 2, 178, 211–12, 214, 216, 230, 234–35
Muradiye Mosque Complex, 106
müşaherehoran, 124, 132
Muslihiddin (*munajjim*), 263n50, 263n52
music (science of), 44, 47
Mustafa (geomancer), 134
Mustafa (*munajjim*), 278n28
Mustafa III (sultan), 235
Mustafa Hatti Efendi, 42
Mustafa Zeki (*munajjim*), 255n5, 256n39, 260n4, 275n66
müteferrika, 115, 123–4
Muvakkit Mustafa, 70, 91, 156, 184–85, 230
Müʾeyyedzade (ʿAbdurrahman), 35, 116, 129, 132, 139–47, 149, 160–61
muʿabbir (dream interpreter), 210
Muʿtazilites, 145
muwaqqit. *See* timekeeper
mystical (or esotericist) astrology, 217–18

Nallino, Carlo, 44
natal astrology (genethlialogy), 48–9, 51, 56
natural philosophy, 22, 47, 88, 97, 100, 104, 106, 116, 125–27, 129, 138, 157, 160–63, 168–69, 177, 228
Navaʾi (ʿAli Shir), 157
Necmeddin b. Seyyid Muhammed (*munajjim*), 69, 70, 181, 263n50
Needham, Joseph, 25
Neugebauer, Otto, 26, 28
Nevʿi Efendi, 247n19
new humanism, 27
newspaper horoscope columnists, 18
Nisaburi (Nizam al-Din al-), 67, 167
Nowruz (or *Nevruz*), 16
Nücumi el-Rumi, 125, 262n35
nujum (*ʿilm al-*), 39, 44–47, 71, 85, 126–27, 203
numerology, 29, 33, 211
Nurbakhshi (Sufi order), 257n57

oblique ascension, 51
observatory. *See* celestial observation; Istanbul observatory; Maragha observatory; Samarkand observatory
occult sciences, 4, 8, 16, 25, 27–30, 32, 37, 47, 78, 148, 186–87, 189, 191, 215. *See also* divination; esoteric knowledge; magic
Old Mosque in Edirne, 69, 181
Orientalism, 26–28, 83
ʿOsman (ruler), 93, 95, 208–9
Ottoman advice literature, 12
Ottoman decline (narrative), 2, 13, 20–23, 36, 38, 120, 155, 163, 226, 232–33
Ottoman palace library inventory (of 1502–4), 47, 55–7, 102, 122, 125, 134, 139. *See also* ʿAtufi.
Ottoman Turkish (as a language), 11, 35, 42, 80, 82, 89, 95, 108, 154, 163, 168, 183, 186, 235
Ottomanization, 154
ownership statements, 36, 67, 99, 108–9, 141, 147, 155, 167
ʿÖzri Çelebi, 133–34

paratexts (or microscripts), 36, 155
Paris observatory, 42, 232, 235
patronage. *See under munajjim*
Persian calendar, 60
Persianate, 35–36, 46, 53, 64, 72, 79, 82, 86, 89, 97–98, 103, 106, 110–12, 116, 123, 132, 139–40, 153–54, 157, 165, 229
Pfeiffer, Judith, 142
philosopher-king, 35, 77, 116, 125, 137–39
physiognomy, 47, 173, 189
Pingree, David, 26
plague, 11, 76, 94, 168, 172, 174–75, 177
planetary conjunctions. *See qiranat*
Plato, 132, 139, 178
Pontus Trebizond, 118
Pope Nicholas V, 119
post-classical Islam, 21, 46–47, 80, 144, 225, 243n52
post-Mongol, 29, 65, 93, 110
prestige, 15, 72, 181, 184, 186, 203
printing press, 23, 91, 113, 121, 226
prognostication: 16; astrological, 43, 58, 60, 62, 68, 118; difference from prophecy 28, 31, 188–89. *See also* almanac (with prognostications)
projection, 10, 17, 188–89
prophecy, 29–31, 37, 152, 154–5, 173, 187–88, 190–91, 208–10, 213, 216–17, 220–24
Prophet Muhammad, 30, 64, 78, 82, 93, 104, 106, 210, 215, 220, 222–23
prorogators, 61, 103, 146
pseudepigraphy, 173
pseudo-science, 25
Ptolemy, 43, 46, 55–56, 63, 89, 103, 120, 135, 143, 167, 197

al-Qabisi, 231
Qadi-yi Baghdad, 127
Qasrani (Yaʿqub b. ʿAli al-), 57, 70, 133, 250n74
Qazvini (Hakim Shah Muhammad al-), 157
qibla, 44, 69
qiranat, 49, 53–55, 76, 104, 177
qiyas (analogical reasoning), 63
Qizilbash, 96
quadrants, 72, 132–33, 151, 176–77, 184
Quigley, Joan, 246n7
Qurʾan, 64, 100, 189, 206, 222
Qurʾanic exegesis (science of), 67, 159–61, 223, 229
Qurtubi (Maslama al-), 271n86
Qusta ibn Luqa al-Baʿlabakki, 135
qutb (axis mundi), 149, 183, 219

Ragep, Sally, 67
raml (*ʿilm al-*). *See* geomancy
rasad. See celestial observation
Rasulids, 138
rational sciences, 67
Razi (Fakhr al-Din al-), 45, 47, 57, 143, 167
Razi (Shahmardan b. Abi'l-Khayr al-), 57, 143
Reagan, Nancy, 246n7
Reagan, Ronald, 137, 246n7
Receb (student of Khitabi), 108
Renaissance polymaths, 121, 142
Renan, Ernest, 244n70
Rhodes, 134, 157
rijal al-ghayb, 183–4, 211
ritual specialists, 193, 205
Riyazi (ʿAli), 36, 148, 155, 158–60, 163–81, 185–87, 196–208, 210, 212, 214, 218, 236, 263n50
Roman antiquity, 179
Rudolph II, 232
Rumeli Hisarı, 119
Rumelia, 7, 102, 143–4, 148
Rumi (Mawlana Jalal al-Din al-), 189
Rumi (Qadizada al-), 62, 67, 122, 130, 138, 142, 160–61, 230
Ruʾus records, 9
ruzname, 33, 171

saʿati. See timekeeper
Saʿati (Yusuf b. Ömer el-) (*munajjim*), 91, 108–10, 124–25, 156, 168, 172, 180–84, 198, 263n50
Sabians, 145
Sabra, Abdelhamid, 67, 71, 251n102

Sabuncuoğlu Şerefeddin, 140
sacred kingship, 30
Safavids, 8, 29, 96, 127, 140, 148, 161, 165, 170, 218, 220, 229
Safi (Mustafa), 206
sahib-qiran, 119, 132, 149, 219, 222
Sahl ibn Bishr, 250n74
Şahin, Kaya, 204
Sahn madrasas, 69, 88, 107, 127, 162
Saliba, George, 41, 69
Salih Zeki, 236
Samarkand, 7, 87, 98, 122, 130, 138, 160
Samarkand observatory, 40, 42, 54, 65, 87, 89, 98, 107, 122, 138–39
Samuel Abulafia, 129
Sarton, George, 27–28
Saruhani (Ilyas b. ʿIsa), 213, 216–18
Sassanian (astral tradition), 80
Saturn, 49, 52, 54, 104, 200, 219
Sayılı, Aydın, 65
sayyid munajjim (as an honorific), 82, 89, 104–7
Sayyid Munajjim. *See* Khitabi
Sayyid Munajjim (of Timurids), 106–7, 164
science of letters, 6, 29–31, 37, 47, 148, 152, 154–55, 216–20, 222, 224
science-versus-religion dichotomy, 2, 20, 120, 233
sciences of the ancients, 66
Scientific Revolution (narrative), 21, 225
seers, 155, 190, 208
Selim I (sultan), 85, 91, 115, 124, 130, 132–33, 140, 142, 156–58
Selim II (sultan), 170, 22, 277n13
Seljuqids, 93, 118
Semiz ʿAli Pasha, 169–71, 229, 236
Şemseddin Sami, 42, 247n15
Serpent Column, 179
seven-climes theory, 97
Şeyh Hamdullah, 140
Seyyid Battal Gazi, 150
Seyyid Lokman, 227, 229
Sezgin, Fuat, 27
Shah Ismaʿil, 96, 127, 148, 157
Shah Tahmasb, 170, 220–21
al-Shaqaʾiq al-nuʾmaniyya fi ʿulama al-dawlat al-ʿUthmaniyya (Crimson peonies among the scholars of the Ottoman state), 88, 142, 158, 160. *See also* Taşköprülüzade
Sharf al-mawaqif (Commentary on stations), 141, 144. *See also* Jurjani
Sharia, 41, 63, 161, 210
Shaybanid Uzbeks, 220
Shayhuniyya madrasa, 229
Shi'i imams, 166, 211
Shiraz, 7, 45, 98, 141–2
Shirazi (Qutb al-Din al-), 46, 67, 144
Shirazi (Sayyid Niʿmatullah), 89–90, 106
Shirvani, 76, 125
Shirwani (Fathallah al-), 138
Si fasl (Thirty chapters), 55, 100, 102, 122, 134, 136, 143. *See also* Tusi (Nasir al-Din al-)
al-Sijzi, 250n74
Silk Road, 140
Sinan Pasha, 131, 262n35
al-Sirr al-maktum (Hidden secret), 143, 167. *See also* Razi (Fakhr al-din al-)
Şirvani (Sadreddinzade), 247n19
Sivasi (Pir Muhammed el-), 256n38
sıvış yılı, 75
Socrates, 88
Sokollu Mehmed Pasha, 212, 230
solar eclipse, 94, 104, 157–58
solar year transfer, 16, 48, 58, 99, 110, 183, 207. *See also Nowruz*; Spring equinox
soothsayers, 42, 190
spasmatomancy, 92
Spring equinox, 58, 219
St. Augustine, 267n151
street astrology. *See* lay astrology
sublunary world, 30, 45–46, 62–63, 76–77, 100, 145, 168, 175–76
substantive law (science of), 159–60
Şücaʿ Efendi, 211, 214
Süleyman I (sultan): 4, 7, 36–37, 106, 115; and messianism, 151–52, 154, 190–91, 219–24; and *munajjims* at his court, 124–25, 134, 184–86; nativity of, 200,

207; relations to Haydar the geomancer, 204–5, 218; relations to Riyazi, 165–66, 168, 170, 172–75, 196, 199, 203
Süleyman Sudi Efendi, 42, 236
Süleymaniye madrasas, 68
Süleymaniye Mosque: 1, 68, 168, 173, 196–98, 201, 204–7; *waqfiyya* (endowment deed) of, 71
Sultan Ahmed Mosque, 206
Sultan Selim Mosque, 70
Sun, 47, 53, 58, 96, 110, 193, 200, 219
Sunnitization, 2, 161, 239n6
superstition, 26–28, 43
Surgatmishiyya madrasa, 229
Suyuti, 206
Syrian calendar, 60

Tabakat-ı Müneccimin (Classes of *munajjims*), 42, 236. *See also* Süleyman Sudi Efendi
Tabriz, 7, 54, 97
tacit knowledge, 12
Tacizade Caʿfer Çelebi, 119, 132–3
al-Tadhkira fi'l-hayʾa, 122, 135, 143–44. *See also* Tusi (Nasir al-Din al-)
al-Taftazani, 159
Tahrir al-Majisti (Recensions of the Almagest), 135, 167. *See also* Tusi (Nasir al-din al-)
tajriba (experience), 62–3, 128, 157, 176
taliʿ. *See* ascendant
talisman, 48, 127–28, 143, 168, 172–79, 184, 213–14, 236
tanjim, 43, 145. *See also ahkam al-nujum* (*ʿilm al-*)
Tankalusha, 202, 274n42
Taqi al-Din, 2, 14, 23, 64, 91, 164, 171, 178, 185, 212–14, 225–26, 228–31, 233–36, 277n6, 277n8, 277n13, 278n15. *See also* Istanbul observatory
taqwim. *See* almanac (with prognostications)
Tár (movie), 248n36
Taşköprülüzade, 45, 67, 88, 130, 142, 158, 160, 247n19, 253n131
Taurus, 34
tax farming, 188
taxonomies (classification) of science, 10, 41, 44–6, 76, 128
Teberrük (*munajjim*), 256n49
terrestrial realm. *See* sublunary world
Terzioğlu, Derin, 115
Tetrabiblos, 43, 46, 56. *See also* Ptolemy
Teucros. *See* Tankalusha
Tezcan, Baki, 94, 235, 256n48
Thamara. *See* Centiloquium
theology (science of), 22, 34, 41, 67, 120, 141, 143–44, 146, 159, 168–69
Thomas, Keith, 245n79
Thorndike, Lynn, 26, 245n79
Timaeus, 139. *See also* Plato
timekeeper, 16–17, 69–72, 113, 124, 181–82, 184–5
timekeeping, 69–71, 180, 184–85, 194
Timur, 95, 107
Timurids, 7–8, 36, 42, 61, 89, 98–99, 102–3, 106–7, 111, 116, 138–39, 153, 164, 229
Tokat, 100, 106
Topkapı Palace, 106–7, 164, 172, 192, 222
Trabzon, 7, 156
transmitted sciences, 66, 68,
tropical signs, 200
true longitudes of planets, 51, 53, 61
Tursun Beg, 119
Tusi (Nasir al-Din al-), 42, 46, 53–56, 67, 76, 87, 89, 100, 102, 104, 122, 126, 134–36, 141–44, 151, 166–67
twelve astrological houses, 33, 49, 52, 58, 61, 100, 146, 149, 192, 220
twin problem, 145
Tyson, Neil DeGrasse, 243n51

ʿulamaʾ, 1, 18, 41, 83–84, 88, 96, 106, 110, 161, 165, 213–14, 233–36
Ullman, Manfred, 27
Ulugh Beg, 54, 87, 98–99, 104, 107, 110, 131, 133, 138–39, 142, 151, 164, 167, 177
ʿUmar II (al-Ashraf) (Rasulid ruler), 138
ʿUmar b. al-Farrukhan, 250n74

Umayyads, 93
Unan, Fahri, 162
upper Mid-Heaven (*Medium Coeli*), 51
Uraniborg, 232
usul al-fiqh. *See* legal theory

van Dalen, Benno, 51
Vatican, 120
Venetians, 126, 149, 238
Venice, 129, 143
Venus, 149, 176, 193, 206
Vettius Valens, 133, 250n74
Vienna observatory, 42
Virgo, 157
vocabulary of sovereignty, 32, 132

Wabkanawi (Shams al-Din Muhammad al-), 42, 54–55, 96, 108, 110, 141
walaya, 217
Wehr, Hans, 39
Wellcome Institute Library, 61
Whewell, William, 26–27
wretched subjects, 26, 28
WWI, 5, 20, 114

Yada stone, 175
Yahya (*munajjim*), 262n35
yaqini (apodictic), 62, 128
Yazdanbakhsh (*munajjim*), 85–88, 123–4, 141
Yaᶜkub Çelebi b. Germiyan, 94
Yaᶜqub (Aqquyunlu sultan), 215–16
Yirmisekiz Mehmed Çelebi, 42
yıldızname, 33

Zacuto, Abraham, 129, 143. See also Almanach Perpetuum
al-Zamakhshari, 223
Zarathustra, 103
Zati, 260n3
zayicha. *See* birth horoscope
zij. *See* astronomical handbook with tables
Zij-i Ilkhani, 42, 52–55, 87–88, 96, 99, 104, 122, 135, 142, 166–67, 183. *See also* Tusi (Nasir al-din al-)
Zij-i Jamiᶜ-i [A]bu Saᶜidi, 98–99, 105. *See also* Amuli (Rukn al-)
al-Zij al-jamiᶜ, 108–9. *See also* Kushyar ibn Labban
al-Zij al-muhaqqaq, 54, 96, 108, 110. *See also* Wabkanawi (Shams al-Din Muhammad al-)
al-Zij al-mujmal, 86, 87. *See also* Yazdanbakhsh
Zij-i Sultani (*or Zij-i Ulugh Beg*), 62, 87–88, 98–99, 104, 110, 131, 133, 142, 151, 164, 167, 177, 183. *See also* Ulugh Beg
Zodiac, 49, 51–52, 55, 58, 61, 100, 145–46, 200, 202–3, 211, 217–18

STANFORD **OTTOMAN WORLD** SERIES

Critical Studies in Empire, Nature, and Knowledge

The Stanford Ottoman World Series showcases cutting-edge interdisciplinary scholarship in Ottoman history from the thirteenth to the twentieth centuries. Books in the series are concerned with three major themes—empire, nature, and knowledge—and the connections among them. The books in this series foster ambitious and innovative scholarship and open new paths in Ottoman studies and beyond.

ROBERT G. MORRISON, *Merchants of Knowledge: Intellectual Exchange in the Ottoman Empire and Renaissance Europe* **2025**

NIR SHAFIR, *The Order and Disorder of Communication: Pamphlets and Polemics in the Seventeenth-Century Ottoman Empire* **2024**

UĞUR ZEKERIYA PEÇE, *Island and Empire: How Civil War in Crete Mobilized the Ottoman World* **2024**

ELIZABETH R. WILLIAMS, *States of Cultivation: Imperial Transition and Scientific Agriculture in the Eastern Mediterranean* **2023**